Complex Systems: Chaos and Beyond

Springer
Berlin
Heidelberg
New York
Barcelona
Hong Kong
London
Milan
Paris
Singapore
Tokyo

Physics and Astronomy   ONLINE LIBRARY

http://www.springer.de/phys/

Kunihiko Kaneko   Ichiro Tsuda

Complex Systems: Chaos and Beyond

A Constructive Approach
with Applications in Life Sciences

With 111 Figures

Springer

Professor Kunihiko Kaneko
Department of Pure and Applied Sciences
University of Tokyo
Komaba, Meguro
Tokyo 153-8902
Japan

Professor Ichiro Tsuda
Department of Mathematics
Graduate School of Science
Hokkaido University
Kita-10 jo, Nishi-8 chome
Sapporo 060-0810
Japan

Title of the original Japanese edition:
Fukuzatsukei no kaosu-teki shinario
© 1996, Kunihiko Kaneko and Ichiro Tsuda
English translation © 2000, Kunihiko Kaneko and Ichiro Tsuda
All rights reserved.
The English edition rights are granted by Asakura Publishing Co., Ltd., Tokyo.

Library of Congress Cataloging-in-Publication Data.
Kaneko, Kunihiko. [Fukuzatsukei no kaosu-teki shinario. English] Complex systems : chaos and beyond :
a constructive approach with applications in life sciences / Kunihiko Kaneko, Ichiro Tsuda.
p.cm. Includes bibliographical references and index.
ISBN 3540672028 (alk. paper) 1. Chaotic behavior in systems. I. Tsuda, Ichiro. II. Title.
Q172.5.C45 K34 2000   003'.857–dc21   00-030752

ISBN 3-540-67202-8 Springer-Verlag Berlin Heidelberg New York

Cover picture: Part of a schematic representation of chaotic itinerary. From: K. Kaneko, Diversity Induced by Chaos. © May 1994, 34, Nikkei-Science, (in Japanese)

Springer-Verlag Berlin Heidelberg New York
a member of BertelsmannSpringer Science+Business Media GmbH

© Springer-Verlag Berlin Heidelberg 2001
Printed in Germany

The use of general descriptive names, registered names, trademarks, etc. in this publication does not imply, even in the absence of a specific statement, that such names are exempt from the relevant protective laws and regulations and therefore free for general use.

Typesetting: Data conversion by LE-TEXJelonek, Schmidt & Vöckler GbR, Leipzig
Cover design: *design & production*, Heidelberg
Printed on acid-free paper   SPIN: 10652079   55/3141/di   5 4 3 2 1 0

Preface

We first had the idea to write a joint book fourteen years ago. First we had intended to write a book on chaos, and then our vision expanded to include the significance of chaos in all fields of science, which exploded to a plan for a dozen-volume series. Since this was not realistic, it did not materialize. Then we made a reduced plan about a decade ago, with explicit outlines, but this did not work out either. In the meantime, our interest has shifted to the study of complex systems, based on, but beyond, chaos. Kaneko made a departure to a 'jungle tour' to search for universal structures in high-dimensional chaos, based on the study of coupled map lattices and globally coupled maps. With concepts and phenomenal he discovered, he started to study life from the viewpoint of a complex system, where life can be viewed as complementarity dynamics itself between the whole and parts. Tsuda proposed a hermeneutic study of the brain and also a dynamic aspect of the brain, taking the information structure of chaos into account. There, he started brain research in order to epistemologically understand chaos, whose essence is regarded as "descriptive instability" and some sort of undecidability.

After this long detour, we finally published the Japanese version of the present book in 1996 with Asakura Publ. Inc. This book is based on the Japanese version, but the contents are updated.

This is neither a textbook on chaos nor a textbook on complex systems. It is not intended to transfer established knowledge, either. Rather, we intend to provoke discussions on complex systems, by showing what we have achieved and what we aim at accomplishing in the field of 'complex systems'.

We take a standpoint that chaos is essential to a practical and philosophical study of complex systems, although the complex systems are such that we must postulate some other concepts beyond chaos. Here we reexamine what concepts in chaos are relevant to the study of complex systems. From this reconsideration, we take the following three standpoints for the study of complex systems: a constructive approach, a many-to-many relationship, and descriptive instability. The last issue is, in particular, based on the observation problem, where we intend to consider seriously what a description of a system means, how descriptions or observations impinge on a system, and then what condition is imposed on a system for stability against descriptions. A clear answer to this issue has not yet been obtained, but we try to provide a basis for future studies.

As people from a variety of fields show interest in the studies of complex systems, there will be diverse ways in reading the present book. Hence we give some possible guidance for readers, so that they are not lost in the jungle of complex systems.

Those who are interested generally in complex systems are recommended to read Chap. 1, and then choose some chapters depending on their interests. Chapter 1 is written so that it can be read as an independent monograph, which we hope will be understandable to the general readership, including nonscientists.

It is recommended that all readers choose some chapters depending on their interests. For example, those who are interested in brain and cognition problems will probably read Chap. 6, but it is hoped that they will also read Chaps. 2–4; while Chap. 5 provides a bridge between these chapters and Chap. 6.

Those who work in the field of statistical physics, mathematical physics, and dynamical systems will find several research topics they should study in the future in Chaps. 2–6. Some topics in these chapters are not yet mathematically refined, but it is hoped that they will be formalized and established as a concrete concept or framework.

Physicists and chemists in general may find Chaps. 2–4 most interesting.

Those who try to understand dynamics in a biological system can find relevant parts in Chaps. 2 and 4 and in a part of Chap. 3, depending on their interests. After scanning these chapters, they will find Chaps. 5 and 6 interesting.

Those who are interested in the engineering applications of chaos and biology-oriented systems, and the search for novel basic ideas, can find some relevant concepts or phenomena throughout Chaps. 2–6.

It is our regret that the translation of this book from Japanese has taken so much longer than we had imagined. Indeed we first looked for a good translator, but in vain. Finally we decided to translate it ourselves. This was a painful decision for us, but there have been some merits, we believe. We have succeeded in updating the contents throughout the book. We revised some vague expressions in Chap. 1 to be more understandable. Chapters 2 and 3 in the Japanese version are joined to form a new Chap. 2, where excessive mathematical discussions have been eliminated. In Chap. 3, we have added some paragraphs on the significance of coupled map lattices, and recent examples of its applications are discussed at length. There are quite recent developments on the collective dynamics in globally coupled maps, which are also included in Chap. 4. Of course, the core of complex systems lies in a biological system, and our recent studies focus on this direction. In Sect. 5.6, we have outlined recent studies on theoretical (cell) biology along the lines of Chap. 1. In Chap. 6, we deleted some lengthy explanations that appeared in the Japanese version, and instead added a concise description. Recent studies about a temporal coding were additionally introduced and their references were updated. Besides these major ones, minor revisions have been carried out throughout the book.

Although we do not list all the names as is usually done, we would, of course, like to give sincere thanks to all the collaborators and those who have shared interest in the topics of this book for their stimulating discussions. We heartily express our gratitude to Dr. Frederick Willeboordse who kindly took care of a critical reading of the mansucript and the correction of the English.

Tokyo and Sapporo
July 2000

Kunihiko Kaneko
Ichiro Tsuda

Contents

1. Necessity for a Science of Complex Systems

1.1 Introduction

Why is a science of complex systems necessary?

When we talk about the study of a complex system, which is an extremely vague and far too general designation, it is fair, we believe, to first state what we are going to explore under the name 'complex system'. The viewpoint in this book might somewhat be biased by the generally prevailing thoughts in the science of complex systems. However, since the present standpoint is based upon important concepts and discoveries which we have proposed, we feel that the general statements made can be considered as fairly universal.

Nowadays many scientists profess the study of complex systems. Nevertheless, this study is still in its infancy. Some scientists insist that it is better to study individual phenomena without defining what a complex system is, since they believe that defining a complex system by some formula might prevent further progress in their study. Nevertheless, we hope to clarify the contours of complex systems through a series of books, starting with this first volume. We feel that our view of the world, acquired through our own studies of chaos, differs much from that provided by conventional science. Furthermore, by trying to expand this difference, we have come firmly to believe that there is a flow in science of which our study is a part. All in all, we do not believe that anything that looks complex or complicated is actually a complex system. We are certain that some requirements are necessary in order to designate a system as complex.

The word 'chaos' has already been used above. We believe that it is impossible to discuss complex systems without mentioning the concept of chaos. The emergence of the concept of chaos is one of the biggest scientific achievements this century, and has inspired interest in complex systems by radically changing many fundamental beliefs in science. We wish to clarify why the study of complex systems is essential today by presenting concepts derived directly from chaos, and by examining theories connected to chaos.

First, the traditional reductionist approach taken in the natural sciences is questioned by the study of complex systems. For more than two hundred years, many natural scientists, including physicists, have employed a method of study which reduces complex natural phenomena to several simple processes, and applies a simple theory to each process; comprehension of the whole phenomenon relies on the superposition of these 'part' processes. This strategy is often referred to as "Ockham's razor". Ockham's razor is a way of thinking that favors a minimum number of assumptions, the smallest number of variables and parameters, and the smallest number of equations when formulating hypotheses. Of course, it is not possible to reduce all phenomena to a system with a small number of degrees of freedom. When the reduction is performed, the parts which deviate from that reduction are recognized as "noise", and probability has been introduced in order to deal with those parts as random, unable to be expressed by determinism. However, the origin and reason for the introduction of the "probability" is not answered so easily, giving rise to the scientific discipline known as ergodic theory. Modern probability theory is based on established mathematical principles. However, whether or not probability theory provides a basis for phenomena is an important issue in applied mathematics which several scientists have just begun to study.

What we question here is not just what could be called 'primitive' reductionism. With primitive reductionism we mean, for example, statements such as "if you understand the theory of an elementary particle, you understand every natural phenomenon" or "if you understand DNA, you comprehend all biological phenomena". If such reductionism would hold, one could understand any level of nature in principle by understanding the nature of each element and the characteristics of the interactions between the elements. Hence both the numbers of variables and equations could radically be reduced, and, as a result, Ockham's razor would perfectly be applied.

Since it is generally accepted though that understanding at the lowest level does not necessarily imply an understanding at the higher levels, it is not necessary for us to refute this 'primitive' reductionism here. Nevertheless, it may be meaningful to offer some simple illustration. As mentioned above, primitive reductionism is based on the belief that we can understand the whole by understanding its elements. This belief is based on the principle of superposition. The principle of superposition states that if $x_1(t)$ and $x_2(t)$ are the solutions of an equation, $C_1 x_1(t) + C_2 x_2(t)$ is also a solution. While this is true for linear systems, it is naturally false for nonlinear systems. (Obviously, $(x_1(t) + x_2(t))^2 \neq x_1(t)^2 + x_2(t)^2$). In this sense, the existence of chaos illustrates a typical flaw in reductionism.

What we really question here is the type of nonprimitive reductionism employed in statistical mechanics. There a reduction to the lowest number of degrees of freedom is often done, typically by the introduction of "order parameters". Generally speaking, in statistical mechanics, we study a system with a large number of degrees of freedom (for example, a large number of particles). We describe the system by the actions of a small number of macroscopic order parameters which are reduced from the system, while the rest, which is the superposition of many complex microscopic degrees of freedom, is regarded as "noise". We can justify the separation of the degrees of freedom in this way because the number of degrees of freedom (for example, the number of particles) is large. This fundamental structure is common to all equilibrium statistical mechanics, linear response theory, and other studies of nonequilibrium systems at a macroscopic level.

As a typical way of thinking in reductionism, the concept of a "mode" has often been effective. An arbitrary motion or pattern (or the solution of an equation) is described by decomposing it into several modes and superposing them. This approach is perfectly justified in a linear system by the principle of superposition. The most established method is the Fourier transformation in which solutions and motions are described by the superposition of sinusoidal waves having different frequencies. It formulates a way of understanding by representing motion as the superposition of periodic components. Perhaps this is what human beings have done since ancient times. It might arise because we human beings have evolved in a cyclic environment including the rotation and revolution of the earth. In addition, as seen in a phrase such as "the sound is high/low", expressions regarding frequencies might also have evolutionary origins.

As will be shown, the existence of chaos leads to a revolution in the concept of reductionism. In chaos, no matter how small the differences between orbits are, they increase exponentially with time. Thus it becomes impossible to separate microscopic (lower) orders from macroscopic (higher) orders in variables. If such an amplification of small differences could be described by a simple stochastic variable, hence replacing the lower-order variables by a probability, we could apply the conventional method where the transformations of the macroscopic order parameters are described by assuming that all motions of lower orders can be regarded as "noise". Since chaos, however, is generated from deterministic rules, such a replacement cannot always be carried out.

As mentioned, the superposition principle may not be effective in a nonlinear system. In particular, a description of a system by superposition of (a finite number of) individual modes is generally not effective in chaos. For example, when trying to understand a chaotic dynamics, we often observe the time series and then transform it into frequency space by using a Fourier transformation. This is a standard procedure for measuring chaos experimentally. However, even in a system with only three variables, the power spectrum can contain an uncountably infinite number of frequencies. Indeed, the existence of this kind of continuous spectrum is often a first experimental step for confirming the existence of chaos. Here, the correspondence between the Fourier modes and the three original variables is essentially intractable. Now reduction to a small number of (Fourier) modes is impossible in the study of chaos.

Here, it should be pointed out that chaos is observed in equations with only three variables. In this sense one may think that 'reduction' to an extremely small number of degrees of freedom is carried out here. The most important point here is that the limit of the reductionist concept manifests itself even in a system reduced to a small number of degrees of freedom. In other words, the denial of reductionism in chaos does not mean simply the return to holism.

In the statistical mechanical analysis of a system with a large number of degrees of freedom, we often separate the system into those with macroscopic degrees of freedom and those with microscopic degrees of freedom, and obtain equations only with averaged macroscopic degrees of freedom. In those examples, the averaged equations can generate chaos. Then, a microscopic change is amplified to a macroscopic level by chaos. This implies that the reduction to a small number of degrees of freedom is essentially difficult.

1.2 Chaos

Since the studies by Poincaré [1899] from the end of the 19th century to the beginning of the 20th century, the concept of chaos has been an exotic field at the forefront of academic research. Celestial motion, regarded as the mathematical realization of harmony in the universe, was found to yield complex orbits whose behavior is impossible to reduce to modes when more than three celestial bodies are involved. However, Poincaré's theory had long been forgotten. In the 1930s, van der Pol, an electrical engineer, discovered chaotic motion in a nonlinear electric circuit, but his discovery did not lead to a systematic study. In the 1960s, Ueda, Kawakami and others, also electrical engineers, discovered chaotic motion in the Duffing equation and performed intensive studies [Ueda 1994]. In the former Soviet Union, Kolmogorov, Arnold, Moser, Chirikov, and others defined the major characteristics which distinguish chaotic from regular motion in Hamiltonian dynamical systems [Arnold 1963; Arnold and Avez 1967/1972; Chirikov 1979; MacKay and Meiss 1987; Lichtenberg and Lieberman 1983]. Chaos in Hamiltonian systems has also been studied in detail by Saito *et al.* in Japan. However, the current study of macroscopic chaos is not a direct descendant of these studies. The rise in the study of chaos itself was not realized until the 1960s when Lorenz [1963] analysed a simple model equation for the earth's atmospheric motion. In his work, chaos was rediscovered and systematic studies were performed. However, the epochal study by Lorenz did not have a scientific impact for a decade. In the mid-1970s, the study of chaos progressed into various areas only after chaos was discovered in an experimental laboratory.

Chaos is a type of unpredictable motion generated by deterministic equations (differential equations or difference equations). Rules which generate chaos are sometimes called chaotic dynamical systems. The characteristics of chaos are described using various indices. The space used to describe a system is often called the phase space. In the phase space of chaotic dynamical systems, two orbits slightly separated from each other will separate exponentially with time. The degree with which two infinitesimally separated orbits move away from or approach each other is measured by the Lyapunov exponent, and is calculated by the long-time average of the logarithm of the amplification (reduction) rate of the difference between the two orbits. Since the number of directions for the deviation of the two orbits is equal to the number of degrees of freedom in the phase space, the number of degrees for separation is equal to the dimension of the phase space. Thus the number of Lyapunov exponents is the same as the number of degrees of freedom. Chaos is often characterized by a system having at least one positive Lyapunov exponent. In other words, when an initial value is changed only slightly, a later state becomes very different. The system is said to have a sensitive dependence on initial conditions. In addition, when chaos is decomposed into Fourier modes, an uncountably infinite number of modes appears in the form of continuous spectra. This is different from quasi-periodic motions with several main frequencies which are indicated by orbits rotating on a torus in phase space (for example, a motion on a doughnut formed by two frequencies). In this case, the spectrum is not continuous but scattered and expressed by several basic frequencies and their superposition.

The set of chaotic orbits is not simply characterized by geometric objects such as points, one-dimensional lines or two-dimensional planes. For example, in the case of chaos as discovered by Lorenz, orbits are observed to exist on a transcendental structure consisting of an uncountably infinite number of finite two-dimensional domains. When a section of the structure is taken, the orbits resemble a set of curves. However, when one of the curves is enlarged, another set of many curves is found. When, in turn, one of these curves is enlarged, the same structure is revealed again. Indeed, the orbit has an infinitely recurring self-similar structure. This structure is characteristic in the sense that it is self-similar in the direction in which the phase volume decreases. From this fact we can see that the geometric nature of chaos is not characterized by ordinary integer-number dimensions but rather by self-similarity. Mathematically, this characterization is measured by the Hausdorff dimension. Since this dimension cannot be calculated practically, correlation dimensions or fractal dimensions are usually calculated instead. In this way, chaos is characterized as a geometric object having non-integer dimensions.

In some eras, chaos was considered an extremely rare phenomenon, but nowadays it is regarded as a universal phenomenon which is observed in many fields. Indeed, a definite rule exists for generating chaos in phase space. It is the iteration of a simple rule in which a bundle of orbits, which shrink in one direction and expand in the other, is folded by nonlinearity and returns to nearly its original location. Such an algorithm is best illustrated by our daily operation used in mixing flour for making rice cakes, bread, pie, or noodles, or in making syrup. In fact, chaos exists everywhere in our lives. We are tempted to imagine that if chaos is such an ordinary phenomenon, perhaps humans discovered chaos and defined the concept in ancient times. Actually, in many mythical stories chaos is described as a state in which heaven and earth are not divided (a state in which everything is mixed). Chaos is also described as an "energy body" responsible for the creation of heaven and earth. Furthermore, in some classics, chaos has a meaning very close to the modern concept. People in ancient Greece, China and Japan, too, had a concept of chaos by intuition. The following quotations from the classics demonstrate how chaos has appeared throughout history in different disciplines. The concept of chaos in the classical disciplines might provide us with some inspiration.

In the 4th century B.C. in ancient China, Zhuangzi[1] expressed a profound thought concerning chaos in a short story "Hun-tun (chaos)".

> "There was a king in each of the countries of South, North and Center. The king of South was "Shu", and the king of North was "Hu". The king of Center was "Hun-tun (chaos)". Shu and Hu were invited and entertained with much food by the king of Center. Shu and Hu wished to thank Hun-tun, and they noticed that Hun-tun's face was different from the faces of others. His face did not have the seven holes present in the faces of others; it did not have two holes to see, two holes to hear, two holes to breathe, and one hole to eat. They opened one hole a day in Hun-tun's face to express their gratitude. On the seventh day, Hun-tun died."

In this story, the names "Shu" and "Hu" signify human shrewdness or quickness, while "Hun-tun" means natural naivety. It can be interpreted to mean that if we human beings put human sense to chaos without seeing itself, or analyse chaos by human techniques, the essence of chaos is lost. As a matter of fact, even today when the study of chaos has progressed considerably, the definition of chaos itself has not yet been perfected. More importantly, we often have the feeling that we have yet to comprehend chaos fully even after thorough analysis; several indices which characterize chaos exist, but we may still not know the answer. We have to recognize that chaos is a complex phenomenon which cannot be described in one word. Chaos contains something which inherently denies comprehension by analysis. Chaos might be understood by being "watched" or "composed," but not by being analysed. The meaning of "watch" and "compose" will be defined later in this book.

Although chaos is clearly an object of natural scientific study, we tend to believe that it may not be possible to capture its essence completely with the help of the current analytical methods. To define the essence of chaos clearly, it also seems to be necessary to employ an engineering method where one tries to do something that exploits chaos or uses chaos for some computation by regarding chaos as an "information processor". We consider that in this manner we may be able to open up "the potential capacity of chaos".

Chaos contains infinitely many periodic solutions whose periods are characterized by all the natural numbers. Being unstable, those periodic solutions

---

[1] Chuang Tzu is another pronunciation.

are not observable by themselves. This instability of orbits, as mentioned earlier, generates a sensitive dependence on initial conditions and can be measured by the Lyapunov exponent. A philosopher in 4th century B.C. Greece, Epicurus of Samos, understood this nature to be the essence of motion of all things in the universe. While his name is famous as the origin of Epicurian, he is of interest to us as an atomic theorist. His atomic theory is as follows.

Epicurus introduced attributes of size and weight to atoms. He supposed all atoms fall downward and have the same velocity regardless of their attributes. Therefore, if left as they are, atoms do not collide with each other at all, and nothing is generated. To make this concept of "generation" realistic, he introduced the notion "Pareqklisis" that atoms have an attribute of "deviation". In other words, he stated that atoms have a tendency to deviate slightly from their original orbit. Because of this "deviation", the collision of atoms occurs and all things in the universe are generated.

Of course, this atomic theory is incorrect. However, he seems to have guessed correctly about the orbital instability of chaos. In chaos, infinitely many "symbol sequences" are generated by the orbital instability, which leads to an aspect of chaos as an information source.

Then, is it possible to regard chaos as an information channel? A pioneering philosophy on this issue appeared in a theory of Anaxagoras of Ionia in the 5th century B.C. Anaxagoras is famous for giving school pupils for the first time the concept of a "summer holiday" by his own death. For Anaxagoras, all things in the universe have self-similarity. Imagine a stone. According to the theory of Anaxagoras, this stone contains the characteristics of all substances. It happens to exist as a stone only because the stone characteristic is dominant. This kind of thought seems to exist in Buddhism, particularly in the Zen sect. When autonomous motion, begun by this self-similarity, starts to express various characteristics in nature dynamically, this body in motion was recognized to be a life form. Conversely, the pure state which differs from this mixed state was recognized to be the spirit (mind). Such a world-view seems to be the same as the one based on chaos. As will be discussed later, according to the studies by Matsumoto and Tsuda, chaos, which is regarded as an information channel, has characteristics in which information mixes in the binary space of state variables and self-similar structures appear in the information. This characteristic establishes sufficient conditions correctly to transmit arbitrary external information and is common to chaos discovered so far in living systems. In addition, when reversible chaotic dynamics is developed in the positive direction of time, a mixed state is generated. Conversely, when that state is developed in the negative direction of time, a pure state is regained. Thus the idea of Anaxagoras may be scientifically realized.

1.3 Chaos and Complexity

The first problem faced in the study of complex systems is answering the question "what is complexity?". Possible definitions of complexity are being proposed in the fields of brain science, computer science, and chaos theory. Here, we briefly discuss the complexity of chaos in connection with algorithmic information theory.

In algorithmic information theory [Kolmogorov 1965; Chaitin 1987a, b], chaos appears as an algorithm with the largest complexity. The complexity of an algorithm is given by the length of the smallest program as measured in bits which is necessary for the calculation of an object. For example, if we write a program to generate a periodic sequence of numbers $01011010110101101011\cdots$, we write

$$\text{FOR } I := 1 \text{ TO } M \text{ DO WRITE('01011')}.$$

The binary expression of the integer M needs the largest number of bits in this program. Therefore, the complexity of the algorithm is proportional to $\log_2 M$. Although the program mentioned above is not necessarily the shortest, it gives the order of the complexity of the periodic sequence of numbers. Thus, the complexity of a program which produces a periodic sequence of numbers is provided by the order of the logarithm of the output digits, and thus the program is said to be compressed. On the other hand, the shortest program which produces a random, nonperiodic sequence of numbers such as $0100010110101111100101\cdots$ cannot be compressed in the above sense. To have such output we need a program as

$$\text{WRITE('0100010110101111100101}\cdots\text{')}.$$

Then the complexity of the algorithm is provided by the order of N. If a decimal point is put at the left of this sequence of numbers, the sequence corresponds to a binary expansion of an irrational number between 0 and 1. On the other hand, a periodic sequence corresponds to a binary expansion of a rational number. For the binary expansions of irrational numbers such as $\sqrt{2}$ or π, the program that reconstructs such a number can be compressed since an algorithm exists. Yet algorithms do not exist for all irrational numbers. Rather, the existence of a compressed algorithm is rare.

The incompleteness theorem of Gödel [Nagel and Newman 1964] can be expressed as "all the theorems on natural numbers cannot be necessarily proven by finite axioms and inference rules". In addition, this theorem inspired theories of computability after being transformed into a proposition of computation by Turing. The proposition "a general algorithm which decides whether an arbitrary computer program makes the computer stop does not exist" was proven. Concerning this, Chaitin [1987a, b] obtained an interesting result by using an algorithmic information theory. He proved that when the k-th bit in the binary expression of the probability Q of a "completely random program" halting is 1, an infinite number of solutions exists for the Diophantine equation corresponding to that bit, and that when the k-th bit is 0, the corresponding Diophantine equation has only a finite number of solutions. A "completely random program of n bits" means the "program of tossing a coin n times". Since Q is algorithmically random, the Diophantine equation corresponding to a different k value can be logically independent. That is, even if the nature of the solution for an equation of a specific k value is known, the information concerning the solution of an equation of another k value is not acquired. In other words, whether the number of solutions for the Diophantine equation is infinite or finite cannot be proven. Thus, in algorithmic information theory, incompleteness, indeterminacy, and randomness are formulated as equivalent concepts. Furthermore, according to algorithmic information theory, what is algorithmically random can be called chaos: as chaos can be replaced by an initial condition of infinite digits and evolution rules of finite digits in the framework of algorithmic information theory, it can be regarded to be algorithmically random. Therefore, chaos, incompleteness, and indeterminacy are conceptually equivalent. Would this also be a good description of chaos in the natural sciences?

While algorithmically chaos and randomness are certainly equivalent, in physics the complexity of chaos should not be described only by randomness. Let us think of statistical descriptions. When a random variable X and its steady distribution $P(X)$ are given, the first moment is defined as the average of X by $P(X)$, and the second moment is defined as the average of X^2. Generally, an average of X^n is called the n-th moment. For X, $Y(s) = \langle \exp(isX) \rangle$ is a characteristic function, and if all n-th moments are finite, $Y(s)$ can be written in the form of a moment expansion. Here, $\langle \bullet \rangle$ is an average of $\bullet$. If $P(X)$ approaches 0 fast enough with $X \to \infty$ as in a normal distribution (Gaussian distribution), its moment expansion is possible, but in the case of Cauchy's distribution $P(X) \propto 1/(1 + X^2)$, in which the higher moments diverge, or in other cases where a higher moment is larger than the lower ones, the moment expansion is impossible. In fact, in most cases of chaos, since the moment expansion based on a steady distribution $P(X)$ requires complete information on chaotic orbits, moment expansion seems to be essentially impossible. Taking this kind of statistical description into account too, the description of chaos is complex.

From the descriptions mentioned above, we find that chaos is complex, not simply random. Do we know what the most random phenomenon in the universe is? Entropy measures randomness. Since entropy is generally dependent on the base, it may be meaningless to search for the most random phenomenon in absolute terms. Thus, we are able to discuss the degree of randomness relative to the base, if the base for the way of observing the system is defined. In the same sense, we may be able to ask whether we can comprehend the most complex phenomenon or pattern in the universe. To ask that, we must (re-)define complexity.

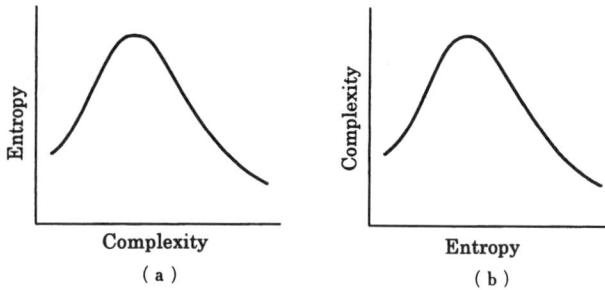

Fig. 1.1a,b. Schematic figure of the relationship between entropy and complexity. (a) Entropy as a function of complexity, (b) complexity as a function of entropy.

Some researchers have proposed new definitions for complexity. Grassberger [1986], Crutchfield and Young [1989], Crutchfield [1994], and others attempted to draw out the structural complexity, and Lloyd, Gell'man [Lloyd and Pagel 1988] and others attempted to make precise the complexity of procedures for problem solving. Ikegami and Tsuda [1992] introduced the viewpoint of "observer", and by formalizing the observer, they defined the complexity relative to the complexity of an observer. In addition, they performed Monte Carlo simulations, calculated the relative complexity for several patterns, and compared their results with the entropy of the patterns. Here entropy is given by a function of complexity (Fig. 1.1a), in contrast with the complexity as a function of entropy by Crutchfield [1994] and Crutchfield and Young [1989]. According to Crutchfield, complexity can be expressed as a function of entropy as shown in Fig. 1.1b. This is a justification for the feeling that a highly entropic state is random and not so complex, and a low entropic state is not complex because of its regularity. However, we believe that this does not necessarily provide true definitions of complexity. Those functional relations only demonstrate that complexity is measured by the scale of entropy so that entropy is the first quantity, and complexity is the second quantity. In contrast, we think that complexity is the first quantity and entropy is in turn dependent on complexity. If we believe that the degree of inconsistency of knowledge and the basic patterns which observers possess is the complexity of the object, from the viewpoint that absolute complexity does not exist and only observers define relative complexity, then entropy itself must be a function of complexity. The result of the calculations by Ikegami and Tsuda is shown in Fig. 1.1a. It shows that the complexity of a highly entropic random state is intermediate, and the state is not noncomplex at all. Extreme complexity and extreme simplicity both exist in the low-entropy state, that is, in a state of regularity. It cannot always be said to be simple even if a system is regular. In chaos, a state with a low degree of randomness shows diverse behavior, which appears highly complex to us with our present level of knowledge. When such chaos appears in the network of agent systems, it appears to exhibit extremely complex behavior for each agent.

Nevertheless, perhaps definitions of complexity will remain incomplete up to the final stage of the study of complex systems.

1.4 How Has Chaos Changed Our Way of Thinking?

With the development of the study of chaos, a new viewpoint of dynamic complex systems has been introduced [Kaneko and Tsuda 1994]. We discuss this new idea based on the two-sidedness of chaos.

1.4.1 Dialectic Method to Overcome the Antithesis Between Determinism and Nondeterminism or Between Programs and Errors

Chaos is formulated by a deterministic equation, but it destroys the basis of the deterministic framework itself. In other words, chaos appears in deterministic dynamics, such as differential and difference equations, and ultimately describes stochastic behaviors. Let us repeat the fact that even a small difference in initial conditions can be amplified into a large difference. Since we are not able to observe a phenomenon with infinite accuracy, we are forced to introduce probability, based on determinism. Thus, chaos provides us with a dialectic method to reconcile the antithesis between determinism and nondeterminism at a higher level of conceptual understanding.

Though this matter has already appeared many times in chaos textbooks, we would like to emphasize here that it is critically significant for the life sciences and engineering where the opposition between the programmed and the unprogammed is often important.

We often encounter the opposition between the standpoint of carrying out programming and the standpoint of taking into account external influences that are impossible to control. A typical example is the antithesis between the standpoints of von Neumann [1956, 1966] and Wiener [1948]. For designing a computer, von Neumann considered a mechanism by which, even if somehow an error had entered an element, a combination of elements would make correct computing possible. In this approach, the concept of error is certainly introduced, but the issue is discussed only in the context of a method to erase errors. There does not seem to be an inclination for coexisting with errors that are inevitable in the actual world. In contrast, Wiener's viewpoint acknowledges the notion of coexistence with noise through feedback.

However, in chaos, "noise" arises from the programmed world. With chaos, an antithesis between the situations is reconciled at a higher dimension. It seems, generally, to be unified to the concept "recipe", which has more flexibility than a written program. A recipe that indicates how to cook is not a complete set of rules for the preparation of that dish. In a recipe, only the minimum information about the ingredients and the procedure is written; depending on the cook and the situation, the exact same dish will not be prepared. But if we follow a recipe, we can cook a specific dish. This demonstrates the necessity of using a recipe for material which cannot completely be controlled, in contrast with assigning a program which completely controls computing and behavior.

This antithesis between programs and errors has continued to influence the study of artificial intelligence and artificial life. In life sciences, this antithesis between programs and errors is widely seen in cell differentiation, aging problems, as well as in neuroscience. Within aging problems, for example, two opposing viewpoints exist; one is that aging is genetically preprogrammed, and the other is that aging is caused by errors accumulated through reproduction [Takagi 1993, 1999].

Chaos may provide a breakthrough for those problems which are still in a stage of opposition. We will see such examples in the rest of this book. We will see that chaos provides a recipe to overcome the traditional conflict between programs and errors and that maintenance of a chaotic state will be relevant to resolve the conflict.

1.4.2 Dialectic Method to Overcome the Antithesis Between Order and Randomness

As discussed in Sect. 1.3, chaos appears complex because it is neither utterly irregular nor completely regular. The word "chaos" might be associated with a total mess, utterly devoid of order. Nevertheless, chaos, as used currently in the natural sciences, is significant because it has broken the traditional dichotomy of order and randomness. Order has been found in chaos, and also, by viewing chaos as random, a statistical mechanical approach has been successful. Chaos, with these two opposite aspects within itself, has an extremely complex structure. In such a transcendental structure, the antithesis between order and randomness is reconciled at a higher level, and the concept of complexity is generated.

1.4.3 Beyond the Antithesis Between Reductionism and Holism

Previously we discussed how chaos challenges reductionism. In chaotic systems, since any tiny difference is amplified to a macroscopic order, tracing one orbit precisely requires infinite precision. In addition, when chaotic motion is decomposed into modes, a continuous spectrum emerges which tells us that chaos contains infinite information. To understand chaos, consideration of the whole description is required. However, it should be noted that this does not necessarily mean holism. Chaos itself, for example, also emerges in a simple differential equation with only three variables and one parameter. In this sense, it is derived from an extremely simple and reduced system at the level of a model. Hence chaos renders the simple dichotomy of reductionism versus holism invalid. It seems to suggest that it is forced to advance to a 'dynamic many-to-many relation'.

This seems to be related to what many people feel when they attempt to quantify chaos using the Lyapunov exponent, dimensions and information content. When we try quantitatively to characterize chaos, we feel that the observed information gained from an orbit is not completely described by such quantifiers. Here we must note that such characteristic quantifiers are measured statistically. This again leads to the antithesis between the deterministic character of one orbit and the statistical character, such as a probability density, of the whole.

1.5 Dynamic Many-to-Many Relations and Bio-networks

1.5.1 The Necessity of Dynamic Many-to-Many Relations

One of the characteristics of chaos is that a tiny difference is amplified into a macroscopic level. Let us imagine a dynamics in which many elements are interrelated. If there is chaos, a tiny change brought about to one element will be amplified and transmitted to other elements. As a result, a weak chain of cause and effect can bring about a large difference. This means that it is difficult to consider each element separately. Then the conventional way of thinking in physics – "to reduce the problem to a superposition of one-body problems and stochastic fluctuations around them" – is no longer possible. Thus, we need to understand situations in which many elements have strong relations which change dynamically (see also Shimizu [1992] and Aizawa [1994] who have similar ideas on this issue).

From what is mentioned above, the situation illustrated by the Japanese proverb "when wind blows, tub shops will earn" emerges. This proverb illustrates how a chain of events with small probabilities generates a large result.

> "When the wind blows, dust gets in one's eyes, then the number of blind people increases; they live by playing the samisen (a three-stringed Japanese banjo), then many cats are captured as their skin is used to make the samisen, then the rat population increases, then increased numbers of rats bite wooden tubs, therefore, the demand for tubs increases."

What we must notice is that the proverb illustrates only one projected line in the flow of events in the world, and that we are able to think of many other possible chains of events, which might result in completely different outcomes. In some cases, different routes of cause and effect may lead to the same conclusion that "tub shops will earn". The proverb "when wind blows, tub shops will earn" has implications for all the possible chains.

Such situations are illustrated in the coupled map lattices (CMLs) and in the networks of chaotic elements to be discussed in Chaps. 3 and 4. In CMLs which consist of chaotic elements, when a small difference (or input) is introduced at one point in space, this difference may become larger and be transmitted to a distant point. There, in some cases, a regular pattern might be formed, but that pattern may be broken by chaos and another pattern is then created. This process can continue indefinitely. In a network of chaotic elements, clusters which oscillate synchronously are often formed and the relations between the elements are formed according to the degree of synchronization. Such relations will change depending on the initial conditions and with time. In some cases, when an input is applied to one element, it greatly changes the relation of the synchronization among the other elements.

This shift to a new view of a "many-to-many" relation provides an understanding of how strong dynamic relations among many elements or modes are formed. Examples of "many-to-many" relations are seen in diverse areas. In chaos, even for systems with a small number of degrees of freedom, when we treat a problem with the conventional notion of modes (for example, the Fourier mode), even if it is expressed as a differential equation with three variables, an infinite number of modes is necessary (this is why a continuous spectrum is observed), and a strong relationship among the modes emerges. This change of viewpoint is becoming important in areas such as traditional solid-state physics, atomic and molecular physics, optics, hydrodynamics, geophysics and others. However, the most profound influence can be found in the study of biological systems, through investigating the network of living organisms, by viewing the process of changing relations as a dynamically complex system. Generally speaking, when a network with many elements is formed in a system that is dominated by nonlinear dynamics through the interaction between elements, a complex dynamics emerges in the whole network, in each element, and in the relationship among elements. We will attempt to understand such successive changes of relations in an integrated manner. Let us discuss briefly several examples of biological networks which should be studied in terms of dynamics.

At first, let us recall that biologists often tend to search for a one-to-one correspondence even in the problem of a network. The "grandmother-cell" hypothesis in neurophysiology may be a typical example, where it is believed that there is a nerve cell which acts according to the object of recognition. In an extreme case, one nerve cell is thought to correspond to one element of a recognized object. However, nowadays, not many scientists believe in such a primitive one-to-one correspondence. Still, many physiologists believe in the correspondence in the looser sense that a certain assembly of nerve cells corresponds to a certain object. The same belief for correspondence is seen in the hypothesis of one antigen-one antibody in immunology, that of one niche-one species in ecology, and that of one gene-one enzyme in biochemistry. However, over the past several years, such a one-to-one-correspondence viewpoint has increasingly been questioned.

1.5.2 Metabolic Systems, Differentiation, and Development

Figure 1.2 shows a schematic representation of only a tiny part of a huge network of chemical reactions in the metabolic system. (Perhaps in the future, after the human genome project is completed, a more complicated map of the genetic network will be drawn.) Then, how can one understand such a reaction network which is intertwined in a complicated way over a large scale? Biochemists have been examining each cycle one by one. However, they still cannot answer why such a complex network functions well as a whole; whether all of those cycles are necessary; whether it functions well when some part of the network is removed; whether there is one vital part without which the whole does not function; how these networks have evolved or been generated, etc.

When trying to answer such questions, one often tends to reduce the problems so that one can discuss them in terms of some established concepts like percolation in a network, chaotic oscillations, or entrainment in each reaction system. However, such concepts cannot provide sufficient tools to answer the essential parts of the questions. We still lack some concepts to tackle these essential problems.

The problems of interactions between cells which carry out metabolic reactions consist of (1) metabolic reactions involving various chemical substances in each cell, (2) the exchange of chemical substances between cells and the medium leading to cell–cell interactions. Here the chain of cause and

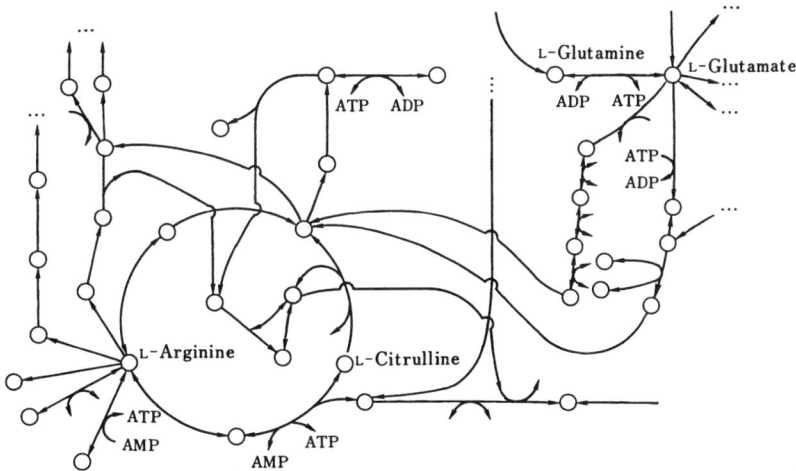

Fig. 1.2. Schematic representation of a metabolic network. Indeed this is only a part of a huge network diagram that requires a few dozen pages.

effect can be seen in the influence of some chemical substance to other cells when the concentration of such a chemical substance is increased slightly. While the effects are described by the change of the chemical concentrations due to the metabolic reactions within each cell, a small difference of chemical concentrations in a cell is transmitted to other cells through the medium, and, consequently, an amplified inter-cellular chain reaction of cause and effect follows. Yet, since the states of the cells change with time, relations among chemical substances change dynamically. If a small difference in the concentration of a chemical component within a cell is amplified by chaos and transmitted to chemical components of other cells, strong interrelations between unexpected chemical reaction parts may appear, and this relation changes dynamically. It is obvious that this kind of problem is commonly seen in the biological networks discussed above. Therefore, a theory which expresses such "relational dynamics" is strongly needed.

Since each cell carries out metabolic reactions, the above problem is related to how cell differentiation occurs and to how morphological organization occurs in a multicellular organism. Differentiation and development are important issues which contain theoretical problems that cannot be solved solely in laboratories. As for differentiation, Kauffman, in his pioneering work [1969, 1993] introduced a model of random Boolean networks with on/off circuits to represent a control network of "on/off switch" genes. In his model, a pattern of 0 and 1 is initially given which then evolves in a random Boolean network to an asymptotic pattern. He demonstrated that in this network many stable steady states coexist which correspond to the types of differentiated cells. In the sense that many possible equilibrium states were taken into account, this study was ten years ahead of a similar study of spin glasses in statistical mechanics. In that sense, Kauffman's study is significant, but it is not clear yet whether such an "equilibrium" viewpoint is appropriate for understanding the problem of cell differentiation.

In differentiation, there are many issues where temporal dynamics is essential, including a division cycle, a metabolic chemical oscillation, and so forth. In a metabolic system, a network consists of many intertwined reaction systems as already discussed. In addition, since a reaction is catalysed by an enzyme, chemical concentrations can show nonlinear oscillations, including chaos, through strong nonlinearity by autocatalytic reactions. Thus, an assembly of cells becomes an interacting system of many dynamically complex networks.

1.5.3 Ecosystems

In an ecosystem, it is recognized that the viewpoint of one niche-one species cannot be well established if interactions between species are taken into account. In an ecosystem, many species form a part of a complex network in which they interact with each other; symbiosis, competition and parasitism are all examples of interactions. Elton [1966] presented data from his observations of forests in England which show that the more complex an ecosystem is, the more stable it appears against an invasion from an outside enemy. On the other hand, May [1973] showed that, for random networks, the larger the number of interacting species is, the more unstable the network is. Generally speaking, it may be difficult simply to correlate stability and diversity. Nonetheless, we must study ecosystems, recognizing the fact that the network which we now see is special in the sense that evolution has made it more complex.

As an example, the formation of a "vital point" may be a key to stability. In an ecosystem, it is often said that a "vital point" called a keystone species exists, which, if taken away, causes a whole ecosystem to break down. Although one might assume that this species is dominant, this is not necessarily the case. It is also possible that some damage to a species that looks unimportant at first glance may lead to the breakdown of the entire ecosystem. We must consider how such a "vital point" is formed.

Tropical rain forests may provide a typical example of a complex network of various species within an ecosystem. For example, by the observation [Wilson 1992] that there are 163 species of beetle in one tree, it is almost impossible to believe that one species occupies one niche obeying a one-to-one correspondence. Also, some researchers have found that a population of species is far from equilibrium in rain forests [Kikkawa 1990; Connel 1978]. In this way, an ecosystem is considered a typical example of a dynamically complex network.

1.5.4 Immune Systems

Jerne [1973, 1974] presented the hypothesis that a network similar to an ecosystem also functions in immune systems. According to his hypothesis, a network similar to a food chain in an ecosystem is formed in an immune system, in which an antibody reacting to an antigen is in turn reacted on by another antibody, which is reacted on by another one, and so forth. In this hypothesis, memories of diseases are stored in the network. This opposes the theory that one antibody responds to a single antigen, and that immunity is determined only by the available concentration of antibodies. A model of such a network was proposed by Farmer et al. [1986], where the populations of a large number of different kinds of antibodies change by an interaction of the "prey-predator" type according to some matching conditions.

Since Jerne's hypothesis, the functions of T-cells, B-cells, and others were discovered in an immune system and consequently it turned out that his network hypothesis could not be used in its original form. Because of this, one may feel that one should abandon such an immune-network viewpoint. This is not true. Although the components of the networks are more diverse than in the original picture by Jerne, the network picture itself is still necessary. We need to revise his immune-network theory to include the diversification and differentiation of cells, and to include the interactions of T-cells and B-cells.

When the concentrations of the antibodies reach a state of near equilibrium, the static aspects of the network could be the main concern of a study. In fact, the model by Farmer *et al.* is related to systems with many meta-stable states as found in spin glasses (see Sect. 1.5.6). In an actual immune system, however, the concentrations of antibodies show various temporal structures, and therefore a logic for understanding dynamic networks is necessary.

1.5.5 The Brain

A similar trend to a dynamic viewpoint can be seen in the brain sciences. For example, as mentioned above, the grandmother-cell hypothesis involves a typical one-to-one correspondence which asserts that one neuron recognizes one object. In contrast, recent studies of neural networks suggest that the firing pattern of a group of neurons corresponds to a certain state, and that memory is allocated to the connection matrix. This case, however, provides only a static picture like the one found in spin glasses (which will be discussed in the following section). On the other hand, theories which treat the dynamic relations of neuron firings have appeared recently [Freeman and Skarda 1985]. Tsuda [1984, 1990a, b, 1991a, b] proposed a dynamic view of the brain, based upon an interpretation of the dynamic activities of neuron assemblies with the help of the concepts of chaotic dynamical systems. Aertsen *et al.* [1992, 1994] studied whether coincidences among neuron spikes change with time, depending on motivated tasks. Also in the study of the visual cortex by Singer, Eckhorn, and others [Gray *et al.* 1989; Eckhorn *et al.* 1988], a dynamic change of a correlation was found, that is, the electric potential (activity) of a neuron assembly shows an approximately 40 Hz oscillation (this is not an exactly periodic oscillation), and the degree of the phase synchronization of the rhythm increases in the parts of simultaneous input. Those data are in accordance with the dynamic viewpoint of the brain as proposed by Tsuda, which can be summarized as follows: a neuron and a neuron assembly are not structured to reveal a single function, but structured such that they can implement multiple functions according to the internal states of the brain and the external environment. Furthermore, these activities should reveal temporally complex behavior which perhaps is related to the chaotic itinerancy that will be mentioned later. Thus, the study of the relational dynamics among the elements involved in the information processing of the brain will be an important issue.

1.5.6 Rugged Landscapes and Their Problems

The static many-to-many relation is often expressed by the term "rugged landscape". Let us consider, for instance, the correspondence of phenotype and genotype. In general, the process of transcribing the information in the DNA to a phenotype makes this correspondence rather complicated.

Here, as an example, let us consider Kauffman's model, which regards a network of on/off switches of genes to be random. In this network, the final state is determined by the initial pattern of 0 and 1, but, even if the initial pattern is only changed a little bit, the final state does not always change smoothly in accordance with the initial change. Generally, a variety of possible final states exists, and there are various sizes of domains (basin sizes) of initial states which lead to each final state. In this model, some function given by the meta-stable states corresponds to the phenotype, and each bit of 0 and 1 in the initial state corresponds to the genotype. Therefore, many 'genes' lead to a single phenotype.

Originally developed for the study of magnetism with impurities, spin glasses provided some basic models for these problems [Mezard *et al.* 1987]. The Sherrington–Kirkpatrick (SK) model, which is the mean-field model of an abstract random magnetic interaction, now serves as a basic model that provides a rugged landscape. This model is given by an energy function with the variable S_i taking the value $+1$ or -1:

$$H = -\sum_{i,j\neq i} J_{i,j} S_i S_j. \tag{1.1}$$

Here, the summation (i, j) is taken over all pairs of elements, and $J_{i,j}$ is randomly fixed to take the plus or minus sign.

If all $J_{i,j}$ are positive, clearly the state of all S_i being 1 or -1 has the least energy. On the other hand, if the term $J_{i,j}$ contains both positive and negative signs, one cannot easily determine which pattern of S_i yields the lowest energy. If turning over the sign of any single S_i raises the energy, that state is regarded as being a metastable state, but it is not evident what kind of allocation of spins is necessary to reach a metastable state. Recent studies of statistical mechanics show that in such a Hamiltonian system, a phase which is called a spin-glass state exists at low temperatures. In a spin-glass phase, many metastable states exist (the number is proportional to $\exp(N)$, where N is the number of spins), and yet many low energy states also exist, whose energy levels are of the same order as the lowest energy.

When the dynamics of spin-glass states at low temperatures are studied, the evolutionary rule "if energy decreases, spins flip-flop with some ratio" is used. Here, starting from some initial condition, the state reaches one of the metastable states described above. In the case of a finite temperature, the flip-flop event of a spin to *increase* the energy can also occur with some probability that depends on the temperature. Then the realized metastable states, as temperature increases, fuse in turn. Thus, many metastable states can be expressed by a branched tree. The energy function has a rugged landscape with respect to the configuration space of spin values, as schematically shown in Fig. 1.3. The thermodynamics and the metastable states in those systems have been analysed theoretically.

Fig. 1.3. Schematic representation of a rugged landscape.

A similar viewpoint is often used in a model of biological evolution. In the study of evolution, the series of S_i are regarded as genes and the energy function above is assumed to give the degree of fitness of the phenotype given by the genotype. Then the evolution according to the fitness is studied following the mutations given by the flip-flop of the 'bit' genes. In addition, it is known that the on/off circuit which Kauffman introduced also has characteristics similar to those of spin glasses.

However, in this viewpoint, the landscape of the degree of fitness is given by the energy function, and each element is adapted independently to the energy minimum, or maximum fitness state. There are several tacit premises here, such as: (1) the energy function or the fitness function to minimize or maximize is given in advance, and (2) the interactions among the elements are only secondary effects on the dynamics of each element. These assumptions are not valid when the interaction among the elements is strong, or when an oscillatory dynamics, rather than the relaxation type to a minimum, exists. For example, even if the degree of fitness is determined by the external environment, with an increase of the population, the competition among each single species gets harder. Then, the species can no longer be regarded as static. Also, even among the same species, differentiation to different types of behaviors can occur, which invalidates the assignment of a single phenotype to each species. Furthermore, if an interaction of "prey–predator type" exists, the population shows an oscillation, and consequently the above assumption cannot be made. The discovery of complex oscillations in the brain and in metabolic systems, as well as other observations, invalidate the viewpoint of the landscape theory.

If two separate time scales exist, the dynamic behavior could in principle be studied based on a static landscape and a dynamic modulation to it. In some cases, such a separation might be possible, but with strong interactions the separation will in general be impossible.

1.5.7 Conclusion

Some conventional network concepts adopted in life sciences are summarized in Table 1.1. In this book, we propose to deal with these systems in the terms of dynamic many-to-many relations. The globally coupled chaotic systems discussed in Chap. 4 provide a basic and abstract framework for such problems. We will show that clustering, chaotic itinerancy, hidden coherence,

Table 1.1. Conventional viewpoints of biological networks.

Viewpoint	1-to-1 map	Many-to-many relation-ship at a "static" level
1. brain science	grandmother cell	equilibrium neural-network model
2. ecology	correspondence of niche and species	random network
3. immune system	correspondence of antigen and antibody	Jerne's idiotypic network
4. differentiation	correspondence of geno-type and phenotype	Kauffman's network
5. abstract model	–	spin glass

homeochaos and so forth will provide us with fundamental concepts needed to understand these network systems.

The dynamics of relationships, discussed in this chapter, are exemplified in the traditional Japanese "renku". In a renku ceremony, many poets participate; the first poet composes his or her short poem which is read, interpreted, and subsequently continued by the next poet. This creative process continues until all have contributed. Since the interpretation of the second poet is different from the thoughts of the previous one, misunderstandings or excess interpretation can occur and be amplified one after another. Through the networks of misunderstandings, the total poem emerges with its new meanings. As Terada [1946] wrote, it often happens that some sentences, used before in the composition of a renku, will be related to other far later sentences. In this manner, in a sense, renku captures a significant aspect in the study of complex systems; that is, a new collective level emerges through interactions between autonomous elements.

1.6 The Construction of an Artificial (Virtual) World

As will be discussed in Chap. 2, a nonlinear system is not always structurally stable, namely, a small change in a model may result in a large change of its behavior. In chaotic dynamical systems, this typically appears as a sensitive dependence on parameters or a sensitive dependence on the number of variables in the model. Furthermore, the behavior of a model may be sensitively dependent on the computational method or the way of observation, a characteristic which Tsuda proposed to call "descriptive instability" (see Chap. 2 for details). The use of the term "descriptive instability" calls for a change in the way how natural phenomena are modeled. In the traditional approach of modeling, one "starts from the most reliable microscopic equations to derive macroscopic equations with the help of appropriate coarse-graining, and then constructs a model using a suitable approximation". If a model is quantitatively unstable with regard to the description, however, the traditional approach is rendered invalid, since the belief that the closer a model is to a "true model", the better it must be quantitatively, is not justified. Even though in such a case a model might nevertheless yield a qualitatively correct result as far as the numerical output is concerned, at the level of "qualitative description", a more powerful method of modeling, that is, a method to create an artificial (virtual) world constructively, needs to be developed.

Examples are the coupled map lattices (CMLs) developed by Kaneko [1984b, 1990a]. Coupled map lattices, which consist of several basic procedures such as chaos in each element, diffusive interaction, and convection, are, unlike in the conventional approach, not derived from basic equations as will be explained in Chap. 3. However, they are capable of describing new phenomenological classes of behavior in complex dynamics, and qualitatively have provided much insight into various complex phenomena. A CML is not a model of a specific system, but is constructed in order to extract qualitative universalities common to various complex phenomena. Because of its descriptive power, a CML can be used as a fairly general constructive model for complex systems, not restricted to a single specific field. In addition, it must be noted that, in complex systems, there is not always a clear quantitative "distance" between a model and the phenomenon it models, due to the descriptive instability. Hence, a descriptive model based on detailed knowledge of a phenomenon does not always fairly and qualitatively approximate the phenomenon. Even if, by chance, a model with a "better" approximation is obtained, there is no method to ascertain whether there is a strict correspondence between the model and the phenomenon.

Constructive models such as CMLs have been criticized for their failure quantitatively to predict experimental measurement values. This, however, is not the main goal of constructive models. The constructive approach is essential to reveal a universal class to which phenomena belong, and thus to grasp the essential factors that underlie complex phenomena. We believe that only through these methods, we will understand *why* certain classes of complex phenomena commonly exist at various levels of nature, and be able to predict what classes of systems can exhibit what type of phenomena.

The approach of constructing a virtual world seems to be indispensable for understanding phenomena such as evolution, which have a historical factor. It has often been much discussed that phenomena which follow a certain historical path cannot be studied by conventional sciences, since specific phenomena which occurred only once in a specific historical path cannot be reproduced, while the conventional sciences are based on reproducibility. In contrast, the construction of an artificial world makes it possible to observe which paths may exist, and thus to provide us with a new way for studying evolution or developmental processes in our cognition.

To summarize, there seem to be characteristics to the understanding of complex systems. A single model which is constructed closely to resemble a phenomenon often does not have universal descriptive powers. In contrast, a model which is moderately remote from reality could have great descriptive power. One may much better *understand* the actual phenomena by studying a virtual world constructed in computers. This fact reminds us of a primary characteristic of a *story*, as often phrased that 'fiction can often tell the truth better'. Why human beings create fictitious stories is interesting as a topical question in the history of evolution. One possible answer is that story telling might be a methodology which human beings have naturally acquired for understanding complex phenomena.

The key question that arises in this 'constructive' approach lies in the relationship between the virtual world and reality. The virtual world should not just be an imitation of reality, but be a sort of abstraction from reality, and be constructed from our side by utilizing some abstracted essential features of reality. Understanding the relationship between the virtual world and reality is a fundamental issue in the study of complex systems with a constructive approach. Obviously, a virtual world must have some interface with reality. To locate such an interface connecting reality and virtuality, we need to have some sort of theory. In fact, in Japanese literature, Monzaemon Chikamatsu (17C), a scenario writer for Kabuki, proposed the theory that the play exists as an interface between fiction and reality.

Let us quote another story from Zhuangzi [4th century B.C.] which relates to our understanding through the construction of a virtual world. The title of this story is "The Dream of a Butterfly".

Once upon a time, Zhuangzi dreamed that he became a butterfly. He felt at ease and very comfortable, so much so that he did not recognize that he was himself. When he woke up, there was Zhuangzi. He did not know whether he became a butterfly in a dream, or whether he dreamed that a butterfly became him. Perhaps therein lies a possible distinction between Zhuangzi and a butterfly. This kind of transformation is called the change of all things.

This story has various implications. First, there is the question of what reality is. The boundary between the real and virtual worlds is not so evident if we think that the entire real world is perceived only by reactions in the brain. Computer science has made virtual reality "real". It is rather natural to think that the real world is also a virtual one, once it is perceived through the human brain. On the other hand, to calculate chaos using computers is impossible, in principle. As will be mentioned in Chap. 2, one characteristic of chaos is that it has uncountably many nonperiodic solutions and countably but infinitely many periodic solutions. Since a computer is a finite automaton, it cannot compute infinity. The result of the computations becomes, after all, a periodic solution (recall Poincaré's recurrence theorem, although the period can be very long indeed). Furthermore, it is known that in many chaotic systems, an approximation of the individual orbits is impossible to obtain. That is to say, what we obtain from computations or experiments is not real chaos but a shadow of chaos. If we can approach chaos only through its shadow, what in chaos do we regard to be real?

Second, this story also implies the concept of "relativity". Although Zhuangzi and the butterfly are two entities which should be distinguishable, the transformation has no causation. The distinction between Zhuangzi and the butterfly is not absolute but relative; change in all things is like this. Since chaos contains all ordered states, one can extract various states of order from chaos depending on one's viewpoint. How chaos is viewed depends on our way of control. If chaos functions in our brains, our recognition of the world depends upon the state of chaos. Our way of control then depends on chaos.

According to Zhuangzi, there is no 'absolute' causal relationship in the change of things, independent of observation. This seems to be correctly applicable to chaos. In deterministic equations such as differential equations and difference equations, without any probabilistic term, the orbits are perfectly determined according to the initial conditions provided. As is well known in Newtonian dynamics, an equation of motion and its initial condition are theoretically independent. However, when an equation generates chaos, the complete prediction of orbits becomes impossible due to the sensitive dependence on initial conditions. In this state, an equation of motion and its initial conditions become interdependent through observation. Since chaos changes its state depending on the way of observation, the causation cannot be absolutely recognized.

1.7 A Trigger to Emergence

In complex systems we try to find characteristics that appear without having initially been programmed into the system. To describe this spontaneous phenomenon we use the term "emergence", although no clear definition of this term exists. We can, however, describe the meaning of emergence by the following condition: the notion of "emergence" requires the condition that a compression of the description into a shorter one to explain the phenomenon is not possible.

Polanyi [1958] was the first to introduce the concept of emergence with this meaning. This concept is nowadays widely accepted because researchers wish to go beyond existing methods in artificial intelligence and neural networks, the logic of which is constructed in terms of top-down versus bottom-up premises. The premise in the methodology of these fields is that a top-level phenomenon is described by factors with a small number of degrees of freedom, while a bottom-level phenomenon involves an extremely large number of degrees of freedom. In this bottom-up case, the characteristics of the top are produced by "order parameters" which coarse-grain the bottom level, and which determine the macroscopic behavior. Conversely, in the top-down case, a small number of instructions from the top determines the behavior at the bottom. Hopefully, emergence plays a role in the bidirectional flow between the two levels. Therefore, the word "emergence" seems to include the nuance that characteristics "which have not initially been built into the program" are spontaneously produced by bidirectional flow without being attributed specifically to either one of the two levels.

How, then, do we define this spontaneous generation? In this context, let us investigate whether emergence can really appear when running a program in a computer. What we had not predicted at the programming stage may often emerge as a result of the simulation. When such unexpected behavior is observed, we are tempted to call this emergence, but it may only mean that our intelligence level was not enough to predict the behavior. In addition, with regard to phenomena in a world of finite states and finite steps, it is logically impossible that "what has not been built into a program" is produced. This discussion then leads to the impossibility of emergence similar to the impossibility of artificial intelligence.

Then, are the recent studies of "artificial reality" searching for 'emergent' properties all in vain? The above discussion is based on the fact that all the states and steps are finite. It leads us to think that it may be possible to use a machine with an infinite tape and an infinite number of states. If we can utilize the tape and have an unlimited amount of time, the possibility of unlimited execution of the program could exist. Unfortunately, whether this is the case or not is impossible to determine as can be seen in Turing's halting problem. Consequently, we might not be able to say that nothing emerges beyond what is initially programmed; this is one possibility.

Furthermore, what happens if the tape has a continuous state? The answer to this question may be found in a study that takes up real-number computations directly. As Moore [1996] proposed, defining inductive logic by integration in real time, and assuming the operator works well, it may be possible to transcend the power of the Turing machine. If we can minimize noise, which unavoidably enters an analog computer, integration can be carried out in finite time by compressing the uncountable state to some finite state in the space of real numbers. Therefore, it is expected that the ability of this kind of machine becomes quantitatively and qualitatively different from a universal Turing machine. Clarifying the relationship between real-number computations [Blum et al. 1989, 1998; Siegelmann and Sontag 1994] and conventional computation theory based on integers may lead us to an understanding of emergence which is difficult to obtain when taking the formal system underlying the latter as a basis.

One of the infinite limits often discussed in physics is the thermodynamic limit, the limit of an infinite number of degrees of freedom (e.g. particles or elements). In the thermodynamic limit in statistical mechanics, important characteristics can be derived by first taking such infinite limit. For example, as with regard to the problem of a phase transition in a thermal-equilibrium system, a singularity in thermodynamic-quantities has appeared only in this

limit. Though some interesting characteristics, such as the divergence of fluctuations, appear due to the limit itself, still essential characteristics of the
phenomena are described by a few order parameters. It does not therefore
belong to the class of "problems which defy simple descriptions" which we
are going to aim our discussion at.

What we point out here is the essentially uncontrollable nature of chaos.
As mentioned previously, in chaotic dynamical systems, in spite of the premise
that "the program for driving dynamical systems is provided", we are not
able to control chaos by that program and, consequently, we are forced to
use "probability". In addition, as will be explained later, chaos can generate "information". Here we emphasize once again that, if chaos is included
in a system, more complex behaviors than those described in the program
could, in principle, appear. This is, however, simply a necessary condition for
emergence in the framework of dynamical systems. To deal with the notion
of emergence more fundamentally, we are concerned with the role of the
observer, that is, the interference of the system with the observations or the
descriptions. This will be an important topic in the future as will be discussed
in Sect. 1.9.4.

Here, we should recall the proverb "if wind blows, tub shops earn". Indeed,
from the fact that the wind blows, the consequence that tub shops earn cannot
be predicted. To explain this proverb, we have to follow and clarify several
steps. Each step is unstable in the sense that it may be moved in a different
direction by some perturbation (therefore, the unexpected causations in "if
wind blows, tub shops earn" appear). Considering all possibilities, the consequence that tub shops earn may be regarded as an "emergence" since there
is no shorter description than the one arising when each step of the sequence
is followed. Here, it can easily be seen that a sequence with descriptive instability is important for emergence. In this respect, the sequence of cause and
effect will become critical for emergent properties in, e.g., elements containing
chaos, the computation theory using chaos, and especially the theory for real-
number (or real-number-time) computation. On the other hand, we can so
far compute chaos in the present discrete computer to some satisfaction. This
suggests that if we use a huge number of states for computations, even if they
are finite, we will be able to see the "shadow" of infinity. Therefore, if there
is a system in which emergence can occur in some infinity, we will have the
possibility of seeing the "shadow of an emergence" using even the current
discrete computer.

1.8 Beyond Top-Down Versus Bottom-Up

The studies of the relational dynamics among many elements described in this
book are based on a new approach which is different from both the top-down
and the bottom-up approaches. In the typical top-down approach seen in the
study of artificial intelligence, the behavior at the bottom level is determined

by the instructions from the top level. In the bottom-up approach, the top level is generated by interactions among elements at the bottom level without instructions from the top level. Particularly, in this bottom-up approach the top level is no longer rigid, in contrast with the top-down approach, since the level is not given in advance but is self-organized. Even when the top level is self-organized, however, the relations between the elements and, after all, the state of the whole system, in general, are fixed. In this sense, the bottom-up approach remains at the level of the framework of traditional statistical mechanics, in which fewer order parameters are generated from a huge number of elements.

Another approach to resolve the rigidity of the top level is introduced by allowing for a feedback between the top and bottom levels. Let us take a simple model of "ants" [Collins and Jefferson 1991] in artificial life, as an example. The model consists of many "ants" that move on a two-dimensional field with various concentrations of a chemical substance (called a pheromone) produced by the ants. In this model, it is assumed that the motion of the ants is affected by the concentration of the pheromone, and that the ants tend to be attracted towards a place with a higher pheromone concentration. In addition, the ants produce the pheromone when they have food. Therefore, when an ant finds food, it produces the pheromone. Then many other ants gather there because they are attracted by the pheromone, and they in turn produce the pheromone. This positive feedback forms a path of pheromone leading to the food. Such a path of pheromone is the "top level", and once this path is formed, the ants move along the path. At this stage, the bottom level (the behavior of an individual ant) is dominated by the top level. However, when the food has gone, the path disappears. Hence it is possible to make a path to the nearest source of food and to erase it in turn.

This approach, however, which involves a simple and direct feedback loop, is not satisfactory for the study of complex systems. First, the motion of the ants is dominated by an environmental factor and the arrangement of food is provided by an outside agent. If food continues to be supplied, and if the path is made to the nearest food source, the path will be fixed. In that sense, there is no autonomy at the bottom level, and the behavior at this level is determined by an environmental factor. This approach essentially does not go beyond the traditional framework of optimization, such as searching for the nearest food source. It should also be noted that the top level is formed by at most one or two paths, and thus is described by a small number of degrees of freedom. Therefore, this approach is not suitable for the complex system studies with which we are concerned.

There are further problems with this type of approach. The relations between the elements are fixed and each element preserves the same features. In contrast, our approach does not suffer from these limitations, and makes it possible to uncover the underlying nature of complex systems. Its main characteristics are listed below:

(1) The top level is not always described by a small number of degrees of freedom. A priori knowledge of whether the top level exists or not is not necessarily available.
(2) Each element at the bottom level has internal degrees of freedom, and displays spontaneous dynamics.
(3) Even if boundary conditions are imposed, the dynamics of each element is neither predictable nor controlled.
(4) Relations among elements are not fixed but change spontaneously.

From these characteristics, one might be tempted to believe that no distinct top and bottom levels exist at all. Indeed, the impossibility of a clear separation between the top and bottom levels is a natural consequence of the inseparability of the macroscopic and microscopic levels seen in chaos, where any small difference is amplified, and thus the microscopic level is connected to the macroscopic level. It is not possible to draw a definite boundary between the top and bottom levels.

Still, we intentionally use these terms, top and bottom levels, because such levels often appear approximately over some time span, and the states during that time span are better understood this way. Even so, generally speaking, it is more relevant to study the dynamics of such phenomena in the light of the generation, collapse, competition, coexistence, and hierarchy of those levels. Furthermore, by deriving contradictions caused by separating the system into levels, an absence of levels can be revealed.

Since our discussion so far is rather abstract, the concrete meaning of each topic might not yet be clear. Examples for each topic will given in later chapters, and it will be shown how the characteristics (1) to (4) naturally emerge in the dynamics of our models.

1.9 Methodology of Study of Complex Systems

If a complex system reveals nonreductionistic features in some sense, a new methodology is required for their study.

Some possible methods are listed below.

(a) observe structural changes from both static and dynamic viewpoints along with dissecting phase space;
(b) reach a general concept, reconstructing immanent abstract structures and relationships from various phenomena;
(c) construct an artificial system by combining several fundamental conceptual elements;
(d) construct a model while observing top and bottom levels from an intermediate level which is neither macroscopic nor microscopic;
(e) construct an adequate language system to understand complex systems, based on a set theory for treating dynamic states of relations and processes;

(f) acquire new intuitions by formulating contra-intuitive events and by experiencing the variety of mathematical phenomena in computers.

The characteristic of complex systems, that their nature is incomprehensible when merely assembling the features of their components, illustrates the limit of the methodology of reductionism. This can typically be seen in the application of the globally coupled map (GCM: to be discussed in Chap. 4) to the traveling salesman problem (TSP) by Nozawa [1992]. The TSP is also an example of a problem characterized by the phrase "construction is simpler than understanding", which is why we briefly discuss it here.

The travelling salesman problem is defined as follows. When a salesman visits each of n cities once, what is the shortest route? This is a typical combinatorial optimization problem which brings about a so-called combinatorial explosion such that the number of routes increases exponentially with the number of cities n. Therefore, when n is large, even if we use a computer, we cannot practically solve this problem by comparing the distances for all cases. Hopfield demonstrated numerically that a solution can quickly be found by means of a neural net that is specially designed for this problem. However, with his method, once a solution is found as a local minimum, the network cannot escape from that local minimum. Nozawa found that by coupling chaotic neurons globally (as in the GCM in Chap. 4), the system is not trapped at a local minimum but wanders around towards a global minimum. At this point, the motion of the system shows the chaotic itinerancy to be discussed later. Nozawa's example typifies a problem that commonly appears in the application of complex system study to problem solving. Although Nozawa was able to construct a system for solving the problem, the dynamics, that is, the mechanism at work here, is difficult to comprehend. As a matter of fact, after Nozawa, many attempts to clarify the mechanism were made, but we still seem to be far from a true answer.

1.9.1 Constructive Way of Understanding

So far we have mentioned the necessity of a new methodology of understanding since we are dealing with a "monster" which cannot be understood by conventional descriptive methods. Indeed, the introduction of a constructive way of understanding for complex phenomena was inspired by the study of chaos. For example, chaos has provided us with a universal framework for understanding the onset of turbulence. The adopted chaotic model is a simple mapping which is not directly connected with hydrodynamics. Still this abstract model has, in a sense, provided deeper insight regarding the onset of turbulence than the conventional study of solving hydrodynamic equations step by step. By extending this approach, one can emphasize the significance of constructive as opposed to descriptive understanding. The coupled map lattice (CML), in a sense, gives an example of such a constructive model for spatiotemporally complex phenomena, as mentioned in Sect. 1.6, while the globally coupled map (GCM) in Chap. 4 will provide another example.

We aim to find common structures of complex phenomena by "constructing the world". As will be discussed in Chap. 3, the constructed world forms a universality class of phenomena at a qualitative level. Here, the criticism might arise that such approaches based on "universality" are almost identical to conventional statistical mechanics, and have no power to deal with the diversity in complex systems. As one possible approach for describing universalities in complex systems without being subject to the above criticism, a "natural history"-like approach could be considered. An example for this can be found in Wolfram's studies [Wolfram 1986] of cellular automata (CA), where he systematically examined CA by scanning classes of rules and proposed a classification for them. Indeed, drawing a phase diagram is an example of such classification. However, when the number of parameters and the observed characteristics are anything but very small, simple phase diagrams cannot be drawn, and we need to adopt some descriptive method, following natural history.

When physicists study complex systems by the ways of "natural history", however, they still tend to search for universalities. It thus becomes necessary to balance the two approaches; the search for a universal structure and the observation of unique events for individual cases of the universality classes. It may be worth recalling that the study of chaos has, to some degree, overcome the dichotomy between universality and individuality. Although, in the mechanism of chaos, there are extremely universal structures such as homoclinicity, individuality can also be found in each type of chaos, for example in the onset of chaos or in the topological structure of a chaotic attractor. This is one of the reasons why the study of chaos has been so rich.

1.9.2 Plural Views

To understand complex systems, plural views as opposed to a "single" view are necessary. For example, to understand the pattern dynamics of spatiotemporal chaos, we need descriptions in both real space and phase space. The dynamical systems viewpoint and the computation-theoretical viewpoint both provided much information concerning the complexity of chaos and cellular automata. In game-theoretic dynamics, descriptions are completed by introducing an algorithm such as a "strategy" which is at a different level than the treatment by dynamical systems theory. To understand the brain, dynamical systems theory, thermodynamics, statistical mechanics, probability theory, set theory, neural-network theory, anatomy, physiology, molecular biology, computer science, and logic are all necessary.

1.9.3 Mathematical Anatomy

The mathematical anatomy of a low-dimensional system provides us with its geometrical structure in phase space. We need something similar for high-dimensional systems. The mathematical anatomy of static complex systems gives the structure of the energy landscape or the fitness landscape. In dynamic complex systems, however, extraction of geometrical structures in phase space is much more difficult, simply because anatomy is originally static. Still, there will be some hope. In recent studies of the dynamics of Hamiltonian systems with large numbers of degrees of freedom, a way of understanding the motion dissecting structure in phase space has been developed [Shinjo 1989; Ohmine and Tanaka 1990; Konishi and Kaneko 1992; Kaneko and Konishi 1994].

1.9.4 The Problem of Internal Observers

The models of complex systems with which we are concerned here, if we may say, are systems containing an artificial reality or virtual reality. To make such models more intelligent, we probably need some internal observers. For example, to understand muscle contractions, we need a detailed knowledge of the interactions in the actomyosin system. The origin of muscle force, at a molecular level, is due to the sliding between two kinds of proteins, actin and myosin, which form the actomyosin system. As myosin moves along the actin, the muscle, as an assembly of such proteins, contracts, leading to some force.

According to recent experiments, this microscopic system can work even at the level of thermal noise. Actin forms a cytoskeleton, and it is known to include structural anisotropy determining the anisotropic potential of the actomyosin system. Theories so far derived a unidirectional sliding motion resulting from the breaking of a detailed balance in the anisotropic potential. But an actual sliding motion of muscles will not be purely stochastic, but much more dynamic. Or, in other words, the motion has some balance between internal deterministic dynamics and interaction with a stochastic heat bath. Some dynamical system mechanism that works in a "thermal sea" should be searched for [Nakagawa and Kaneko 1999]. We need to clarify how internal dynamics work as a kind of 'observer', as is also termed a (Maxwell's) demon. Of course, Maxwell's demon cannot work at a thermodynamic scale, but such mechanism may work over some limited time span. This is not only in the case for muscles. A flagella motor, for example, is seen in bacteria to drive their motion. It has a molecular ratchet with structural anisotropy. In this motor, the inflow of one proton causes the rotation of the flagellum in one direction. This motion could be caused dynamically, not by breaking the detailed balance at the level of thermal fluctuations. A flagellum rotates dynamically in one definite direction, for example, to the left, not turning to the right on average as a result of a breaking of the detailed balance while turning right or left probabilistically.

Also, for models of the brain, Tsuda pointed out the importance of observation from within and introduced the concept of "chaotic hermeneutics" for this purpose. One of the motivations is that if chaos works in the brain, chaos can be a candidate for being an internal observer. As one of the constructive approaches to understand the brain, possible construction of internal observers within a computer has been studied. An effort to construct a dynamics inside a computer which generates a formal neuron by generating the mathematical threshold function has been made [Tsuda and Tadaki 1997; Tsuda and Yamaguchi 1998].

Gödel was the first to point out the importance of an internal viewpoint. He constructed a theory so that a description from outside of a formal system is again contained in the formal system. Here, the internal–external problem of a formal system was formulated, and the complexity of a formal system was clarified. It should, however, be noted that the "descriptive instability" of a complex system and its "observation" are interrelated. Perhaps this is the reason why the notion of an internal observation is inevitable for understanding a complex system. Although an observation from without justifies objectivity, external observations and internal descriptions do not always coincide with each other for some problems in chaos or biological problems. In biology, as Shimizu et al. [1985, 1987, 1992] pointed out, it is important to understand the internal logic and meaning from the viewpoint of the living organisms. These cannot always be understood by observation from without only.

Rössler [1987, 1994] conceptually generalized the notion of an internal observer in formal systems for use in physics, chemistry and biology, and thus proposed a new paradigm called endophysics [Finkelstein 1979]. In a complex system, we are unable to understand the system's dynamic state fully by an external observation only. As will be discussed later, in complex systems, the observation problem is crucial, and it is necessary to describe a system including its observer [Shaw 1984]. In this respect, Rössler's endophysics will be important in the study of complex systems [Gerbel and Weibel 1992].

2. Observation Problems
from an Information-Theoretical Viewpoint

2.1 Observation Problems of Chaos

If chaos could be characterized by one word, it would be "complexity". Even with the greatest of efforts, a finite sequence cannot accurately describe chaos. Chaos can be perfectly represented only in an infinite sequence. Several problems stemming from the discrepancy between this infiniteness and the finiteness of observation and description will be discussed later, but here we briefly touch upon it.

Let us consider a one-dimensional map. A chaotic orbit starting from some initial condition moves in a complicated manner in the interval of definition. Using an appropriate decision point, one may assign '0' for the orbits below the decision point and '1' otherwise. Then, an orbit is, for instance, represented by .0100101110$\cdots$, where a decimal point is introduced for convenience. In this representation, the larger the number of digits obtained, the finer the observation becomes, that is, the digits further to the right express more microscopic states. The operation of cutting such a sequence at some finite digit naturally leads to the introduction of a heat bath (see also Shaw [1981]). In chaotic dynamical systems, an initial condition or its function can theoretically be determined with infinite precision, and the above procedure continued without cutting the sequence at any finite digit. In other words, the information about the initial condition can be obtained through observation, if the function of the initial condition is simple in the information-theoretical sense. Then, in this ideal limit, the initial condition or its function can be uniquely determined. However, in an actual case, we have a finite observation time which may give rise to an uncertainty in the determination of the initial condition. Despite the fact that we follow a deterministic law, we see the natural introduction of a heat bath; thus the system is coarse-grained.

Chaos can appear in ordinary Newtonian equations. One may notice two stages when describing the system with differential equations or difference equations. First, one determines what a state is, and second one represents a time variation of the state in terms of equations. In other words, by differential equations or difference equations, we generally assume that we can describe a time-evolution of the state of the system concerned. Here, we see one of the characteristics of scientific description, namely, first assuming a state and then obtaining an evolution or a flow of the state. Conversely, we can see an inverse relation of state and flow in chaos. Though in chaotic dynamical systems too, the evolution of the state is calculated, starting from some initial conditions, in chaos, the causation between an initial state and a current state becomes vague since the information of the initial conditions decays rapidly. This happens, despite the fact that the causation is ascribed definitely at the level of the equation. Thus, it seems that only the evolution itself is a realistic variable. A state may be defined conversely by an evolution; even an initial condition might be. Though it is difficult in general (see Fig. 2.3), there is a simple example which shows that the initial condition can be determined from the time evolution (Fig. 2.1). This is the special case where the relation between an initial condition and a binary number generated by the evolution rule is linear (more specifically, equivalent).

Furthermore, according to the study of chaos, chaos can contain countable periodic orbits and uncountable nonperiodic orbits, whose infiniteness might justify the evolution.

The strange behaviors in chaotic dynamical systems originate in such infiniteness of chaos, or in the gap between the infiniteness and the finiteness of observation. The noise-induced order (NIO) [Matsumoto and Tsuda 1983] clearly expresses the presence of this gap. The NIO means that chaos diss-apears when applying noise to some chaotic system, and that some ordered motion, which was absent in the original system, emerges. This phenomenon will be discussed again later. What we emphasize here is that this phe-nomenon forces us to reconsider the effect of "observation". The purpose of observation is to understand the nature and structure of the system con-cerned. Suppose we do not know these before observing the system. If the observation inevitably distorts the nature and the structure of the system, what do the observed results mean? Such effects of observation have been discussed mainly in quantum mechanics. In classical systems, the effects also become an important issue if chaos exists there. The issues of observation in dynamical systems are introduced in Sect. 2.6.

To discuss mathematically the observation problem of chaos, the concept of observation will be used in two ways, namely, an external observation distinguished from an internal observation. It is usually assumed that an ob-servation is conducted independently of the system to be observed, implying that the observation never essentially distorts the nature and the structure of the system. We here call this type of observation an external observation, or simply an observation. Contrarily, if an observation inevitably distorts the system, the observation is no longer independent of the system. Then, the observation itself is involved in the object to be described. We call this case an internal observation or simply a participation. It does not solve anything to include the observation that distorts the system in a new system as an object. There seem to be two ways to make an observation from within meaningful. One way is to describe the distorted system with an explicit description of the observation. Another way is to find a new insight or coordinate, in which we

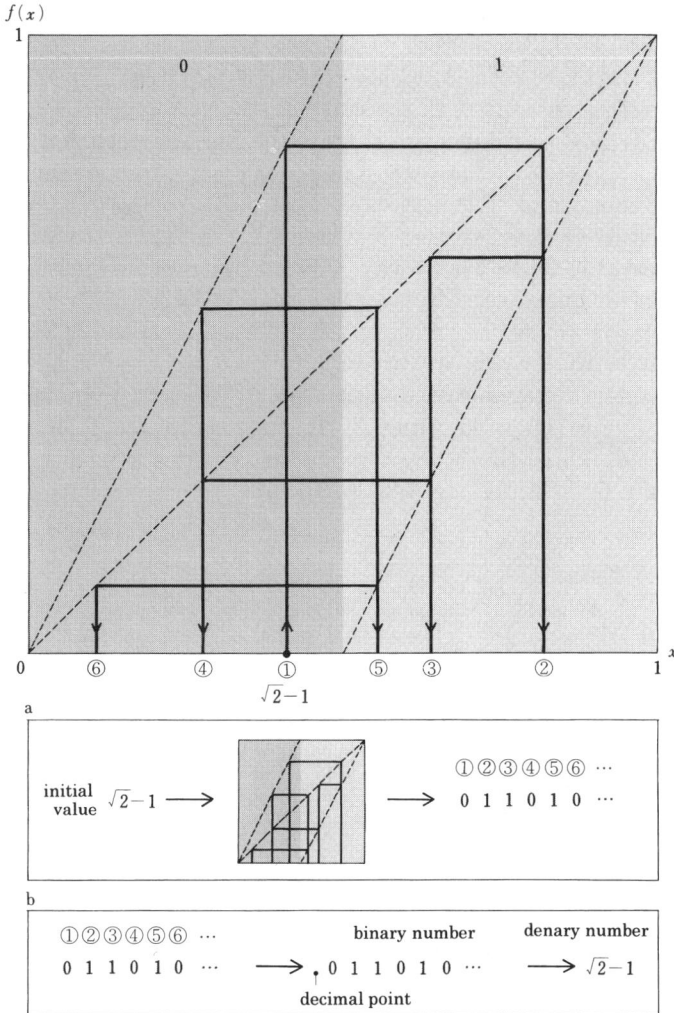

Fig. 2.1a,b. This example is taken from a typical chaotic system called the Bernoulli shift $(X_{n+1} = 2X_n(\mathrm{mod}\,1))$. Encode by "0" all the orbits arriving at the left branch $(X < 0.5)$, and by "1" those arriving at the right branch $(X \geq 0.5)$. (a) For example, for the initial condition $x_0 = \sqrt{2}-1$, the orbit is labeled $011010\cdots$. (b) On the other hand, if it is viewed as a binary expansion of a number s, putting a decimal point at the left side of this label, the temporal evolution of the binary sequence asymptotically approaches $\sqrt{2} - 1$. This holds for any initial condition in this system. Therefore, in the Bernoulli shift, one can have the view that an initial condition is obtained, accompanied by the time evolution of the equation of motion.

can look at a naked system's behavior and structure. The problem is whether or not such an insight exists.

In chaos, there are cases in which a Markov partition (Fig. 2.2) [Ito and Takahashi 1974; Bowen 1979] exists, which fortunately gives rise to the above insight. In a typical one-dimensional map, the Markov partition gives a collection of local one-to-one maps for a nonlinear many-to-one map. With this partition, a sequence of coded orbits is defined, allocating "0" for the orbits on the left from an apex of the map and "1" for the orbits on the right. Each subinterval in finite partitions is labeled by an orbit with finite length, starting from an initial condition within such an interval. If partitioning is performed up to infinity, the width of the subintervals becomes zero, defining points. Then, a codesequence with infinite length corresponds to an orbit with infinite length starting from such a point. Therefore, knowing Markov partitions is equivalent to knowing all the time-evolutions following a given dynamical rule. Thus, the Markov partition provides a means to describe the whole map by compressing Newtonian time. In other words, the Markov

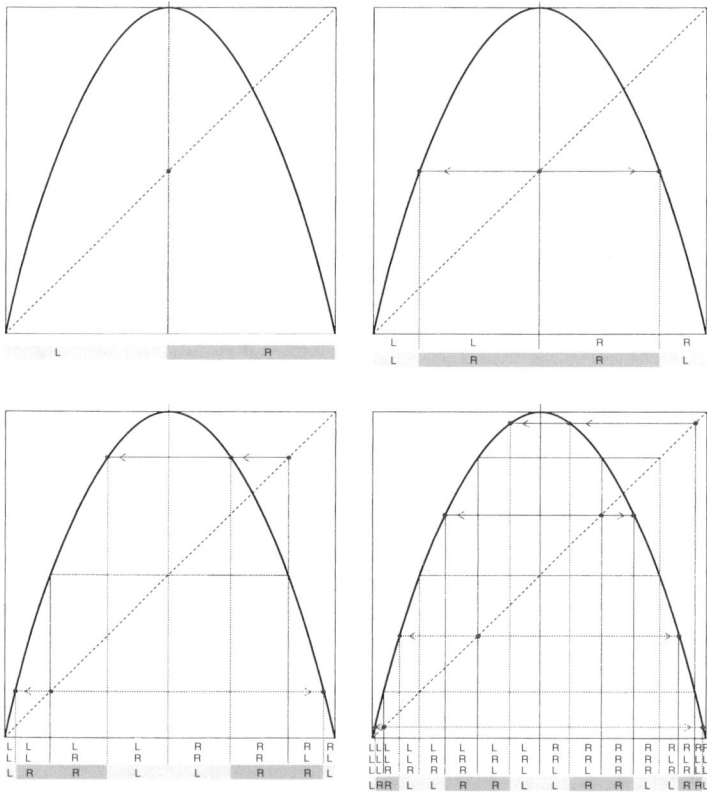

Fig. 2.2. Markov partitions in a one-dimensional map. Several finite partitions are shown. To each cell in a partition a label is allocated.

partition is a means to observe the structure and dynamic behavior of chaos independently of chaos itself, namely without any perturbation. Furthermore, it turns out that the Markov partition also represents chaos itself. Thus, we find an insight in which we can observe chaos from within chaos without any interference.

2.2 Undecidability and Entire Description

The complexity of chaos lies in the diversity of its emergent process, which is not easily observed from without. One way to understand chaos could be to find an insight justifying an internal observation, which gives rise to the problem of total descriptions.

The question arises of whether there is another example where an entire system is described completely. Indeed, Gödel's work on the formal system certainly led to a discovery of this kind of description [Hofstadter 1979; Hilbert and Bernays 1993; Nagel and Newman 1964]. Gödel was able clearly to demonstrate an important characteristic that inevitably appears when a formal system describes the whole of itself. This characteristic is undecidability. He demonstrated the incompleteness of formal systems, recognizing the undecidability problem. The method Gödel used for description was applied to the problem of what cannot be achieved by computation machines, and consequently the decidablity or the computability of machines was defined. In machines which perform only limited computations, one cannot find universal undecidability. On the other hand, undecidable problems exist for machines which treat every problem that can be described with a finite algorithm. From this, an entire description of the system leads to universal undecidability within finite descriptions.

The idea of description from within is not new. The relativity theory and the incompleteness theorem mentioned above are typical examples. Many issues discussed in quantum theory concern the delicate relationship between external observations and the internal quantum world which inevitably is distorted by an observation. Recently, observation problems in classical chaotic systems as in the 'endophysics' of Rössler, and in quantum mechanics as in the 'endophysics' of Finkelstein have been reexamined with a similar idea.

In order to approach an entire description of chaos, we must first find an undecidable problem implicated in chaos. For a small class of chaotic dynamical systems like the one-dimensional map, we can find a tool for an entire description, that is, a Markov partition. However, it is an open problem if there is such a strong description for a general class of chaos.

2.3 A Demon in Chaos

We here call an internal structure that can accurately perform the Markov partition a demon, after Maxwell. Several situations where Maxwell's demon is supposed to work have been proposed. The following is a kind of Maxwell's demon. Let us assume that gaseous molecules are in a box. These molecules are supposed to collide with each other without chemical reactions. Now imagine that there is a partition at the center of the box with a minute door through which molecules can pass. According to the second law of thermodynamics, the molecules that initially gather in one part of the box spread throughout the whole box over time. On the other hand, if there exists a demon who perfectly knows the direction of the molecules' motion, the demmon can gather all molecules which spread throughout the entire box in an early period. If this demon perfectly knows the state of a gaseous molecule, it can decrease the entropy of the ensemble of molecules in time. However, the impossibility of such a demon was proved. More specifically, there cannot be a demon that works for an arbitrarily long time. As Wiener pointed out [1961], however, the demon could work for some finite time in a system possessing a metastable state whose relaxation time to the equilibrium state is sufficiently long.

In chaos, we can discuss a similar problem, though the situation is different from thermodynamics. Here, let us call something that has previously known all the information in a Markov partition, i.e., all the information about the dynamical system, a demon. Suppose we allocate symbols (e.g., 0 and 1) to the cells made by dividing the state of the system into two parts. Since a dynamical orbit passes through each cell as the dynamics develops, a symbol sequence is generated. Then, different sequences with a fixed length are fine divisions of partitions. If an initial condition is uniquely determined when knowing a sequence with infinite length, the entropy associated with the partitions is maximum. Thus, maximum entropy is assured if the demon works.

However, it will be impossible to make such a demon work in a realistic physical system. Knowing Markov partitions precisely is equivalent to observing the orbits precisely up to positive/negative infinite time. Hence, such a demon either has a whole knowledge of the chaotic system in advance or has a "super"-ability of analysis. In either case, the demon must have a mechanism of knowing, memorizing, and calculating information, that is, a brain. To have a brain, the demon must be in some ordered state. Such an ordered state will not appear in thermal equilibrium, but appears in far-from-equilibrium [Nicolis and Prigogine 1977; Haken 1979]. On the other hand, energy dissipation is necessary for far-from-equilibrium. If energy is dissipated in a closed system, the observation by the demon becomes inaccurate because of the perturbation by this dissipation. Hence, a demon is impossible in a closed system.

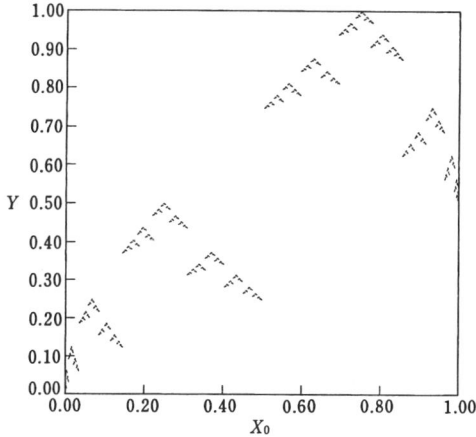

Fig. 2.3. The relation between the initial condition X_0 and the corresponding binary number in the logistic map $X_{n+1} = aX_n(1 - X_n)$ and $a = 4.0$. The orbit starting from the initial condition X_0 is labeled "0" if $0 < X_n < 1/2$ and "1" otherwise. If this sequence is viewed as a binary expansion of Y, the relation between X_0 and Y is plotted.

The following is a mathematical representation illustrating the impossibility of a demon. We place a decimal point at the left end of a symbol sequence consisting of 0 and 1, and regard it as a binary expression of a position in phase space. Suppose the demon knows this binary expression perfectly. Now the problem the demon must solve is to decide the initial value of the orbit only with this information. Figure 2.3 shows the relationship between these two values with finite computations for the case of the logistic map. The graph should be discontinuous everywhere, but here only a part is shown. This is only a part of the function that the demon must calculate. Conversely, such a transcendental relation might represent one side of the characteristics that the demon possesses. Because of this transcendental character one will not be able to make such a demon physically.

2.4 Chaos in the BZ Reaction

Next, we will explain noise-induced order (NIO). NIO was first discovered by Matsumoto and Tsuda in the Tsuda-Tomita model (BZ map) of the Belousov–Zhabotinsky (BZ) reaction [Matsumoto and Tsuda 1983]. Before addressing this problem, we provide a brief review of a part of the early studies of chaos in the Belousov–Zhabotinsky (BZ) reaction, which led to the discovery of NIO.

Prigogine is probably the most influential scientist to emphasize the importance of the BZ reaction [Tyson 1976]. When Prigogine advocated a new paradigm implied by the notion of dissipative structures, the Brussels group led by Prigogine proposed a nonlinear equation called a Brusselator as a typical model that shows a dissipative structure [Nicolis and Prigogine 1977]. Although at first it was considered to be a model of the BZ reaction, it was pointed out that the equation was not appropriate for a chemical-reaction model because it contained a third-order nonlinear term. In chemical reactions, the collision of two bodies often occurs, but the collision of three bodies seldom occurs. In spite of this criticism, by studying the model's limit cycles and the spatial patterns obtained by the introduction of the diffusion term, the Brusselator was shown to be a useful model for the study of dissipative structures. Tomita and Kai discovered chaos and the period-doubling bifurcations leading to chaos in the forced Brusselator [Tomita and Kai 1978; Tomita 1982b].

The construction of more realistic models of the BZ reaction was attempted. Noyes proposed a model called the Oregonator. Based on this model, Tomita *et al.* [1977] proposed a simpler model which does not spoil the essence of the Oregonator. The BZ reaction is an oxidation reaction of malonic acid catalysed by cerium or iron ions. Tomita and others considered reactions of malonic acid and the constancy of the total concentration of the catalyst. Furthermore, Tyson [1978] studied intensively the Oregonator model. However, no model ever demonstrated chaos. Under these circumstances, Schmitz *et al.* [1977] reported that chaos was observed in the BZ reaction. This study produced an atmosphere in which model studies demonstrating chaos and further experimental studies were necessary.

Tomita and Tsuda, with the help of Ueda and his staff, discovered 'chaos' in the simulation by analog computer of a modified Tomita–Ito–Ohta model [Tomita and Tsuda 1979]. Immediately after this discovery, the Bordeaux group led by Vidal experimentally verified the existence of chaos of the same type that Tomita and Tsuda found [Roux *et al.* 1980]. However, the analysis of the Tomita–Tsuda model by digital computer did not show chaos.

At the same time, Hudson *et al.* [1979] performed precise experiments and discovered specific periodic and chaotic behaviors, where the periodic and chaotic behavior appear one after the other in parameter space. On the other hand, Noyes attempted to verify his belief that chaos should not exist in chemical reactions. In fact, he obtained high-order periodic solutions in a more precise model made from the Oregonator, which were very similar to the chaos discovered by Hudson *et al.* [Showalter *et al.* 1978].

Tsuda made the return map concerning the local extremum in the time series of the concentration of Br^-, and found a very precise one-dimensional map. Furthermore, through the plots (so-called Lorenz plots), for several cases of the control parameter used in the experiment, Tsuda found that the parameter in the maps is simply shifted. In Fig. 2.4, two Lorenz plots of chaotic time series in the case of different velocities of flow of chemicals are shown. What is understood by superimposing these two maps is that the

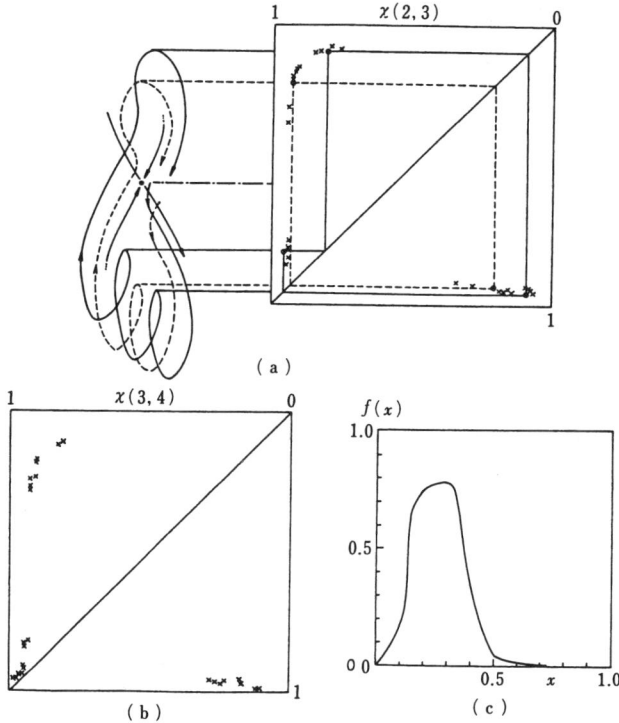

Fig. 2.4a–c. One-dimensional plot (z_n, z_{n+1}) concerning the local minima z_n of the time series of the concentration of Br$^-$ found in Hudson's experiment. Two maps in (a) and (b) are obtained for two different flow rates of chemicals. From these plots, the functional form of the map was determined as a gamma function. (c) Compared with two empirical maps, the functional form is the same and simply shifted vertically. Thereby the bifurcation parameter is determined [Tomita and Tsuda 1980].

functional form of the map does not change but only shifts vertically. This empirical map is not continuous and breaks at many points. Tsuda considered that this discontinuity must stem from the presence of an extremely low density of orbits in an originally continuous map, and then approximated the supposed map by the following one-dimensional map with a bifurcation parameter of the shift type:

As $x \mapsto f(x)$,

$$f(x) = \begin{cases} [-(0.125 - x)^{1/3} + 0.50607357]\exp(-x) + b & (x \leq 0.125) \\ [(x - 0.125)^{1/3} + 0.50607357]\exp(-x) + b & (0.125 \leq x \leq 0.3) \\ 0.121205692[10x\exp(-10x/3)]^{19} + b & (0.3 \leq x). \end{cases}$$

$$(2.1)$$

Tsuda investigated the bifurcation structure of this one-dimensional map in the parameter space b, and could reproduce all the periodic and chaotic be-

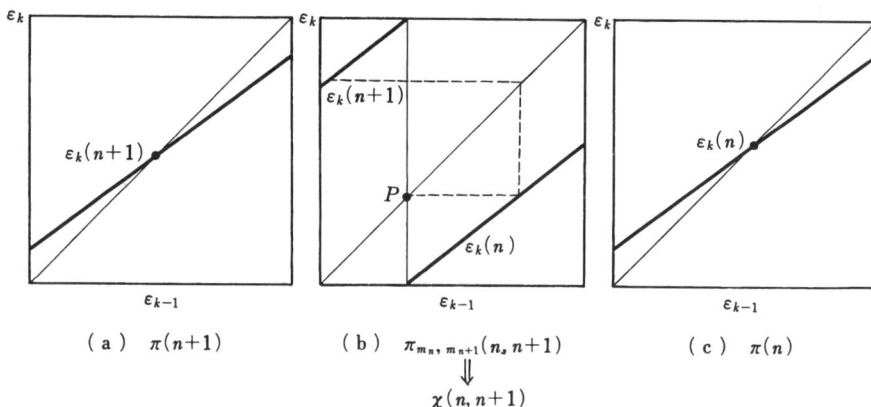

Fig. 2.5a–c. Approximation by the piece-wise linear map of the supposed one-dimensional Poincaré map of the supposed three-dimensional flow in the BZ reaction system. The bifurcation parameter is the shift. This map does not have observable chaos, but allows the existence of chaos. (a) The period $n + 1$, (b) the period $M_n n + M_{n+1}(n + 1)$, where M_n and M_{n+1} are integers, (c) the period n [Tomita and Tsuda 1980].

haviors which Hudson had discovered in the experiment. Tsuda also predicted numerically the detailed bifurcation structure of the BZ reaction system possessing a flow term as the bifurcation parameter [Tomita and Tsuda 1980]. In fact, Hudson experimentally reconfirmed that the solutions exist as predicted [Hudson and Mankin 1981].

Furthermore, Tomita and Tsuda analysed the possible structure of the vector field, and constructed a shift map of piece-wise-linear type by defining an appropriate Poincaré map [Tomita and Tsuda 1980]. This is shown in Fig. 2.5. By analysing this map, the periods of the periodic solutions which must appear in the BZ reaction were clarified, together with the Lesbegue measure of such solutions in parameter space.

The shift map of the piece-wise-linear type which Tomita and Tsuda investigated in detail was mathematically analysed by Hata [1982, 1998] independently. According to Hata, in this map chaos exists on a Cantor set in parameter space, though what type of chaos it is is still an open problem.[1] This map was also shown by Nagumo and Sato [1972] to be equivalent to Caianiello's equation, which is a mathematical model of a neuron (see also

---

[1] As can be seen in Fig. 2.5b, there is a discontinuity in the map. Only in the neighborhood of this point an orbital instability can appear, and hence the measure of the contribution for orbital instability is zero. On the other hand, if topological chaos is defined as the existence of a positive topological entropy, the chaos existing in this map is not literally topological chaos. If the discontinuous point, however, is viewed as a third branch with an infinite slope, a positive (nonzero) topological entropy could be obtained.

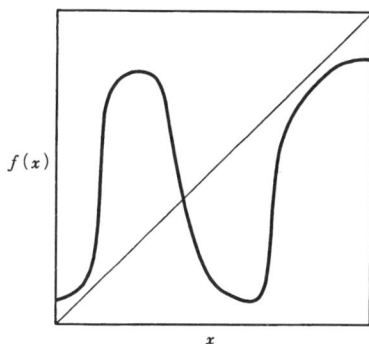

Fig. 2.6. One-dimensional map of the BZ reaction system over all domains of flow. Introducing the shift parameter, all solutions found in experiments are reconstructed (the extended experiment of this type was first conducted by M. Horai (Master's Thesis, Osaka University, 1983)).

Sect. 6.5). After these studies, chaos and bifurcations in the BZ reaction were studied and analysed by Swiney and others in Texas.

We must note one thing here. Equation (2.1) was obtained empirically, referring to Hudson's experiment. This equation however describes only half of the structure of BZ chaos. According to Tomita and Tsuda's theory [1980], which was based on the simulation results by analog computer of the model on a three-dimensional vector field, there must be two steady states, each of which appears in high or low concentration of Br^-, and in between the bifurcation is symmetric. Taking into account this point, the map for the whole structure of the BZ reaction will be as shown in Fig. 2.6, where the bifurcation parameter is a shift.

2.5 Noise-Induced Order

Noise-induced order (NIO) was first found by Tsuda in the Tsuda–Tomita model of the BZ reaction (2.1). Matsumoto and Tsuda investigated NIO in detail from various directions, culminating in the information-theoretical study of the plasmodium of slime mold (by Matsumoto) and the brain (by Tsuda). Here, let us briefly explain NIO.

The nature of systems can be changed by observations. The word observation, as used here however, is not yet strictly formalized. One of the directions for possible formalization is discussed in Sect. 2.6. Here by the word observation we simply denote seeing a response of the system to some stimulation given from outside the system. In the microscopic world governed by quantum mechanics, the nature of the system changes by receiving electrons or photons which are emitted from the measurement apparatus. On the other hand, in classical mechanics, people believed that this type of essential changes could not appear, and that 'perturbational' changes could simply be observed, namely the relation between the changes and the perturbations is linear. In the macroscopic world, however, it turned out that state changes by observations can occur. One of the most remarkable examples is NIO.

One can consider classical systems to be always exposed to a continually varying environment because of thermal fluctuations and the finite precision of the measurement apparatus. To model this situation, a noise term is introduced in the right-hand side of (2.1). Since the behaviors of the system are simulated here with a computer, pseudo-random numbers generated by the computer can be regarded as "noise". Here, we use a uniform noise. In Fig. 2.7, the NIO is demonstrated by three quantities: (a) the change in the power spectrum, (b) the change in the Lyapunov exponent, and (c) the change in the Kolmogorov–Sinai (KS) entropy. The upper figure in (a) denotes the power spectrum in the case without noise, and the lower one the case with noise whose amplitude is 1% of the system size. As can be seen in this figure, the power of a certain frequency component rises sharply due to noise, and that component is not dominant in the original chaos. Thus, this can be interpreted as the generation of order by noise out of chaos.

As can clearly be seen in Fig. 2.7b, the Lyapunov exponent changes from a positive to a negative value at a noise level of around 0.1%. This also indicates the appearance of a certain order out of chaos. However, one should note that the Lyapunov exponent is well defined only when the system's development is expressed as an orbit. It is certain that applying noise, more or less, breaks the orbits, even if the noise plays a role of making the orbits aggregate rather than segregate. A randomized piece of orbit cannot be viewed as a dynamical orbit. Thus, for the appearance of order out of chaos in the system with noise, the Lyapunov exponent simply suggests the transition. Additionally, Paladin *et al.* [1995] recently introduced the notion of a Lyapunov exponent for noisy systems in the study of NIO. They emphasized that the orbital instability is still present for different orbits produced by different noise series.

The KS entropy shown in Fig. 2.7c is a good indication even in this noisy situation. The KS entropy is defined as follows. Let us consider a partition by a diffeomorphism f of the phase space Ω, where f is defined on Ω. Then, let the partition $\alpha \vee \beta$ be the most coarse one among the subdivisions common to α and β. For $A^{(n)}$ such as $A^{(n)} = f^{-n}\Omega \vee f^{-n+1}\Omega \vee \cdots \vee f^{-1}\Omega \vee f^0\Omega \vee f^1\Omega \vee \cdots \vee f^{n-1}\Omega \vee f^n\Omega$, we specify elementary partition cells as $A^{(n)} = \bigcup_{i=1}^m a_i^{(n)}$. Provided the probability $p_i^{(n)}$ of the orbit arriving at the i-th cell is known, the entropy $H^{(n)}(A^{(n)}) = -\sum_{i=1}^m p_i^{(n)} \log p_i^{(n)}$ is defined. The asymptotic form of the increment of this entropy, that is, $h(A, f) = \lim_{n\to\infty} H^{(n)}(A^{(n)})/n$, is the entropy of the partition. Taking the supremum over all ways of partitioning, the KS entropy is defined, namely $h(f) = \sup_A h(A, f)$. This is invariant for dynamical systems, hence it does not change under homeomorphisms.

In actual computations, we perform the following procedure since one cannot take the 'supremum'. In general, we use n equal partitions, calculate

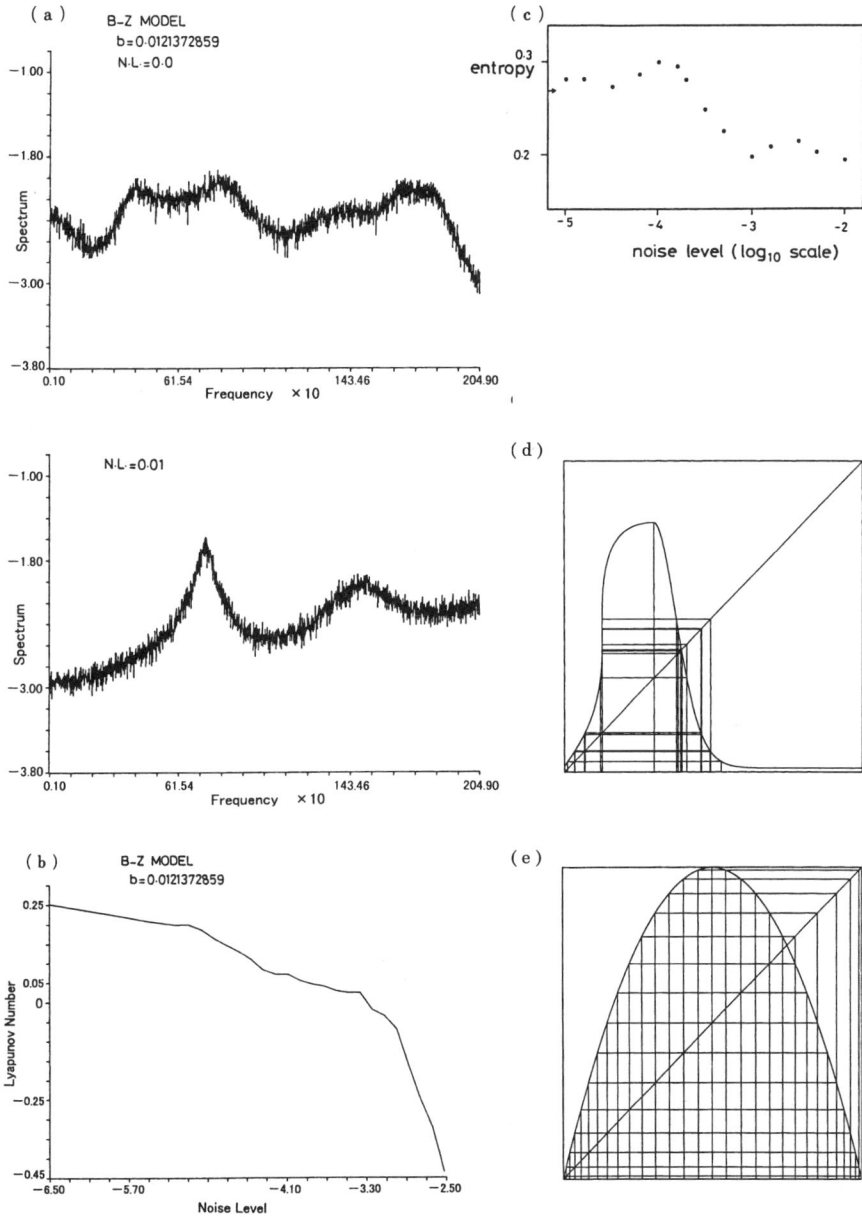

Fig. 2.7 a–e. Three indices of noise-induced order. (**a**) Power spectrum, (**b**) Lyapunov exponent, (**c**) KS entropy (reprinted from K. Matsumoto and I. Tsuda: *J. Stat. Phys.* **31** (1983) 87, with the permission of the publishers), (**d**) Markov partition of BZ map, (**e**) Markov partition of logistic map. Both partitions are the same depth of 5.

the entropy in each, and find a convergent value when increasing n. Specifically, in one-dimensional maps, there are the cases, in which we easily find Markov partitions [Ito and Takahashi 1974; Bowen 1979] (see below for the definition of Markov partition). The entropy calculated for the n-th order Markov partition is higher than the entropies for any other partitions. Thus, the Markov partition is the most efficient way for calculating the entropy. Actually, in the present case of the BZ map, the Markov partition was used for the computation of the KS entropy. Namely, the entropy was calculated for the case that the map possesses Markov partitions, and the noise-dependence was investigated. As shown below, the observation problem can be discussed using Markov partitions.

For the n-th order Markov partition $(a_{n_1}, a_{n_2}, \cdots, a_{n_l})$, $H = -\sum p(a_{n_1}, a_{n_2}, \cdots, a_{n_l}) \log p(a_{n_1}, a_{n_2}, \cdots, a_{n_l})$ is the quantity to be calculated. This gives, however, a slow convergence in the calculation. Then, actually, the KS entropy is obtained through the calculation of $H = -\sum_{ij} p_i p_{ij} \log p_{ij}$, by using the transition probabilities p_{ij} of the Markov process. Matsumoto and Tsuda investigated the noise-dependence of such an entropy, where at each noise level the entropy was calculated for each of the 0-th to 12-th order Markov partitions. Concerning the BZ map, the entropy rapidly decreases at the 4-th order and does not change for the higher orders. Thus the 12-th order calculation is reliable also in case of the presence of noise as long as its amplitude is not too high. Additionally, the convergence of the calculation is better in the case with noise than without noise. The result calculated in this way is shown in Fig. 2.7c.

The arrow at the left side of the figure indicates the value of the case without noise. Although the application of a small amount of noise contributes to raising the value of the entropy a bit, the entropy abruptly decreases when the noise level is around 0.01% of the system size. It reaches its lowest level from approximately 0.1% to 1%. Since a high-entropy state indicates disorder and a low-entropy state indicates order, a relative decrease in entropy due to noise shows the formation of order out of chaos with the help of noise. Since the KS entropy is an indication of orbital separation by division of space, it is evident that an extremely large amount of noise will increase the entropy. However, an abrupt decrease of the entropy due to (not too) small noise was unexpected.

The reason for the occurrence of NIO can be explained intuitively as follows. The dynamical entropy expresses an increment in the number of permissible orbits which are separated by partitions of phase space. Here, the role of noise is twofold. Separating orbits in different cells by noise increases the number of permissible orbits. This contributes to the increase in the entropy. On the other hand, an aggregation of orbits coming from different cells into a certain cell decreases the number of permissible orbits. This contributes to the decrease in the entropy. If a map possesses an extremely small slope, as is the case for the BZ map, the latter contribution can be higher than the former. This is particularly clear when the Markov partition is considered.

Let us define the Markov partition for the case of a one-dimensional map, where the definition is recursive. Let the n-th partition be $A^{(n)} = \{(a_1), (a_2), \cdots, (a_n)\}$. Then, the $n + 1$-th partition is defined by $A^{(n+1)} = \{(a_1 a_1), (a_1 a_2), \cdots, (a_m a_m)\}$, where $(a_i a_j) = (a_i) \cap f^{-1}((a_j)) \subset (a_i)$, and $(a_i a_j) \neq \emptyset$. The Markov partition of the BZ map has remarkable characteristics (see Fig. 2.7d, e). In partitions of the same order, cells of extremely different sizes (which are here the lengths of subintervals) coexist. The Markov partition is viewed as an inherent observation window, and as mentioned above, this partition is a partition by which the entropy is maximized. Hence, an inherent character of the system is preserved only by the observation with the Markov partition. In the case of the BZ map, where the size of the subpartitions is extremely different depending on their position in phase space, the inherent observation is complex. Namely, the observation that preserves the inherent character of the system must be made with a different precision depending on the position in phase space. On the other hand, applying noise is one way of observation, which can be viewed as an external observation, contrary to an inherent observation viewed as an internal observation. This type of external observation fixes the observation window, in particular, in the case of uniform noise, the precision for the observation is independent of the position in phase space. Thus, in the BZ map, noise destroys its inherent character. In fact, in the BZ map, many Markov partitions with a width smaller than the noise amplitude exist. Information on these parts is completely lost. One should note that such a phenomenon is not limited to uniform noise. The important point is that the observations break the inherent character of the system, as far as we do not know detailed information on the Markov partitions of the system, prior to observation.

Furthermore, another characteristic of NIO is that the noise generates an order which was not originally conspicuous. The generated order is different from a noisy limit cycle. This was demonstrated by the investigation of the change in the dominant frequency of the power spectrum, depending on the change of the bifurcation parameter. In the case of a noisy limit cycle, the dominant frequency does not change up to the next bifurcation, whereas in the case of NIO the dominant frequency continuously changes as the bifurcation parameter changes.

2.6 Could Structural Stability Lead to an Adequate Notion of a Model?

Following the discovery of NIO, Tsuda considered whether the concept of structural stability needs to be extended in order to understand dynamical

complexity better. The notion of structural stability was formulated by Andronov and Pontryagin [Arnold 1982] on the basis of the belief that a model
for a natural phenomenon must preserve at least the topological character
under perturbations. Taking into account a whole set of dynamical systems
(diffeomorphism), a dynamical system is called structurally stable in the C^r
sense if any other dynamical systems in the C^r neighborhood of such a dynamical system is topologically conjugate (C^0 conjugacy) with the dynamical
system concerned. Many one-dimensional maps such as the logistic map and
the BZ map are not structurally stable, since the structure of infinitely many
bifurcations is embedded in any neighborhood of the chaotic solutions in
parameter space.

However, the above discussions cannot be directly applied to systems with
noise, since generally a noise term is not a dynamical system. Here, it might
be useful to describe briefly a skew-product transformation, which will be
discussed again in Chap. 6. If one regards a noise term as produced by some
chaotic dynamical system, a system with noise can be expressed by a new
dynamical system defined in an extended phase space through a skew-product
transformation. The parameters of the dynamical system expressing the noise
term are, for example, designated as follows. The amplitude of the noise is
designated by the connection strength to the original dynamical system, and
the type of the noise by the bifurcation parameter of the dynamical system.
In such a way, structural stability can be discussed also in a dynamical system
with noise.

Although the BZ map is in the class of structurally unstable systems, taking into account the bifurcation structure (see also Sect. 2.4), the topological
character does not seem to be changed by a slight change in the connection
strength of the extended dynamical system. This is also numerically observed
in many chaotic dynamical systems.[2] This seems to indicate that the original
BZ map keeps a sort of 'structural' stability. Hence, it will be concluded that
the concept of structural stability is too narrow to describe sufficiently the
effects of perturbations. On the other hand, in NIO, the stability by noisy perturbations whose amplitude is not too small must be solved, in relation to the
partitions in phase space. Therefore, it appears difficult to define the stability
in the present case by extending the structural stability. Tsuda emphasized
the necessity of formulating a new concept of (in)stability to capture the
above problems, and proposed the notion of *descriptive (in)stability* [Tsuda
1987a/b, 1990a; Tsuda and Tadaki 1997].

Let us here briefly touch upon what we mean by descriptive instability.
What we are dealing with is the stability of systems under observations. This
was also the case for Andronov and Pontryagin, but the formulation was

---

[2] A similar issue has been pointed out in, for example, the Lorenz system. The
numerical simulation of the Lorenz system shows an invariance of the structure of
chaos in a wide parameter range, though mathematically it should be structurally
unstable.

not necessarily sufficient. The observation in the case of NIO is formulated by the Markov partition. The nonuniformity of the Markov partition leads to nonuniformly observing the states. This nonuniformity brings about a large difference in observation between the cases with and without noise. We would like to deal with the stability problems of the systems under this type of observation. Let us consider the observation by separating it into two ingredients: *measurement apparatus* and *measurement action*. Then, it would be natural to think that a measurement apparatus can be described by a classical dynamical system. The action would be defined by describing the states of the system consisting of the original dynamical system and the dynamical system of the measurement apparatus. In the case of NIO, the measurement apparatus is noise, or another chaotic dynamical system generating noise, and the measurement action is the partition of the extended phase space. Thus, in NIO, it turns out that the measurement action determines the system's behavior.

In the following, let us provide examples where the measurement apparatus changes the system's behavior drastically, despite the fact that it is structurally stable. The first example was given by Moser [1969], and the second one by Rössler *et al.* [1995].

(The model by Moser)

$f : (x, y) \rightarrow (Ax, by)$, where $x \in T^2$ (that is, x is a vector on a two-dimensional torus), $y \in \mathbf{R}$, A is a unimodular matrix, and $b < 1$.

The orbit of this map asymptotically approaches a smooth invariant torus. The following perturbation is added:

$$g : (x, y) \rightarrow (Ax + \varepsilon p(x, y, \varepsilon), by + \varepsilon q(x, y, \varepsilon)), \tag{2.2}$$

where p, q are periodic functions and ε is small. Then, an invariant torus exists. Concerning the smoothness, the following propositions are proved.

(1) The eigenvalues of A are λ and λ^{-1} since A is unimodular. Supposing $|\lambda| < 1$, a smooth invariant torus exists when $0 < b < |\lambda|$. Generally, an invariant torus is r-times differentiable when $0 < b < |\lambda|^r$, where r is an integer.
(2) When $b > |\lambda|$, an invariant torus is nowhere-differentiable, and s-order Hölder continuous, where $s = \log(b) / \log(|\lambda|)$.

In other words, an invariant torus is smooth if it is attracted sufficiently fast, but if not, it loses differentiabilty. Furthermore, two dynamical systems with and without perturbation are topologically conjugate, thus they are transformed into each other by homeomorphism, preserving their topological structure. Therefore, differentiable and nondifferentiable manifolds coexist in a structurally stable class.

This example indicates that structural stability is a wider concept. On the other hand, however, if we use diffeomorphisms, the classification is too fine since each point in the space of diffeomorphisms is classified as being a different dynamical system. More concretely, in the classification by diffeomorphism, dynamical systems with a different linear term are viewed as different dynamical systems. A new method will be necessary to classify dynamical systems producing differentiable and nondifferentiable invariant manifolds.

In our present context, the above example implies that the invariant manifold can be described as being different by the use of the measurement apparatus p or q, independently of the measurement actions. The "fractalization" of the torus described here was independently found by Kaneko [1984a, 1986a]. Furthermore, such an attractor can be represented by nowhere-differentiable functions. This issue and its relation with chaos were mathematically investigated in several references [Hata and Yamaguti 1984; Yamaguti *et al.* 1993].

(The model by Rössler *et al.*)

$$R_0 : (x, y, z, w) \to (9x(\mathrm{mod}\,1), 0.3y - 0.7\cos(9x), 0.3z + 0.7\sin(9x),$$

$$0.3w - 0.7\sin(18x)), \tag{2.3}$$

where $(x, y, z, w) \in \mathbf{R} \times \mathbf{R} \times \mathbf{R} \times \mathbf{R}$.

The chaotic component is the x variable only, and it drives and perturbes the three-dimensional torus. This system is an Axiom A dynamical system, and hence it is structurally stable. Furthermore, the motion on the cross-section which is taken such that the chaotic component is invisible, that is, $x =$ constant, reveals a fractal structure. This way of taking the cross-section yields the observation of orbits in contraction space. In this three-dimensional space, the attractor is represented by a singular-continuous (that is, continuous on a Cantor set) but nowhere-differentiable function, $w = f(y, z)$. In this case, it is implied that an invariant torus is observed as a singular-continuous nowhere differentiable attractor when chaos is used as a measurement apparatus.[3]

These two examples show that the concept of structural stability is too broad to discriminate a smooth and a nonsmooth manifold, and a smooth and a discontinuous manifold. Thus, a new concept of stability different from structural stability is necessary to describe stability under observation in the sense mentioned in this section. For this, we proposed the notion of descriptive stability (see Tsuda and Tadaki [1997] for a possible definition of descriptive stability with a similarity to pseudo-orbit tracing properties).

2.7 Information Theory of Chaos

Information theory of chaos was begun by Oono [1978] and Shaw [1981], and was further developed as a theory by Crutchfield and Packard [1982], by

---

[3] A similar dynamical situation has been found also in neural-network models. See Tsuda [1996] and Tsuda and Yamaguchi [1998].

Matsumoto and Tsuda [1985, 1987, 1988], and by Grassberger [1986, 1989]. Applications of information theory of chaos to the brain sciences started in the works of Nicolis [1982, 1991] and Nicolis and Tsuda [1985, 1989]. Information theory for CMLs was developed by Kaneko [1986b]. Here, we describe the fundamental framework according to the works of Matsumoto and Tsuda. We deal with the case of the one-dimensional map, but it is not difficult to extend the theory to high-dimensional dynamical systems and even flow.

Let us consider a map f defined on the interval J, $f : J \to J$. The fundamental quantity is the following Kullback divergence:

$$I_K(p) = \int p(x) \log \frac{p(x)}{q(x)} dx, \tag{2.4}$$

where $x \in J$. This indicates a relative information content of $p(x)$ to $q(x)$. The evolution rule of the state variable x is given by the dynamical system f, and that of the distribution p is given by the following operator F, where the equation is derived from the law of conservation of distribution:

$$Fp(x) = \sum_{y=f^{-1}(x)} \frac{p(y)}{|df(y)/dy|}. \tag{2.5}$$

This operator F is called the Frobenius–Perron operator.

Let us define the information flow by the difference of the information content before and after applying this operator:

$$\Delta I_K = I_K(p) - I_K(Fp) = \sum_i \{I_K(p_i) - I_K(Fp_i)\}. \tag{2.6}$$

Here, by p_i on the right-hand side is denoted the distribution restricted to the i-th subinterval $p_i(x)$. The partition is appropriately chosen. By simple calculations, the following equation is derived:

$$\Delta I_K = \int dx\, p_0(x) \sum \rho_i(x) \log \rho_i(x)^{-1}, \tag{2.7}$$

where $\rho_i(x) = Fp_i(x)/Fp_0(x)$, and $p_0(x)$ is a stationary distribution, namely $Fp_0(x) = p_0(x)$. Furthermore, the following equation holds if $p_0(x)$ is absolutely continuous with respect to the Lesbesgue measure:

$$\Delta I_K = \int dx\, p_0(x) \log |df(x)/dx|. \tag{2.8}$$

As one sees easily, the right-hand side is just the Lyapunov exponent. Hence, the Lyapunov exponent represents the average information loss per unit time of initial distribution.

Here, our aim is to establish a framework that allows us to understand details of the information structure of chaos. Then, in order to investigate the

observation window

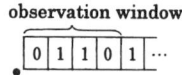

Fig. 2.8. Virtual register. In the treatment of the information content as a statistical quantity, an ensemble of the register is considered.

information flow in detail, let us consider a virtual computer register (see Fig. 2.8). The right direction indicates the microscopic digits of the variable x, and the left direction the macroscopic digits. As shown in the figure, we consider the 'observation window' which is designated by a certain finite number of digits starting from the most macroscopic digit. Our observation should be done only within this window. In the case of chaos, information flows on average from microscopic to macroscopic digits. What we want to calculate, however, is not an average quantity like this but all information going out of the window. Thus, the fluctuations of the information flow should be taken into account. These fluctuations are calculated by the mutual information:

$$I(i;j) = \sum_j p(j) \log p(j)^{-1} - \sum_i \sum_j p(i)p(j/i) \log p(j/i)^{-1}. \qquad (2.9)$$

This is the shared information content of the subintervals i and j. Here, for simplicity, $p_i(x)$ is denoted by $p(i)$, and so on. $p(j/i)$ is a conditional probability of j under a given i. Actually, as shown in the following, it turns out that the fluctuations of the information flow can be calculated using (2.9).

If for $x \in J$

$$F(p_0(x)\chi_i(x)) \sim p(j/i)\chi_j(x), \qquad (2.10)$$

where $p_i(x) = p_0(x)\chi_i(x)$, then

$$\Delta I_K \sim \sum_i \sum_j p(i)p(j/i) \log p(j/i)^{-1} \qquad (2.11)$$

holds.

Condition (2.11) means that the information flows mainly from microscopic to macroscopic digits, namely that the fluctuations of the information flow are small. Then, the information flow is approximated by the second term of the mutual information. In other words, when the fluctuations of the information flow are small, the information on the information structure obtained from the calculation of the mutual information is equivalent to that obtained from the Lyapunov exponent. On the other hand, when the fluctuations are large, the mutual information reveals not only the information flow ΔI_K but also information structures other than ΔI_K, namely the fluctuations of the information flow.

Actually, the local motion of the information flow can be described as follows. A sharp distribution is flattened where the local orbital separation is large, thereby the local information flow directs microscopic digits

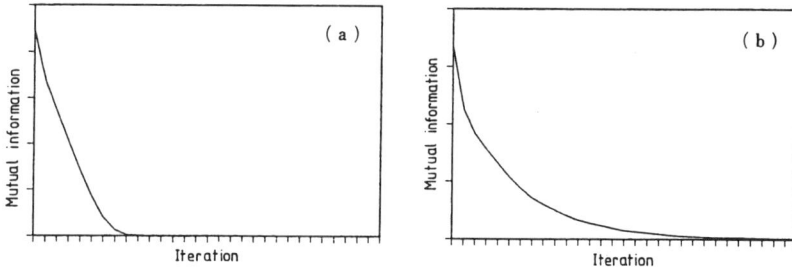

Fig. 2.9a,b. Temporal change of the mutual information. Typical cases of (a) a small and (b) a large fluctuation of the information flow are shown. (Reprinted from K. Matsumoto and I. Tsuda: *J. Phys. A: Math. Gen.* **18** (1985) 3561, with the permission of the publishers.)

to macroscopic digits. On the other hand, if the local orbital separation is small, a flat distribution is sharpened, thereby the local information flow directs macroscopic digits to microscopic digits. Thus, the fluctuation of the information flow is determined by the distribution of the local orbital separation. The information content that this distribution possesses is obtained by calculating the shared information of states. In such a way, the structure of the fluctuation of the information flow can be calculated by the mutual information.

In order to illustrate this point concretely, let us define the time-dependent mutual information:

$$I^{(n)}(i;j) = \sum_j p(j) \log p(j)^{-1} - \sum_i \sum_j p(i) p^{(n)}(j/i) \log p^{(n)}(j/i)^{-1}, \quad (2.12)$$

where n is the discrete time. As shown in Fig. 2.9a, in the case of small fluctuations of information flow, the time-dependent mutual information decays linearly in time, except for the asymptotic form which always possesses an exponential tail. Here we neglect a tiny asymptotic behavior since we extract an essential information structure. In general, if the information on the initial condition decays linearly in time, the same content of information is lost in each time step. This means that the fluctuation of the information flow is small. On the contrary, in the case of a large fluctuation of the information flow, the time-dependent mutual information decays slower than linear, for example by an exponential or a power. This is shown in Fig. 2.9b. In the case of exponential decay, the same ratio of information content is lost in each time step. This is easily derived as follows. If $I^{(n)} = I^{(0)} \exp(-\gamma n)$, $\Delta I^{(n)}/I^{(n)} = $ constant. In the case of a power decay, the same statement holds for the logarithmic time scale. In these slow decays, in the time-development, information in each digit contains information of other digits. This is called information mixing.

This feature is particularly important in a coupled chaotic system. A perfect transmission of input information is allowed because of the information-

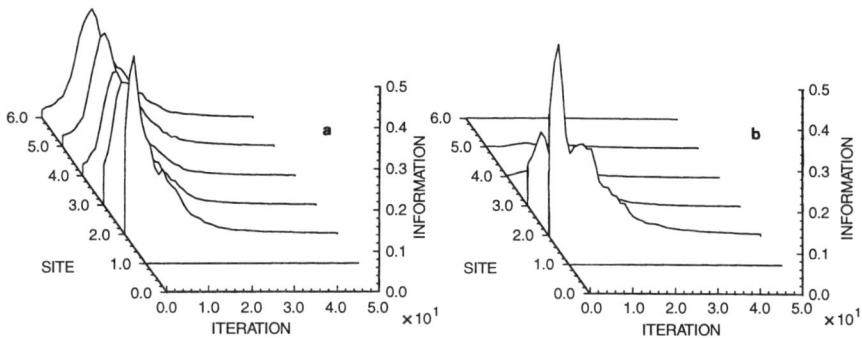

Fig. 2.10 a,b. The transmission of the input information in unidirectionally coupled chaotic maps (open flow). Two typical cases, where elementary chaos possesses (a) a large and (b) a small fluctuation of the information flow are shown. In (a), the BZ map is used. The bifurcation parameter b of elementary map denoted by the site number i $(= 0, 1, \ldots, 6)$ is 0.150, 0.149, 0.148, 0.147, 0.146, 0.145, and 0.144, respectively. In (b), the logistic map is used. The bifurcation parameter a of elementary map denoted by the site number i $(= 0, 1, \ldots, 6)$ is 2.00, 2.50, 3.00, 3.20, 3.40, 3.50, and 3.60, respectively. A coupling strength is fixed to 0.12 in both cases. The change of coupling strength brings about the change of transmission speed of information. (Reprinted from K. Matsumoto and I. Tsuda: *Physica D* **26** (1987) 347, with permission from Elsevier Science.)

mixing property if at least one digit survives, even if the information contained in most digits is lost. This is shown in Fig. 2.10, where the transmission of information fed from outside in a linear chain of chaotic maps is clarified. In the case of coupled chaotic systems, for each element which does not have such a mixing property, the information does not transmit to other connected chaotic elements since the information decays fast in each element.

In Fig. 2.11, the creation of virtual connections between registers for the case of unidirectional coupling of chaotic systems is schematically drawn. Because of information mixing, many-to-many (virtual) connections between digits are created in spite of the one-to-one (real) connection between the chaotic elements. This can be verified by a detailed analysis of the information in terms of the bit-wise mutual information [Matsumoto and Tsuda 1988]. Regarding each bit of a chaotic element as a formal neuron, one may see the formation of multi-layered neural networks with interlayer connections. Then, since each layer is a single chaotic element, in M-coupled chaotic systems, each element which possesses the mixing property of information within an N-bit window can be a simulator of an M-layered neural network with N formal neurons in each layer. It should be noted that the presence of chaotic elements assures an enhancement of the information-processing abilities by a fine-precision computation. The information transmission by information mixing was verified also in neural networks. Thus, we think that a dynamic preservation of information in the brain can be achieved by the presence of chaotic activities in neurons. Related to this issue, see Sect. 5.3 and Sects. 6.7–6.10.

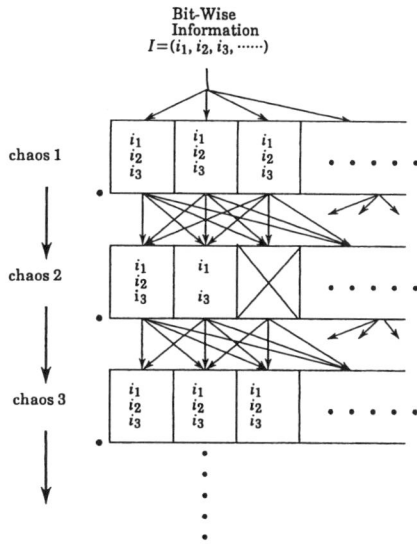

Fig. 2.11. A schematic figure expressing the mechanism of perfect transmission of information in unidirectionally coupled chaotic systems.

3. CMLs: Constructive Approach to Spatiotemporal Chaos

3.1 From a Descriptive to a Constructive Approach of Nature

In physics, we are used to adopting a descriptive approach of nature. Often, one might assume that the macroscopic level of nature can be understood based on an understanding of the microscopic level. For example, the following chart flow is often applicable.

Microscopic law (e.g. Newton's law for molecular dynamics) → coarse-graining → macroscopic equation (e.g. Navier–Stokes equation) → some approximation (e.g. reductive perturbation etc.) → tractable model equation.

Here one might implicitly have assumed a one-to-one correspondence between a model and a natural phenomenon, and examine the predictive power of the model by investigating whether the setting of some parameter to some value will indeed yield the expected results. The term 'control parameter' comes from such an assumption.

However, the above flow chart is often based on 'dangerous' approximations. Especially if a system has descriptive instability, such lucky one-to-one correspondence may no longer hold. By changing the model or a parameter only slightly, an enormous quantitative difference could result.

Still, one might claim that natural phenomena can be reproduced by large-scale computations of a model at the microscopic level. However, this approach is marred by significant fundamental problems.

First, a microscopic model is not necessarily established without the knowledge of the macroscopic level (as a typical example, one should recall that quantum mechanics is constructed such that observed data are consistent with thermodynamics, the macroscopic description of nature).

Second, the microscopic details are amplified to a macroscopic level, and often such microscopic models have descriptive instability. Indeed this is one of the reasons why we need a macroscopic viewpoint. We try to exclude the descriptively unstable level from our consideration of nature in order to understand her. For example, we disregard the individual orbits of molecules and focus on the macroscopic (thermodynamic) level so that the description is stable. In this sense, the description at the macroscopic level is more important 'fundamentally'. Still, there may be the possibility that when chaos exists at the macroscopic level, and thus lead to descriptive instability, that one needs to go up to a higher level of description. In order to avoid such an elevation of the model hierarchy, we adopt a constructive, rather than a descriptive approach of nature.

Third, a further problem of the microscopic description is the 'paradox of computational physics'. This paradox means that the better one succeeds in reproducing the phenomena, the closer one comes to the level of the observation of the phenomena themselves. Even if one traces the orbits of a huge number of particles, it may be impossible to utilize all such information, and indeed, generally, some form of coarse-graining is applied based on existing assumptions and expectations. After such a reduction, we again have to focus on some macroscopic quantities. Then, even if one succeeds in reproducing the phenomena at a macroscopic level, one does not know what is essential to the phenomena.[1] Hence, the descriptive approach derived from microscopic levels, even when concluded successfully, may not lead to the *understanding* of such systems.

Rather, we aim at a constructive approach here. Combining several fundamental processes, we try to 'reconstruct' the phenomena, and from such reconstruction we obtain a better understanding of the phenomena themselves.

Indeed the third problem raised against the microscopic approach is also applicable to other levels. Conventionally, a model equation in physics is believed to have a one-to-one correspondence with the phenomenon under consideration. For example, if one tries to model turbulence, one often adopts a model with a tight connection to the phenomenon studied. If one succeeds in reproducing the phenomena from the model equation, then what can one learn? One dangerous trap in computational physics is that, when successful, what one might obtain as the result is just that the equations are correct or reasonable, without gaining any *understanding* of the complex phenomena. For example, this might be the case for some of the successful simulations that employ the Navier–Stokes equation.

Let us recall a lesson from low-dimensional chaos. If one conducts a splendid numerical simulation on a set of equations with a velocity field and temperature (e.g., Navier–Stokes with buoyancy and heat), one possibly can get the same oscillatory behavior of rolls as in convection experiments. Does this success further our intuition on the origin of this strange oscillation any more than a simple chaotic dynamical system would?

In the studies of CMLs to be presented here, we would like to put forward the point of view that significant understanding can be gained by applying the constructive approach to analysing natural phenomena. By employing CMLs we try to discover novel qualitative universality classes which characterize complex behavior in nature.

---

[1] Of course, there may be the merit that in computer experiments the 'experimental' conditions might be better controlled.

3.2 Coupled Map Lattice Approach to Spatiotemporal Chaos

3.2.1 Spatiotemporal Chaos

The surprise in the research of chaos lies in the fact that a system with a few degrees of freedom, as long as it is nonlinear, exhibits complex behavior. However, the complexity encountered in natural as well as artificial phenomena is more likely to be attributed to a combination of nonlinearity and the fact that the number of involved degrees of freedom is large. Although an important step towards complexity is taken by low-dimensional chaos, it is not the last step.

Roughly speaking, there are four reasons for the emergence of low-dimensional chaos in experiments and real data.

(1) The system has a few degrees of freedom from the beginning (e.g., a pendulum or a Josephson junction with an external force).
(2) Parameters are chosen or tuned around the onset of chaos, so that only a few modes are excited.
(3) Spatial degrees of freedom are suppressed: examples are chemical chaos in which a system is fully stirred or Rayleigh–Bénard convection with a small aspect ratio, where only a few rolls are allowed.
(4) The system is self-organized towards low-dimensional chaos. This possibility may be important in biological systems if low-dimensional chaos yields a better ability of adaptation or fitness in the evolution than high-dimensional chaos, or other dynamical states (periodic or static states, which possibly are not powerful to face the complex "environment").

Reasons (1)–(3) seem to be so restrictive that we have to admit that low-dimensional chaos is not so common in nature. Reason (4) has neither been denied nor confirmed so far, although in an example of an evolution model, the possibility is answered in the negative. Regarding some physiological data such as the heart rhythm, there are evidences for low-dimensional chaos. We note, however, that these examples are often found in sick states, while the dynamics in a healthy state may be too high-dimensional to be analysed with present-day techniques for chaos. Chaotic dynamics in neural activity also seems to be high-dimensional in most cases, except in abnormal states like epilepsy. In short, high-dimensional chaos also seems to be essential for the complex dynamical behavior encountered in biological systems.

Most nonlinear nonequilibrium states tend to exhibit high-dimensional chaos. (Temporally) high-dimensional chaos with spatial degrees of freedom is called *spatiotemporal chaos* (STC) (see Cross and Hohenberg [1993] for a review). It covers turbulent phenomena in general, including Bénard convection, electric convection in liquid crystals, boiling, combustion, MHD turbulence in plasma, solid-state physics (Josephson-junction arrays, spin-wave turbulence, charge-density waves and so on), optics, chemical reactions with spatial structures, and so on. It is also important in biological information processing with nonlinear elements like neural dynamics. In STC, we assume that the involved degrees of freedom are large and increase with the system size. Here, let us elaborate a little on some of these phenomena that provide STC.

- Fluid dynamics:
  Many fluid experiments exhibit spatiotemporal chaos. Rayleigh–Bénard convection is one of the most famous examples in low-dimensional chaos. When a small aspect ratio is adopted to restrict the number of rolls (see Fig. 3.1a), the motion can be described by a set of only a few variables which govern the dynamics of the roll. In fact, such an experiment provides a test-bed for the theory of low-dimensional chaos. When the aspect ratio is large, however, the convection leads to a large number of rolls, whose dynamics involves chaos in space and time (see Fig. 3.1b,c). As the temperature gradient increases, the convection shows turbulent behavior, with a transition in spatial patterns. Related examples can be found in Taylor–Couette experiments (fluid in a rotating cylinder) and Faraday experiments (surface waves). Open flow (like pipe flow, boundary layers, and air jets) also provides us with rich examples for spatiotemporal chaos.
- Chemical reaction–diffusion systems:
  Chemical oscillations are known to be a rich source of chaotic dynamics. In a well stirred system, they have provided some important examples for low-dimensional chaos. If a reaction system is not stirred, it often gives fascinating structures in space and time. Spiral structures and waves are well known to occur in the Belouzov–Zhabotinsky reaction. In a more general situation, chemical turbulence is observed.

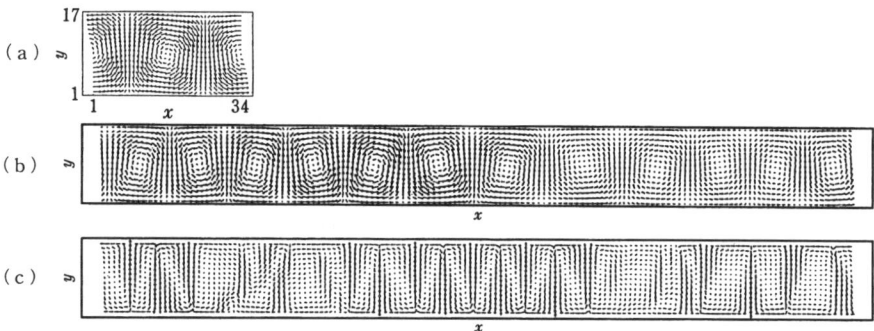

Fig. 3.1 a–c. Snapshot of a velocity pattern in Bénard convection. Computed by the CML model in Sect. 3.5. The horizontal axis has periodic boundary conditions. (a) Two rolls, (b) 12 rolls in the formation process, (c) chaotic oscillations with a larger number of rolls. (With the courtesy of Tatsuo Yanagita.)

- Solid state physics:
  Many phenomena in solid-state physics involve the interaction and propagation of local excitations. If a solid-state system is externally pumped by some fields or currents, it is in general governed by nonlinear dynamics, that can be described by the interplay of local dynamics and spatial interactions. Examples include spin wave turbulence in magnetic materials, arrays of Josephson junctions, charge density waves, nonlinear transmission lines such as optical fibers, coupled systems of nonlinear optical elements, video feedback, chains of electronic nonlinear oscillators, and electron–hole plasmas. The electric convection in liquid crystals gives one of the most important examples for spatiotemporal chaos.
- Climate and geophysics:
  Important examples of spatiotemporal chaos can be found in geophysics. Indeed, a study of the dynamics of the atmosphere is the origin of the field of chaos [Lorenz 1963]. However, the atmosphere can never be low-dimensional, but involves spatial pattern dynamics and turbulence. Some other examples of spatiotemporal chaos in geophysics include mantle convection, long-term climate dynamics, ocean dynamics, and so on.
- Biological networks:
  Biological systems provide a rich source of interacting nonlinear oscillators, besides the above-mentioned neural dynamics. Heart rhythm and capillary oscillation can be modeled as a coupled system of local nonlinear dynamics. A metabolic reaction network involves an ensemble of nonlinear oscillators which can show chaotic behavior. The population of antibodies and antigens, forming an immune network, changes chaotically in time, and provides an example for high-dimensional chaos. On a more macroscopic level, population dynamics for ecological and evolutionary systems often produces high-dimensional chaos, where spatial variation is important.
  Of course a neural network is a coupled dynamical system with a complicated connection. For this problem see Chap. 6.

3.2.2 Introduction to Coupled Map Lattices

To start a novel paradigm, we often need a standard model. In low-dimensional chaos, there are some well-known dynamical equations (e.g., Lorenz equation, Rössler equation, logistic map, Hénon map, and so on) that provide this role.

For high-dimensional chaos, we have some candidates for a standard model. The traditional approach (for spatiotemporal chaos) adopts partial differential equations. There are some drawbacks to this approach, however, when we try to study heuristically the phenomenology of high-dimensional chaos, as will be discussed. A cellular automaton is another possible candidate, but it has again some drawbacks that will be mentioned later.

The coupled map lattice (CML) was introduced to answer the needs for an extensive and intensive study of spatiotemporal chaos [Kaneko 1984b, 1986a;

Waller and Kapral 1984; Crutchfield and Kaneko 1987]. It is presented as a dynamical model for the evolution in time of a spatially extended system, and as a tool to explore the behavior of such a system. Studies in CMLs have expanded not only in the field of spatiotemporal chaos and pattern formation but also in biology, mathematics and engineering.

A CML is a dynamical system with discrete time ("map"), discrete space ("lattice"), and a continuous state. It usually consists of dynamical elements on a lattice which interact ("coupled") with suitably chosen sets of other elements.

The essence of a CML lies in the reductionism in procedure, not in the level of the elements. Starting from a suitably coarse-grained description, we introduce reductionism on a macroscopic level. The construction of a CML is carried out as follows:

(A) Choose a (set of) field variable(s) on a lattice. This set of variable(s) is not on a microscopic, but on a macroscopic level (e.g., temperature, fluid-velocity field, concentration of some chemical substances).

(B) Decompose the phenomena into independent units (e.g., convection, reaction, diffusion, and so on).

(C) Replace each unit by simple parallel dynamics ("procedure") on a lattice: the dynamics consists of a nonlinear transformation of the field variable at each lattice point and/or a coupling term among suitably chosen neighbors.

(D) Carry out each unit dynamics ("procedure") successively.

Let us take a simple example (see Fig. 3.2 for a schematic illustration). Assume that we want to study a phenomenon (in a fluid), created by a locally chaotic process, and by diffusion. A simple reductionist might start from a

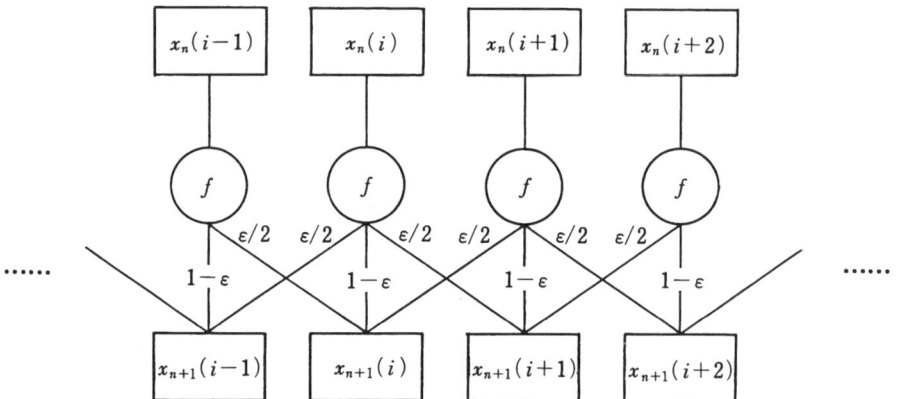

Fig. 3.2. A schematic representation of a CML. The variable $x_n(i)$ at time n and lattice site i evolves according to the two procedures, the transformation leading to local chaos $x \to f(x)$, and the diffusion coupling.

microscopic model, like molecular dynamics or lattice gas cellular automata. Some others might believe only in equations on a coarse-grained level like the Navier–Stokes equation. Our CML approach is clearly different from both of the two; reductionism in procedure. We reduce the phenomena into local chaos and diffusion. Then we choose a suitable lattice model on a coarse-grained level for each process. As the simplest choice we can adopt some one-dimensional map for chaos, and a discrete Laplacian operator for the diffusion.

The former process is given by

$$x'_n(i) = f(x_n(i)), \tag{3.1}$$

where $x_n(i)$ is a variable at time n and lattice site i, and $x'_n(i)$ is introduced as the intermediate value. The discrete Laplacian operator for diffusion is given by

$$x_{n+1}(i) = (1 - \epsilon)x'_n(i) + \frac{\epsilon}{2}\{x'_n(i + 1) + x'_n(i - 1)\}. \tag{3.2}$$

Combining the above two processes, our dynamics is given by

$$x_{n+1}(i) = (1 - \epsilon)f(x_n(i)) + \frac{\epsilon}{2}\{f(x_n(i + 1)) + f(x_n(i - 1))\}, \tag{3.3}$$

where n is a discrete time step and i is a lattice point ($i = 1, 2, \cdots, N =$ system size).

The mapping function $f(x)$ is chosen to depend on the type of local chaos. For example, one can choose the logistic map, $f(x) = 1 - ax^2$, for typical chaos emerging through period-doubling bifurcations.[2] The dynamics of this logistic map shows bifurcation from fixed point to period-2 cycle, then period-4, 8, $\cdots$. This period-doubling bifurcation accumulates at $a = 1.4011 \cdots$, above which chaos appears. Several windows of stable cycles exist among the parameters for chaos.

In the model (3.1), the independent procedures of (B) are local transformation and the diffusion process, which are separated parallel procedures. The model consists of the sequential repetition of these two procedures. This argument leads to the following equivalent form with the above model: $y_{n+1}(i) = f((1 - \epsilon)y_n(i) + \epsilon/2(y_n(i + 1) + y_n(i - 1)))$.

If we adopt different procedures, we can construct models for different types of dynamical behavior of spatially extended systems. For problems of pattern formation, it is useful to adopt a map with bistable fixed points (e.g., $f(x) = \tanh(\beta x)$) as a local dynamics. Examples with a different type of coupling ("convective coupling") are also possible. The extension to a higher-dimensional space is quite straightforward. One only has to use a higher-dimensional discrete Laplacian operator.

---

[2] The form $f(z) = rz(1 - z)$ is often adopted. With the transformations $a = r(r - 2)/4$ and $x = 4(z - 2)/(r - 2)$ these two maps are identical.

Another easy extension is a spatially asymmetric coupling: instead of the diffusion procedure, we often encounter a spatially asymmetric interaction in nature. A typical case, to be discussed later, is open fluid flow, where the coupling from up-flow to down-flow is larger than the one in the opposite direction.

3.2.3 Comparison with Other Approaches

Our CML approach is of a constructive nature since we try to model a system by combining procedures and to find the phenomenology of complex spatiotemporal behavior. We believe that the constructive approach is essential for the understanding of complex systems. If this approach gives a novel class of notion, which reproduces some natural phenomena, we can *understand* the complex behavior, even if (or *because*) the model equation itself is not completely derived from a microscopic level. This approach is based on the belief in the existence of universality classes in physics. A model cannot be exactly same as nature herself anyway, and we have to assume that there is some universality of phenomenology, independent of the details of the modeling.

In CML studies, we search for novel qualitative universality classes, without bothering with the details of the phenomenology. Through this approach, we understand how such phenomenology appears, in what class it is commonly seen, and what the essence of the phenomena is. Only through this approach we can understand why some type of complex behavior is common in nature, irrespective of the details, and then we can predict what class of systems leads to such behavior.

Of course, the traditional approach to STC is the use of partial differential equations (PDEs). In the context of physical problems, the use of partial differential equations is standard.

In PDEs, state, time, and space are all continuous. The other extreme limit towards discretization is the so-called cellular automaton (CA), originally introduced by Ulam and von Neumann as a model for self-reproduction and the computer architecture of the present-day type. From the side of physics, the CA has recently been used as a simple simulator for statistical behavior, like the Ising model-type CA or a lattice-gas automaton for fluid dynamics.

Basic structures of the three models are summarized in the following Table 3.1.

Besides the significance of having a constructive approach, the merits of CMLs over PDEs and CAs can be summarized as follows:

(1) Since the model is numerically efficient with parallel computation, we can carry out a heuristic study. One can discover novel phenomena that are not observed in the experiment yet, and propose a new concept that can be tested in experiments.

Table 3.1. CA, CML, and PDE.

Model	Space	Time	State
Cellular Automaton	D	D	D
Coupled Map Lattice	D	D	C
Partial Differential Equation	C	C	C

D = Discrete,   C = Continuous.

(2) Since the model is constructive, it is easy to generalize the observed phenomena and propose a new concept underlying several phenomena in nature. By adapting the meaning of the procedure to each specific problem, one can see which part of the process is essential for the phenomena.

(3) One important feature of a CML lies in its semi-macroscopic (mesoscopic) description. Each variable at a lattice point represents not a microscopic but a semi-macroscopic state. This is in contrast with a CA or a spin system of Ising type. In order to simulate physical problems, a CA requires a huge number of cells, while a moderate number is sufficient for a CML.

(4) Since the model has dynamical variables and parameters, dynamical systems theory is applicable. This is in strong contrast with a CA. Since there is no inherent continuous parameter in a CA, it is rather difficult to see a change of a state with a bifurcation parameter. Applications of dynamical systems theory are difficult in a CA. Also, by adopting a map studied in low-dimensional chaos one can understand some features of spatiotemporal chaos in terms of dynamical systems. This is not necessarily easy in the PDE approach.

(5) Due to the discreteness in space–time, one can apply some statistical mechanical argument developed in lattice systems.

(6) By changing the procedures in a CML, one can easily construct a model for dynamical phenomena in space–time, where the essence is abstracted.

3.3 Phenomenology of Spatiotemporal Chaos in the Diffusively Coupled Logistic Lattice

3.3.1 Introduction

Novel qualitative universality classes in spatiotemporal chaos are discovered from extensive numerical simulations of coupled map lattices. In the present section we give a rough introduction to such phenomenology by discussing the coupled logistic lattice. Here we restrict ourselves only to the qualitative nature, while the original papers can be consulted for a detailed quantitative analysis.

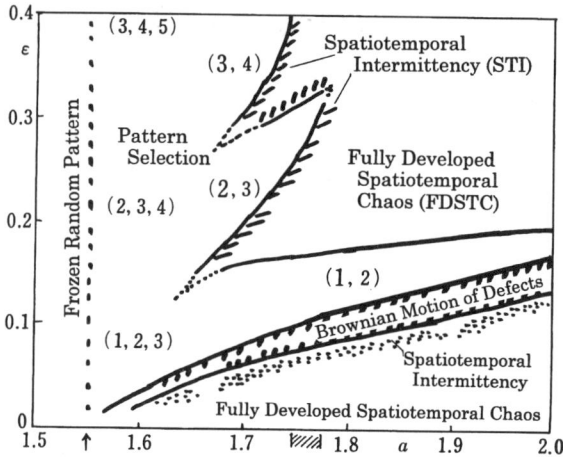

Fig. 3.3. Phase diagram of the CML (3.1). (From K. Kaneko, *Physica D* **34** (1989a) 1, with the permission of the publishers.)

In the logistic lattice, as will be shown below, we can observe the following main succession of patterns: doubling of kinks, frozen random pattern, pattern selection, traveling waves, spatiotemporal intermittency, and fully developed spatiotemporal chaos. Since the model has only two parameters, the nonlinearity a and the coupling strength ϵ, one can draw a phase diagram in the (a, ϵ) plane (see Fig. 3.3). Note that this class of successive changes is found in a wide range of CMLs, as well as in a class of spatially extended dynamical systems, supporting our assumption of qualitative universality.

3.3.2 Frozen Random Patterns and Spatial Bifurcations

As the logistic map shows period-2^n oscillation, it is relevant to study the phase of oscillations with respect to period-2 motion. For example for the parameter corresponding to the period-2 motion, each lattice point shows period-2 motion, but the phase of the oscillation is not necessarily identical. Several domains are formed where elements oscillate in-phase within, and out-of-phase with the neighboring domain. Generally speaking, for the period-2^k region, all lattice sites in a domain have the same phase of oscillation. The domains are separated by kinks that develop at sites whose amplitudes are near unstable periodic points of 2^{k-1} period. Just like in the case of the isolated logistic map, the coupled logistic lattice (1) also exhibits a period-doubling of kinks as the nonlinear parameter a is increased [Kaneko 1984b]. By these doublings, domains of various sizes are formed.

After the cascade of doublings, the system exhibits chaotic behavior. Due to the sensitive dependence on initial conditions that is characteristic of chaos, a homogeneous state becomes unstable in our CML with chaotic components.

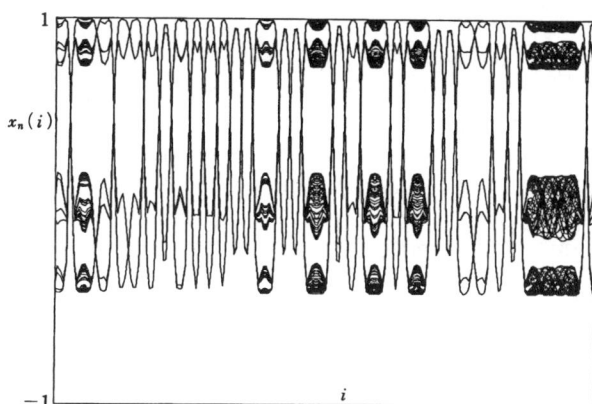

Fig. 3.4. Pattern of $x_n(i)$, overlaid from $10\,000$ to $10\,250$ steps. $a = 1.42$, $\epsilon = 0.4$, $N = 200$. Depending on the domain, the phase of the oscillations is different.

The domain structure is spontaneously created by chaos, even if we start from an almost homogeneous initial condition.

Domain structures are formed, as regions in which the site values are correlated up to some specific degree in space and time. The correlation here is partial coherence with respect to the period-2^k oscillation, instead of a complete phase coherence. These domain structures remain even if the patterns are chaotic. Examples of chaotic states are shown in Fig. 3.4, where the patterns through 16 time steps are shown after the transients have decayed away.

A noteworthy feature for the spatially-extended systems is the *spatial bifurcation*. Even if the dynamics itself is homogeneous, the local state can differ from lattice site to site. In the example of Fig. 3.4, the motion is chaotic in a large domain, while it is almost period-8 at smaller domains, period-4 for much smaller domains, and period-2 for the smallest ones.

A description of the spatial bifurcation can be developed as follows. As already noted, the kinks pass through saddle points which are separatrices of two out-of-phase regions. That is, the kinks connect two domains through locally unstable fixed or periodic points. The dynamics in a domain can be approximated by the dynamics of a small system with fixed boundary conditions at both ends; specifically, fixed at the value of the unstable points. The narrower the domain, the more highly constrained the dynamics and the simpler the behavior. Thus as a function of domain size, we have a bifurcation sequence to more complex behavior to chaos.

Of course, the replacement of a domain by a finite system is an approximation. In principle, the kinks may move or the domains may affect one another. Nonetheless the approximation appears valid, at least if the coupling is not too strong.

Another interesting point is the hierarchical domain structure. Within each domain structure with the π phase difference, there are often sub-domains that are separated by the $\pi/2$ phase difference (i.e., with respect to the period-4 structure). Within such sub-domains, again smaller-scale domains with a $\pi/2^k$ difference may exist successively (see Fig. 3.5). As one looks at a lower level of sub-domain, they scale both with a and the domain size decreases successively. In fact, this scale change can be calculated by extending Feigenbaum's renormalization group [Feigenbaum 1979] method for the period-doubling cascade [Kuznetsov 1986, 1993].

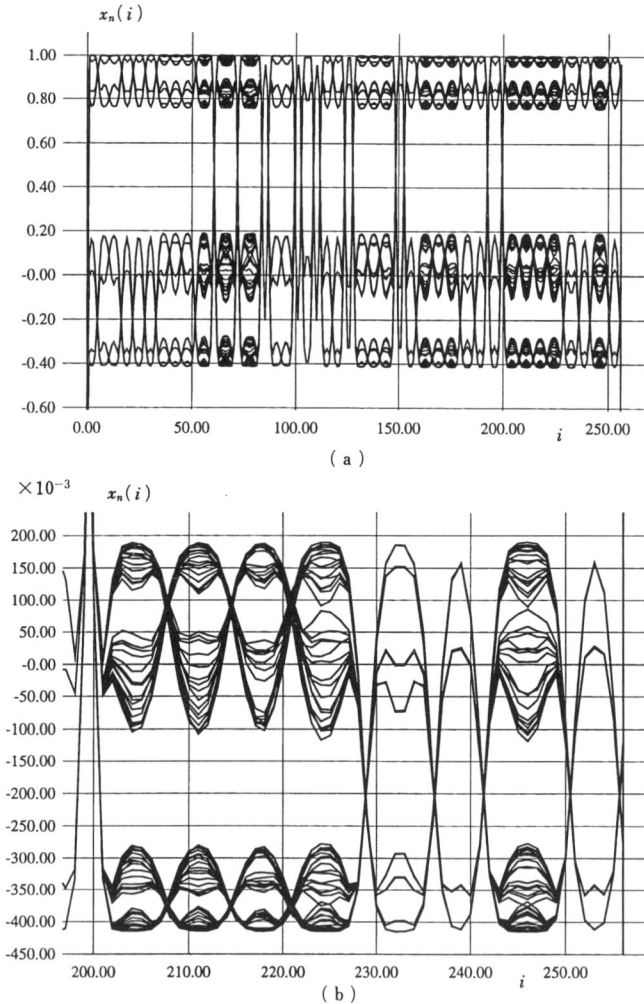

Fig. 3.5. (a) Pattern of $x_n(i)$, overlaid from 10 000 to 10 250 steps. $a = 1.41, \epsilon = 0.2, N = 250$. (b) Expansion of a part of (a).

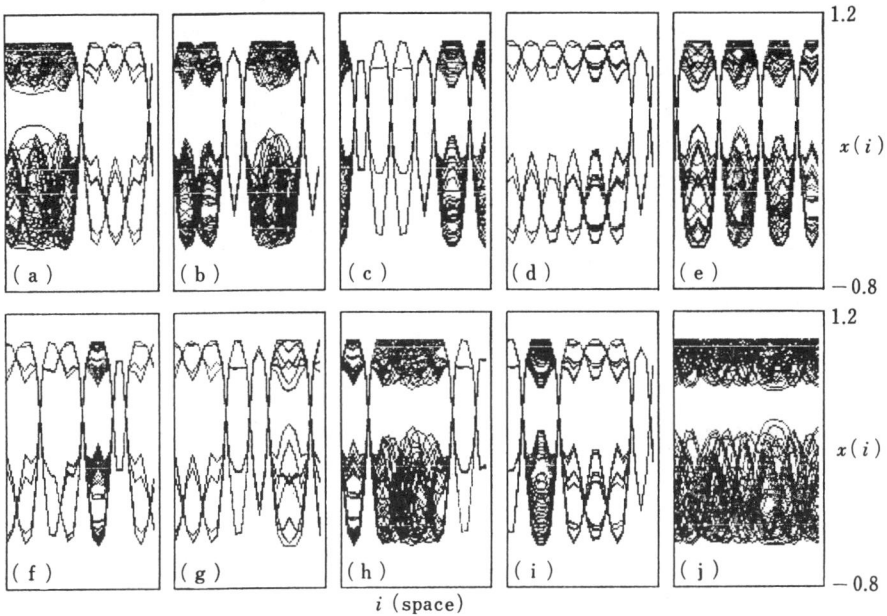

Fig. 3.6 a–j. Pattern of $x_n(i)$ overlaid from 10 000 to 10 250 steps. Examples from 10 randomly chosen initial conditions are shown in (**a**) to (**j**). $a = 1.5$, $\epsilon = 0.3$, $N = 25$.

The distribution of domain sizes can depend on the initial conditions. We can choose initial conditions so that attractors have an arbitrarily large domain. The number of attractors is huge, which is expected to increase exponentially with the system size. In Fig. 3.6, we have plotted final states starting from 10 arbitrarily chosen random initial conditions. Note that they fall on different attractors with different domain structures.

When the variable of one lattice point is perturbed externally, the perturbation is amplified by chaos and propagated to other lattice points. Thus it is possible to switch from one domain structure to another, by injecting an input at some lattice point. In this case, 'sub-domain' structures are more easily destroyed, and the switch can be 'hierarchical'. Also, the amplification and propagation rates depend on the degree of chaos in each domain. Hence the effect of the input depends on the domain structure itself. Relevance of the present domain structure to hierarchical information processing will be discussed in Chap. 5.

3.3.3 Pattern Selection with Suppression of Chaos

As the nonlinearity is increased further, larger domains start to be unstable and split into smaller domains (see Fig. 3.7). Initial conditions are no longer

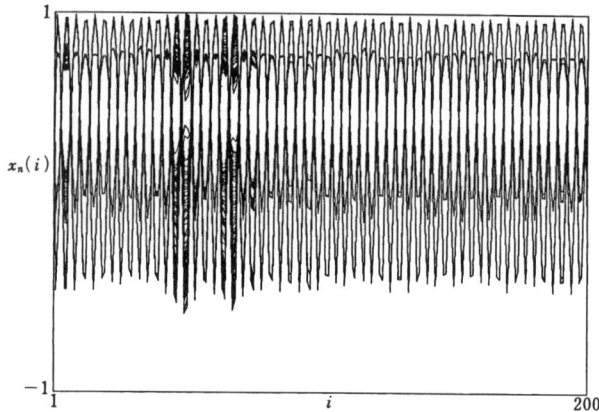

Fig. 3.7. Pattern of $x_n(i)$. Overlaid from 10 000 to 10 250 steps. $a = 1.71$, $\epsilon = 0.4$, $N = 250$. (From K. Kaneko, *Physica D* **34** (1989a) 1, with the permission of the publishers.)

preserved. Through the transient process, domains of a few special sizes are selected.

After the selection, the pattern of domains is frozen and does not move in space. Selected are domain sizes such that the dynamics therein is less chaotic, that is, a motion with shorter periods. In the frozen random pattern, chaos is suppressed only in domains of small sizes.

The diffusion tries to homogenize a system, while the chaotic motion makes the system inhomogeneous due to the sensitive dependence on initial conditions. These two tendencies conflict with each other. In a large domain the chaos is so strong that it splits into smaller domains. Since the neighboring domains oscillate out-of-phase (with π phase difference), the chaotic instability is smeared out. Once a domain structure is formed with suppression of chaos, the conflict is resolved and the domain structure is stabilized. Still, we have no clear mathematical foundation to support the hypothesis of pattern selection that suppresses chaos.

The suppression of chaos often leads to quasiperiodic behavior. In nonlinear systems, quasiperiodic states are commonly observed. Many dynamical systems exhibit a bifurcation sequence to successively more complex behavior that follows the path

Limit cycle (Hopf bifurcation) $\rightarrow$ quasiperiodic state (two-dimensional torus) $\rightarrow$ locking (devil's staircase structures) $\rightarrow$ chaos or high-dimensional torus $\rightarrow$ development of chaos.

3.3.4 Brownian Motion of Chaotic Defects and Defect Turbulence

The simplest example of pattern selection is that of domain size 1 (i.e., a wavelength of two lattice sites), with (almost) period-2 temporal oscillations.

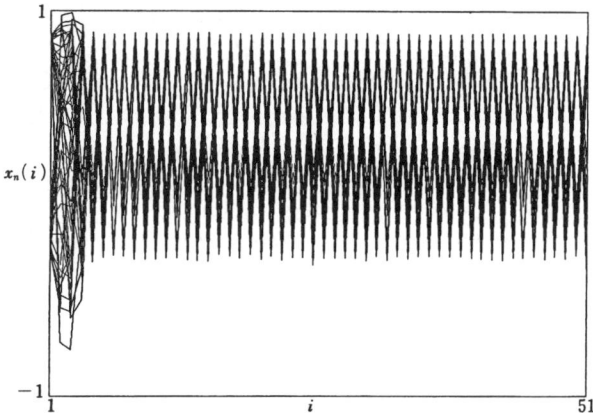

Fig. 3.8. Pattern of $x_n(i)$. Overlaid from 10 000 to 10 250 steps. $a = 1.83$, $\epsilon = 0.1$. Since $N = 51$ is odd and the boundary condition is periodic, a defect remains and moves around (in the figure it is located around $i \approx (1 \sim 4)$). (From K. Kaneko, *Physica D* **34** (1989a) 1, with the permission of the publishers.)

The (period-2) oscillation of the zigzag pattern is formed as shown in Fig. 3.8. There can be two regions of the zigzag pattern with a different phase of oscillation. In one domain, the oscillation is large-small-···, while for the other it is small-large-··· (i.e., the phases differ by π). These two regions are separated by a defect. Although in the zigzag pattern chaos is suppressed, this is not the case in the defect region. Thus the dynamics is chaotic at the defect position. This defect, changing its waveform chaotically in time, also moves in space (see Fig. 3.9). In fact this motion in real space can be interpreted as Brownian motion. Note that in contrast with the chaos in 'phase space' (the N-dimensional space of variables $x(i)$), the Brownian motion here is in 'real' space. Here the chaos in phase space is the trigger of the Brownian motion. The diffusion constant of the Brownian motion changes with a, almost proportionally to the degree of chaos (Kolmogorov–Sinai entropy of the defect, measured from the sum of positive Lyapunov exponents).

If the nonlinearity a is small, these defects pair-annihilate by collision. Thus the number of defects decreases with time to form a zigzag pattern with a single domain. For a larger nonlinearity a, the defects start to be formed spontaneously, although they still pair-annihilate by collision (see Fig. 3.9c). Thus the system repeats the formation and collapse of defects. This type of phenomena is studied as 'defect turbulence' in convection of liquid crystals, and belongs to the following class of 'spatiotemporal intermittency'.

3.3.5 Spatiotemporal Intermittency (STI)

As the nonlinearity a is increased further, the domain is no longer stable and the pattern starts to collapse. As shown in Fig. 3.10, each lattice point alternates irregularly between the ordered states with patterns (to be called laminar regions) and spatially disorganized, temporally chaotic regions (to be called burst regions). In the study of low-dimensional dynamical systems, intermittent chaos between ordered (laminar) states and chaotic bursts was studied in detail. When measured at a lattice point, our phenomena show similar switching between the two states. Here the intermittent switching is seen also in space, as we move from one lattice point to another. STI is generally observed at the transition from an ordered state to a completely disordered turbulent state and has now become one of the most thoroughly examined topics in spatiotemporal chaos.

In STI, both laminar motion and turbulent bursts coexist in space–time. Laminar motion is characterized by periodic or weakly chaotic dynamics with a spatially regular structure, while turbulent bursts have no regular structure in space–time. In fact, two types of STI are known.

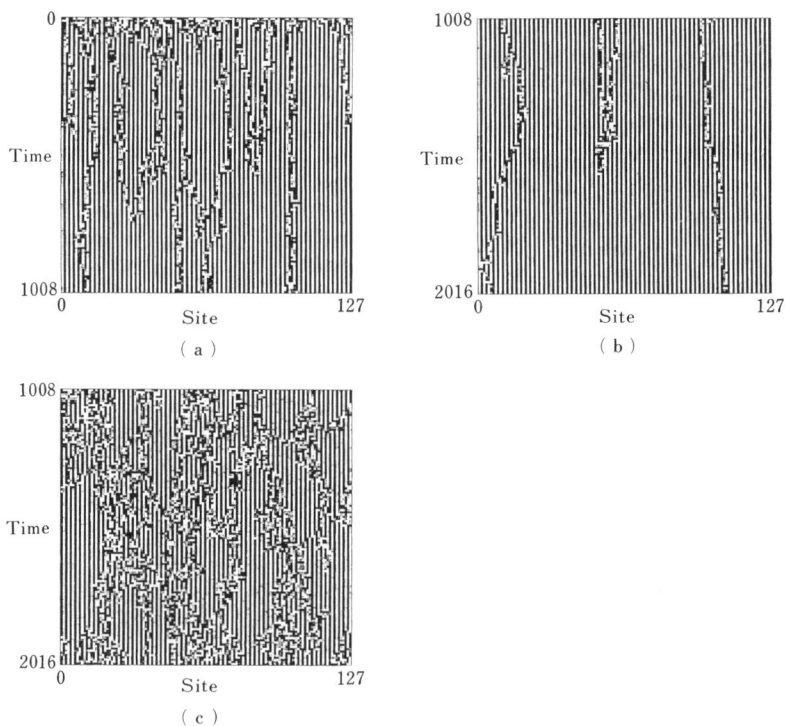

(a)

(b)

(c)

Fig. 3.9 a–c. Space–time diagram of the evolution of $x_n(i)$. The value of $x_n(i)$ is plotted as a gray-scale plot where each pixel corresponds to (i, n), with the horizontal axis representing space i, and the vertical axis representing time n. The pattern is plotted every 8 time steps ($N = 128$, $\epsilon = 0.1$). (a) $a = 1.85$, (b) continued from (a), (c) $a = 1.895$.

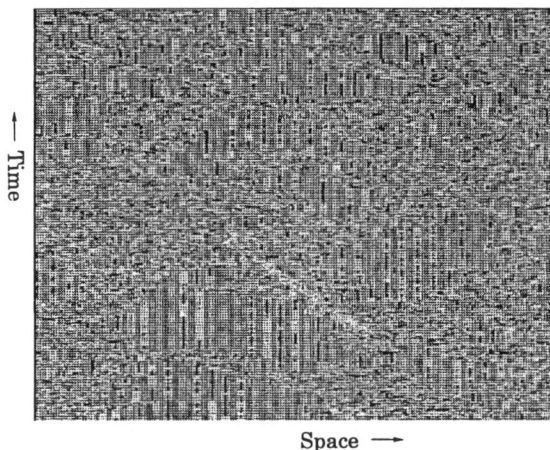

Fig. 3.10. Space–time diagram of the evolution of $x_n(i)$. Gray-scale plot is used that represents the value of $x_n(i)$. The pattern is plotted every 8 time steps ($N = 200$, $a = 1.75$, $\epsilon = 0.3$).

In the first type of STI [Kaneko 1984b, 1985a; Chaté and Manneville 1988], there is no spontaneous creation of bursts. If a site and its neighbors are laminar, it remains laminar at the next step. Before the onset of STI, a spatially homogeneous state is stable, which is temporally periodic.

This first type of STI is for example at a CML with a weak coupling, when the local dynamics includes topological chaos, but has a periodic attractor. A typical example is a CML with local dynamics with the period-3 window of the logistic map, where the laminar state is a homogeneous period-3 domain. It is discussed [Chaté and Manneville 1988; Grassberger and Schreiber 1991] that this STI transition is related to (directed) percolation [Obukhov 1980; Pomeau 1986].

In the second type of STI, spontaneous creation of turbulent bursts exists [Keeler and Farmer 1986; Kaneko 1989a, b], as long as some coarse-grained reduction of states is adopted. There is some probability that bursts are created, even if all the states of a site and its neighbors are laminar. It is necessary to introduce (infinitely) many states between "laminar" and "burst" so that no spontaneous creation from "laminar" is possible.[3]

---

[3] If this partitioning is possible with a finite number of states, the distinction between the two types of STI may be superficial. In typical cases, we suspect that such finite-number partitioning may not be possible. Then, the STI here cannot be represented by a cellular automaton or a related percolation model. For example, all the lattice points may come close to a regular pattern, and stay there for many time steps. Then bursts start to be created spontaneously at some lattice points. The mechanism of creation is possibly due to an instability associated with a riddled basin [Sommerer and Ott 1993] and on–off intermittency [Fujisaka and Yamada 1986; Yu *et al.* 1990]. Many initial conditions (with a finite

Indeed, the example given in Fig. 3.10 belongs to this type-II STI. This type of STI is observed as a transition from local to global chaos, and even before the onset of STI there is a spatial structure given by the selection of patterns. The STI here is more robust than the type-I STI, and the intermittent behavior is seen in a larger range of parameter regimes.

In STI, the temporal change corresponding to the selected pattern has a very long memory. For example, we have measured the residence-time distribution at laminar states in Fig. 3.10. In particular the distribution of the lifetime $P(t)$ of laminar states measured at each lattice point obeys the power law

$$P(t) \approx t^{-\beta} \tag{3.4}$$

with $\beta \approx 1.7$. Correspondingly, the power spectrum obtained from the Fourier transform of the spatiotemporal pattern shows a characteristic behavior. Let us adopt the power spectrum in frequency and wavelength domain as

$$S(k,\omega) = \left| \sum_j \sum_n x_n(j) \exp(2\pi\mathrm{i}(kj - \omega n)) \right|^2. \tag{3.5}$$

In Fig. 3.11, we have plotted $S(k,\omega)$ by fixing k at 0, 1/4, 3/8, and 1/2 corresponding to the STI for Fig. 4.9c. Only when k is close to k_p the wavelength of the selected pattern (in the present case $k_p = 1/2$), $S(k,\omega)$, shows a power-law behavior with respect to ω. In fact $S(k,\omega) \approx \omega^{-1.9}$ for $k \approx 1/2$ in Fig. 3.10. This selective flicker-like noise characterizes the long-term motion of the pattern, as is already shown in the power-law distribution of the laminar time. On the other hand the motion with other wavenumbers corresponds to the chaotic burst motion, and is governed by a short time scale. This is the reason why the Lorentzian spectrum is observed in Fig. 3.10a, b.

STI has recently been observed in various examples at the transition from regular states to STC. It is studied in a variety of CML, PDE systems such as the Ginzburg–Landau equation [Chaté and Manneville 1987]. It is also found in several experiments [Ciliberto and Bigazzi 1988; Diaviaud et al. 1989, 1990; Nasuno et al. 1989; Michalland et al. 1990], including Bénard convection with a large aspect ratio, Faraday instability of waves, 2-dimensional electric

---

measure) are attracted to the regular pattern, while some unstable directions from the attractor remain, from which the orbits escape out of the regular pattern state. Although it has not been confirmed, we believe that this state is a Milnor attractor (or close to it) [Kaneko 1997]. Here the Milnor attractor, following the first argument by Milnor, is defined as a state which is an invariant set by the dynamics, and has a finite measure of attracting orbits to it, but there are orbits that go out of it in some neighborhood arbitrarily close to it [Milnor 1985]. If it is proven that the regular pattern within type-II STI corresponds to the Milnor attractor, the spontaneous creation of bursts naturally follows. (See also Hata et al. [1996] for stability of multiple attractors in a CML.)

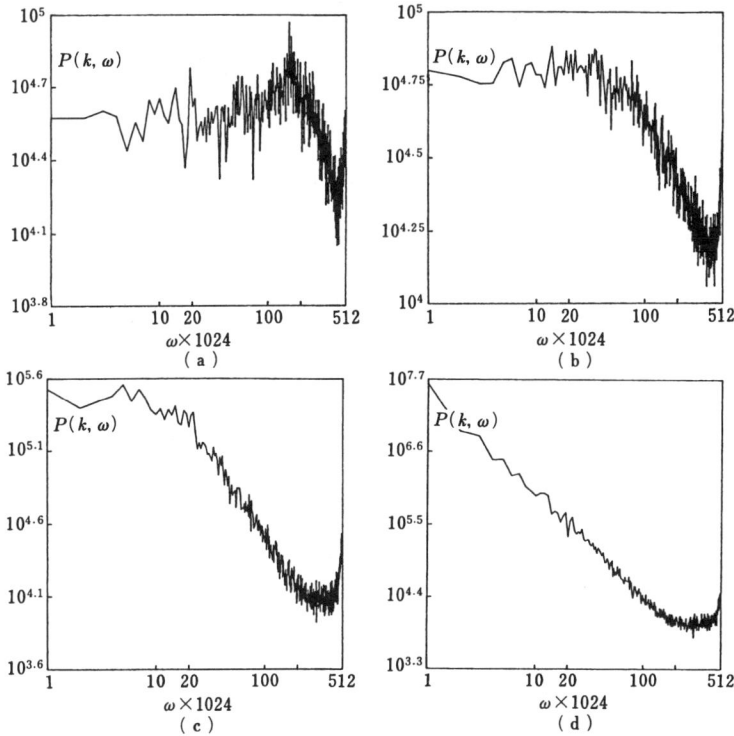

Fig. 3.11 a–d. Log–log plot of $S(k,\omega)$ as a function of ω ($N = 256$, $a = 1.89$, $\varepsilon = 0.1$). (a) $k = 0$, (b) $k = 1/4$, (c) $k = 3/8$, (d) $k = 1/2$. These power spectra are calculated from the data of $x_{2n}(i)$ for 512×2 time steps, after 10 000 steps of transients, and from 50 sequential samplings. Random initial conditions were used. (From K. Kaneko, *Physica D* **34** (1989a) 1, with the permission of the publishers.)

convection in liquid crystals, and viscous rotating fluid. Indeed the power-law distribution of residence time at laminar states, as well as the selective flicker noise in $S(k,\omega)$, are confirmed therein. It may be also interesting to note that all the experimental observations so far seem to correspond to the type-II STI.

3.3.6 Stability of Fully Developed Spatiotemporal Chaos (FDSTC) Sustained by the Supertransients

As the nonlinearity a is further increased, STC has no ordered spatial structure. This state can be interpreted as a direct product of local chaos at each lattice point. Here the spatial correlation decays exponentially, and two lattice points whose distance is farther than this correlation length are regarded as independently changing chaotically.

This FDSTC has stability against parameter change. To see this recall that in the single logistic map, chaos is structurally unstable (see also Sect. 2.6). In the logistic map, small periodic window structures are interspersed in any parameter regime. In any neighborhood of the parameter space leading to chaotic motion, a window exists where the attractor is periodic. Accordingly, the Lyapunov exponent, plotted as a function of a, shows several 'valleys' (negative values) as shown in Fig. 3.12a. By decreasing the increment of a in the plot, the number of such valleys increases. Indeed the number increases to infinity as the increment decreases to zero, corresponding to the existence of a countably infinite number of windows.

On the other hand, we have not observed any window structures in FDSTC, for any random initial conditions. Indeed, the maximal Lyapunov exponent (Fig. 3.12a) and the Kolmogorov–Sinai entropy (computed by the sum of positive Lyapunov exponents) changes smoothly with the parameter a in FDSTC, in contrast to the (infinitely) many drops in the Lyapunov exponent of the single logistic map. The absence of window structures is rather surprising, since in a coupled system the homogeneous state with the stable cycle for the window is linearly stable. If x^k $(k = 1, 2, \cdots p)$ is a stable periodic solution of the logistic map, then the homogeneous solution $x(i) = x^k$ independent of i is a linearly stable solution of our CML. This can easily be confirmed by recalling that our CML consists of the processes of local chaos $x \to f(x)$ and the diffusion process. First, if $x(i)$ is homogeneous, the diffusion cannot give any change, and our dynamics reduces to a single logistic map. Hence $x(i) = x^k$ is a periodic solution of our CML. Then, if we start close to this homogeneous solution, the diffusion works to homogenize the system, and the local dynamics forces the orbit to approach $x(i) = x^k$ since the solution is linearly stable in the single logistic map. Thus as long as we start from an initial condition in the vicinity of a homogeneous state, the windows should exist.

Then why are such windows not observed? In fact, the time to be attracted to such a homogeneous state $x(i) = x^k$ is practically infinite. Here we mean by the term 'practically', that the transient time necessary to be attracted to such state increases exponentially with the system size, as will be discussed in Sect. 3.3.8 as a "supertransient". As long as the system size is not too small (e.g., larger than 10 lattice sites), it is in practice impossible to wait and see whether the system really falls onto the homogeneous attractor. The stability of FDSTC is sustained by this supertransient.

Of course, there are some initial conditions that are attracted to the homogeneous state within a few time steps. However, the volume of such initial conditions that lead to the homogeneous state decreases very rapidly (exponentially) with the system size.

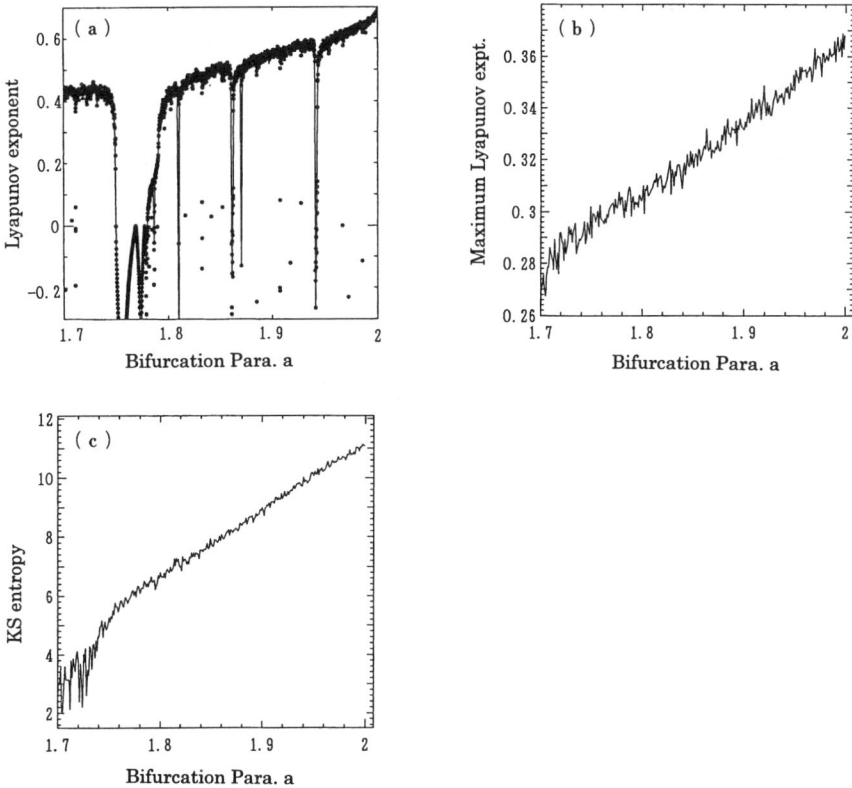

Fig. 3.12. (a) Lyapunov exponent of the single logistic map $x_{n+1} = 1 - ax_n^2$ for $1.7 < a < 2.0$. The solid line is obtained by incrementing a by 0.001, while the points ● are obtained by using an increment of 0.0001. (b)The maximal Lyapunov exponent of the CML $x_{n+1}(i) = (1-\varepsilon)f(x_n(i)) + (\epsilon/2)\{f(x_n(i+1)) + f(x_n(i-1))\}$ for an increase of a by 0.002. (c) The plot of the KS entropy measured as the sum of positive Lyapunov exponents for the data of (b). Here the computation of the Lyapunov exponents is carried out over 3000 steps, and the error around 0.005 remains, which leads to the variation around the smooth curve in (b) and (c). (From K. Kaneko, *Prog. Theor. Suppl.* **99** (1989c) 263, with the permission of the publishers.)

3.3.7 Traveling Waves

When the coupling ϵ is large (e.g., larger than 0.45), the domain structures in Sects. 3.3.2 and 3.3.3 are no longer fixed in space, but can move with some velocity [Kaneko 1992a,1993a]. At the parameter corresponding to the frozen state in Sect 3.3.2, the motion of a domain is rather irregular (see Fig. 3.13). This type of floating domains has some correspondence with the dispersive chaos found in Bénard convection [Kolodner *et al.* 1990].

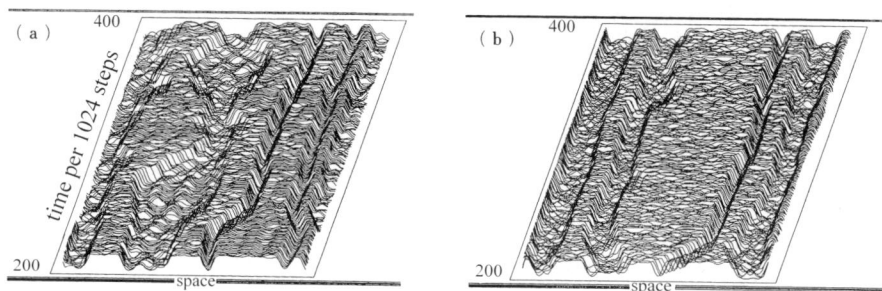

Fig. 3.13 a,b. Amplitude–space plot of $x_n(i)$. 200 sequential patterns $x_n(i)$ are plotted with time (per 1024 time steps), after discarding 20 480 initial transients. Random initial conditions were used. $\epsilon = 0.5$, and $N = 100$. (a) $a = 1.47$, (b) $a = 1.52$. (From K. Kaneko, *Physica D* **68** (1993a) 299, with the permission of the publishers.)

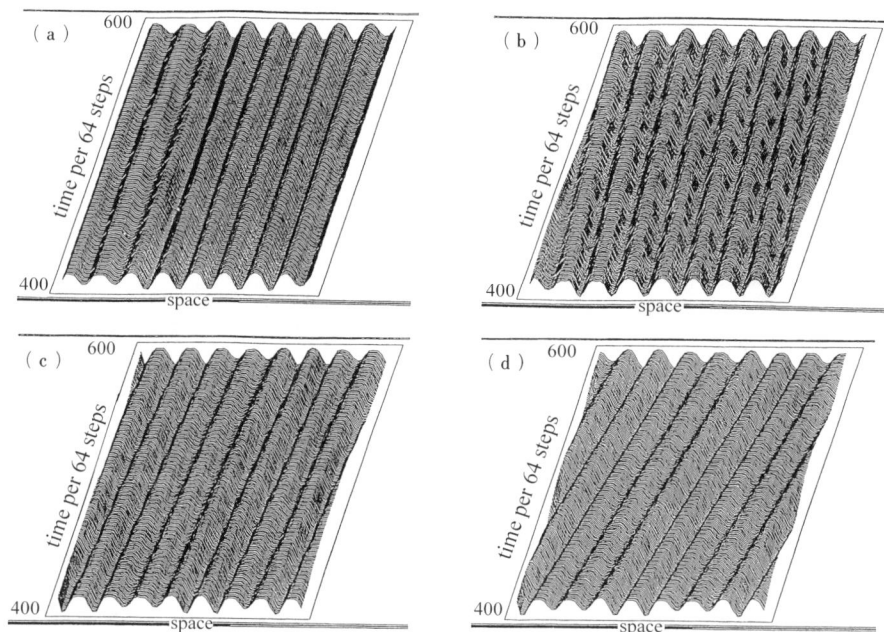

Fig. 3.14 a–d. Amplitude–space plot of $x_n(i)$. 200 sequential patterns $x_n(i)$ are displayed with time (per 64 time steps), after discarding 25 600 initial transients. Random initial conditions were used. $a = 1.71, \epsilon = 0.5$, and $N = 64$. Four attractors from different initial conditions are shown for (a)–(d). (a) is a fixed wave pattern with temporal period 4. (b)–(d) give the traveling wave with (b) $v_p = -v_1$, (c) $v_p = v_1$, and (d) $v_p = v_2$. (From K. Kaneko, *Physica D* **68** (1993a) 299, with the permission of the publishers.)

On the other hand, a regular traveling wave is seen in the pattern-selection regime. Examples of attractors are given in Fig. 3.14. Attractors with different wave velocities coexist. Admissible velocities by attractors lie in a small band located at $0, \pm v_1, \pm v_2(\approx) 2v_1$, $v_3(\approx) \pm 3v_1$, and $v_4(\approx) \pm 4v_1$. For example, $v_1 = 1.3 \times 10^{-3}$, $v_2 = 2.75 \times 10^{-3}$, and $v_3 = 4.1 \times 10^{-3}$, for $a = 1.71$, $\epsilon = 0.5$, and $N = 50$. Numerically, v_k is approximately proportional to k.

To understand the mechanism of this velocity selection, we note that $x_n(i)$ oscillates in time. One can assign a phase of oscillation to a lattice site i relative to $(x_n(i), x_n(i+1))$. When there is phase change of 2π between sites i and $i + \ell$, it is numerically found that this interval unit $[i, i + \ell]$ maintains the traveling wave. For example, in the attractor with velocity v_1 in Fig. 3.15, the oscillation is close to period 4, with slow quasiperiodic modulation. It is possible to assign a phase change $m\pi/2$ ($m = \pm 1$) between a lattice site i and a lattice site $i + j$ in a neighboring domain, according to the order of the period-4-like motion. If the boundary condition is periodic, the total phase change should be $2M\pi$. The velocity is found to be exactly zero for an attractor with $M = 0$. If $M = 1$, there must be a sequence of 5 domains with phases $0, \pi/2, \pi, 3\pi/2, 2\pi$ for successive lattice sites i (Fig. 3.15). This unit gives a (positive) phase slip. A negative phase slip is defined by the mirror-symmetric pattern of the positive one. Numerically it is confirmed that the velocity of an attractor of phase change $2M\pi$ (M equals the number

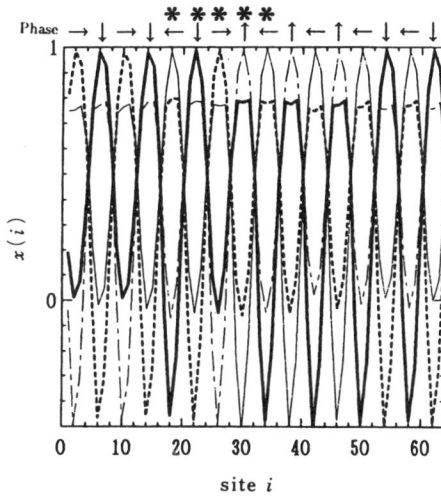

Fig. 3.15. Amplitude–space plot: plotted are 4 sequential patterns $x_n(i)$, in the order of thin line ($n = 10\,001$), thick line ($n = 10\,002$), thick dotted line ($n = 10\,003$), and broken line ($n = 10\,004$). $a = 1.72$, $\epsilon = 0.5$, and $N = 64$. There is a negative phase slip (as shown by the asterisks), and the velocity is $-v_1$. (From K. Kaneko, *Phys. Rev. Lett.* **69** (1992a) 905, with the permission of the publishers.)

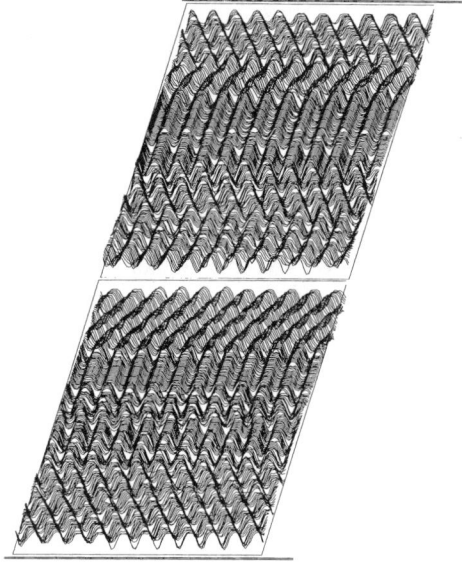

Fig. 3.16. Amplitude–space plot of $x_n(i)$. 200 sequential patterns $x_n(i)$ are depicted with time (per 1024 time steps), after discarding 40 960 initial transients. Random initial conditions were used. $a = 1.69$, $\epsilon = 0.5$. $N = 93$.

of positive phase slips with the negative ones subtracted) falls on the velocity band of v_M.

Since this phase slip is localized in space, one might think that the movement is a local phenomenon like soliton propagation. This is not the case. In the present case, this phase slip must pull all other regions to make them travel, changing the phases of oscillations of all lattice points. Thus a local slip influences globally all lattice points. Our dynamics gives a connection between local and global dynamics. One clear manifestation of the global aspect is the additivity of velocity. In our system, the velocity of the wave is proportional to the number of the phase slips. This proportionality gives a clear distinction between our dynamics and soliton-type dynamics, where, of course, the speed of a soliton does not increase with the number of solitons present.

The global change by the local phase slip is clearly demonstrated by changing the value of one lattice point externally, in order to change the number of phase slips. With this local information, the speed of the global traveling wave changes by v_1. (Of course the speed does not change instantaneously, to satisfy causality. It takes roughly N steps for it.) This suggests that our CML can be used to transform local to global information.

Since our system has a selected wavelength R (≈ 8 in the examples here), the fractional part of N/R is relevant to the nature of the traveling wave. If this fractional part is not close to 0.5, the motion is quasiperiodic for most sizes and parameters. On the other hand, when there is mismatch between the size and the wavelength R (the fractional part of N/R is near 0.5), any pattern with any velocity and wavelength has frustration. This frustration introduces remnant chaotic motion, which leads to spontaneous switching among patterns with different velocities. In Fig. 3.16, traveling velocity spontaneously switches. Here states with different velocities merge, which are disconnected attractors for most other sizes without frustration. This spontaneous switching arises from the chaotic motion of each pattern. The state chaotically itinerates over "attractor ruins", as will be discussed as chaotic itinerancy in Chaps. 4 and 6. We have measured the distribution of the residence time in a traveling-wave state with unchanged speed. This distribution $P(t)$ obeys a power law

$$P(t) = t^{-\beta} \qquad (3.6)$$

with $\beta \approx 1$. This shows the long-term correlation between switches of traveling waves.

The traveling wave by phase slip has also been discovered in an experiment on Bénard convection [Flesselles *et al.* 1995].

3.3.8 Supertransients

In spatiotemporal chaos, we often encounter very long transients. Indeed, in a class of spatiotemporal chaos, the transient time before falling on an attractor diverges exponentially or faster with the system size. In the transient regime, spatiotemporal chaos is quasistationary, and the orbit is 'suddenly' attracted to the final attractor after such long transients (see Fig. 3.17). The quasistationary state is almost indistinguishable from attractors.

As an example, let us consider a CML (1) in the parameter region for a period-3 window. As has been discussed, the homogeneous state with $x(i) =$

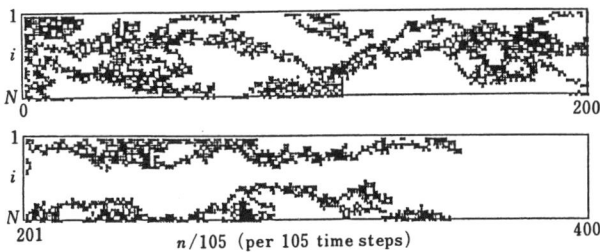

Fig. 3.17. Spatial derivative plots for the coupled logistic lattice (1), starting with random initial conditions. If $|x_n(i+1) - x_n(i)|$ is larger than 0.3, the corresponding space–time pixel is painted black, gray if it is between 0.1 and 0.3, while it is left blank otherwise. Plotted per 105 steps. $a = 1.752$, $\epsilon = 0.001$, $N = 30$. After around 36 000 steps, bursts disappear and the system falls on a spatially homogeneous state with temporal period 3. (From K. Kaneko, *Phys. Lett.* **149A** (1990b) 105, with the permission of the publishers.)

x^k (x^k as a cycle for the single map $x' = f(x)$) is linearly stable. We have made a long-time simulation for a small-lattice system [Kaneko 1990b] to check the transient length before the system falls on such a homogeneous state. After a great number of time steps, we have found that the system falls onto the homogeneous periodic state when the coupling is not very close to zero. The average length of transients before the attraction to the final attractor fits

$$T = \text{constant} \times \exp(N(\epsilon - \epsilon_c)^\nu), \qquad (3.7)$$

where ν (≈ 1) is an exponent characterizing the growth of the correlation length (see Fig. 3.18). The above growth form is rather rapid: unless the system size is very small (e.g., less than 10 lattice sites), it is in practice impossible to wait and see whether the system really falls on the homogeneous attractor or not.

In the transient regime, we have observed spatiotemporal chaos, where the dynamics is quasistationary. We can compute Lyapunov exponents and other quantities for chaos, which converge rather well during these time steps. In these steps, it is almost impossible to distinguish the dynamics from the motion on an attractor.

This supertransient leads to type-I spatiotemporal intermittency on the one hand, and causes the stability of fully developed spatiotemporal chaos on the other hand.

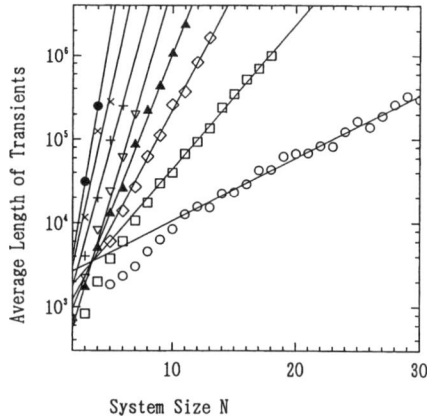

Fig. 3.18. Average transient length versus system size N. The average number of time steps before our system is attracted to a period-3 cycle is calculated from 100 randomly chosen initial conditions. $a = 1.752$. Semi-log plot for $\epsilon = 0.0011$, ($\circ$), 0.0012 ($\square$), 0.0013 ($\diamond$), 0.0014 ($\triangle$), 0.0015 (∇), 0.002 ($+$), 0.003 ($\times$), and 0.005 ($\bullet$). The slopes are 0.074 ($\epsilon = 0.0011$), 0.17 ($\epsilon = 0.0012$), 0.29 ($\epsilon = 0.0013$), 0.38 ($\epsilon = 0.0014$), 0.47 ($\epsilon = 0.0015$), 0.76 ($\epsilon = 0.002$), 1.0 ($\epsilon = 0.003$), and 1.4 ($\epsilon = 0.005$). (From K. Kaneko, *Phys. Lett.* **149A** (1990b) 105, with the permission of the publishers.)

Such long transients are not restricted to the above case. The first discovery of such supertransients was made by applying a CML with a piecewise-linear map by Nagumo and Sato in Sect. 2.4 [Crutchfield and Kaneko 1988]. For example, by taking a local map $f(x)$ with an "extinct" state $(x = 0)$ and chaotic transients, it is expected that the chaotic transients avoiding total extinction last over the time steps whose number increases exponentially with the system size. In this case, during the transient time steps, the dynamics shows spatiotemporal intermittency, with intermittent appearance of "extinct" areas in space–time. Another interesting example is given in Hastings and Higgins [1994], where long transients are studied in a CML model with a longer-range coupling.

3.4 CML Phenomenology as a Problem of Complex Systems

We have discussed the phenomena that are observed in the diffusively coupled logistic map lattice. One can see the diversity of the phenomena that this single model produces. The existence of such a variety constitutes one of the reasons why we emphasize a naturalist's approach in Chap. 1.

On the other hand, the classes of phenomena observed in this single model are also observed in other CMLs, partial differential equation systems, and experiments including Bénard convection. Thus this CML provides a good example to consider the problem between universality and individuality. By studying an individual model suitably abstracted, one can extract universal features. Note that the universal features are extracted since we do not choose a model specific to some phenomena, but construct a model combining fundamental procedures.

In contrast to the universality independent of the local mapping or interactions, some features are specific to each model. For example, there is a phenomenon called 'soliton turbulence', which is an example of spatiotemporal intermittency. An example of soliton turbulence is found in the coupled circle map lattice where each map is given by the circle map $f(x) = x + (K/2\pi)\sin(2\pi x) + \Omega$. Here the ordered state consists of a homogeneous state and a localized structure that propagates like a soliton. These localized structures lead to spatiotemporal chaos by collisions, as seen in Fig. 3.19 [Crutchfield and Kaneko 1987]. Generation of these soliton-like structures through chaotic dynamics and their propagation are also interesting from the viewpoint of information processing. Indeed a cellular automaton that can constitute a universal Turing machine has some common features with this soliton turbulence.

To study the variety of phenomena in CML, one requires multiple viewpoints as mentioned in Chap. 1. For example, we need to study the phenomena both from the 'anatomy' of phase space and from the motion in the real space.

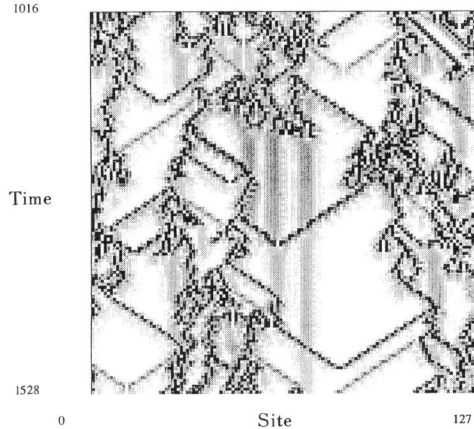

1016

Time

1528

0   Site   127

Fig. 3.19. Soliton turbulence in the CML with the circle map ($K = 1.3$, $\Omega = 0.3$, $\epsilon = 0.2$). Spatial derivative plot. The absolute values of the spatial difference $|x_n(i+1) - x_n(i)|$ (mod 1) are plotted with gray scale in space–time. (From J.P. Crutchfield and K. Kaneko, Directions in Chaos (World Scientific, 1987), with the permission of the publishers.)

For the Brownian motion of defects, the chaotic dynamics in phase space is revealed in real space as Brownian motion. In spatiotemporal intermittency, the interplay between local chaos and global spatial structure is important.

3.5 Phenonemology in Open-Flow Lattices

3.5.1 Introduction

As another example of a CML let us study the asymmetric-coupling case, especially focusing on the extreme case, where the coupling is one-way (only from left to right) [Kaneko 1985b]:[4]

$$x_{n+1}(i) = (1 - \epsilon)f(x_n(i)) + \epsilon f(x_n(i-1)). \tag{3.8}$$

One motivation for the asymmetric-coupling model lies in the study of the behavior of a partial differential equation system with a first-order spatial derivative. A typical example is the open fluid flow system, with a flow from upstream to downstream. Although it is one of the most well-known examples in turbulence, it is not well understood compared with the turbulence in a closed system.

Furthermore, spatiotemporal chaos with uni-directional coupling can be seen in some optical systems and in circuits. As to general studies for networks

---

[4] See Aranson *et al.* [1988] for a similar uni-directional coupling system for a coupled differential equation.

of chaotic elements, it is important to study the simplest case with uni-directional information flow. It will provide a theoretical basis for the study of more complicated cases, like signal transduction in a cell.

In real applications, asymmetric rather than one-way coupling may be more natural. Still, it is important to study the extreme case of one-way coupling as a starting point. Also, the present model provides an example of chaotic modulation from one element to another, discussed in Sect. 3.6 in the context of the observation problem and nowhere-differentiable torus.

Here we choose again the logistic map $f(x) = 1 - ax^2$ for the local map. As a boundary condition the left end $(x(0))$ is a fixed constant, while the boundary condition for the right end is not necessary since there is no coupling from right to left.

3.5.2 Spatial Bifurcation to Down-Flow

One remarkable phenomenon in open-flow lattices is spatial period-doubling. As shown in Fig. 3.20, the period of the motion of each lattice point increases as $1, 2, 4, 8, \cdots$. The spatially period-doubled pattern develops in complexity as follows: at lattice sites $0 < i < i_1$, $x(i)$ is a fixed point, while at $i_1 < i < i_2$, $x(i)$ is a period-two cycle, at $i_2 < i < i_3$ a period-four cycle, and so on. If the nonlinearity a is not large enough, the period of $x_n(i)$ is 2^{k+1} for $i > i_k$ and no more bifurcation occurs. If the nonlinearity a is large, the pattern is turbulent for $i > i_c$ after some finite number of period-doubling bifurcations in space. If chaotic behavior occurs up-flow (at $i = i_c$) for the above model, periodic behavior cannot be observed further down-flow $(i > i_c)$.

Although the final state at the down-flow is determined by the parameter (and the boundary condition), the bifurcation sequence itself (i.e., the value of

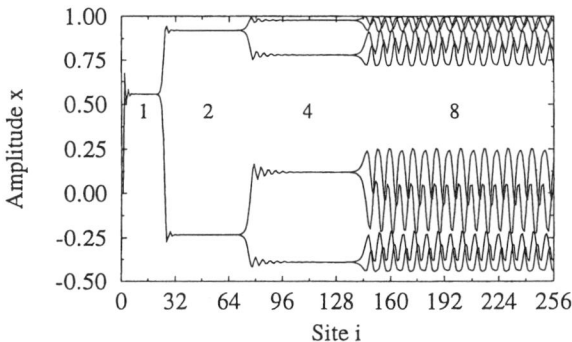

Fig. 3.20. Spatio–temporal bifurcations in the one-way coupled map lattice (3.5). The values of $x_n(i)$ are overlaid. As the lattice point is followed down-flow, bifurcations from the fixed point to periods $2, 4, 8$ occur. $a = 1.45$, $\epsilon = 0.5$. (From F.H. Willeboordse and K. Kaneko, *Physica D* **86** (1995) 428, with the permission of the publishers.)

i_k, $k = 1, 2, \cdots$), is sensitive to small perturbations. In fact, even the adopted floating method in a computer can alter the value of i_k.[5]

Note that, although the intermediate state of period 2 or 4 is unstable the dynamics has to pass through such states to reach stable down-flow dynamics. The dynamics has to experience some history to reach the stable state. To study such a mechanism, the following convective instability is essential.

3.5.3 Convective Instability and Spatial Amplification of Fluctuations

In an open-flow system, it is important to distinguish convective from absolute instability [Briggs 1964; Bers 1983]. If a small disturbance grows in a stationary frame, it is absolutely unstable. Even if the disturbance is reduced in the stationary frame, it may grow in a moving frame (see schematic Fig. 3.21). In this case, the system is called convectively unstable. This convective instability is often seen in open fluid flow, where the perturbation grows as it goes down-flow.

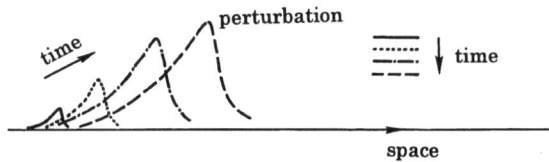

Fig. 3.21. A schematic representation of the development of the perturbation. At each lattice point the perturbation decays, while it is amplified as it moves down-flow.

For a finite-size system, such a perturbation finally decays away, and the final state is stable. However, if there is some noise in the system, it is always amplified down-flow. Thus the down-flow dynamics cannot stay in the same state. Hence even if the system is linearly stable in a stationary frame, the state is unstable against noise when it is convectively unstable. Bifurcation to other structures can occur from the convectively unstable state. In many cases, the gap between linear stability and convective instability leads to interesting phenomena intrinsic to open-flow systems.

---

[5] In fact KK was annoyed by the dependence of the numerical result on the adopted computer. IT has also experienced such dependence on the floating-point method when he studied the one-way coupled map lattice of the BZ map, discussed in Sect. 6.3. This sensitivity is understood by noting that the computational process itself forms a 'small dynamical system'. A difference in the computational process changes the nature of this small dynamical system, which couples each chaotic element, and is amplified. A detailed analysis was recently given by Yamaguchi [1997].

Let us explain this property by taking an example of the one-way coupling model (3.9). First, we consider the stability of a homogeneous fixed-point state, i.e., $x(i) = x^* (= (\sqrt{(1 + 4a)} - 1)/(2a))$. The Jacobi matrix for the model is given by $J_n(i, j) = (1 - \epsilon)f'(x_n(j))\delta_{i,j} + \epsilon f'(x_n(j - 1))\delta_{i,j-1}$. For the state $x(i) = x^*$, the eigenvalues of the matrix are degenerate and are $(1 - \epsilon)f'(x^*)$. Thus, the homogeneous state is stable in the stationary frame if $|(1 - \epsilon)f'(x^*)| < 1$. Let us then consider the evolution equation of the disturbance $\delta x_n(i)$, which is given as

$$\delta x_{n+1}(i) = A\delta x_n(i) + B\delta x_n(i - 1) \tag{3.9}$$

with $A = (1 - \epsilon)f'(x^*)$ and $B = \epsilon f'(x^*)$. If we start from the disturbance $\delta x_0(i) = d\delta_{i,0}$ ($\delta_{i,j}$ is the Kronecker delta), the above equation is solved as

$$\delta x_n(i) = \binom{i}{n}A^{n-i}B^i d. \tag{3.10}$$

The growth of the disturbance at the co-moving frame with the velocity v is obtained by substituting $i = vn$ as

$$\delta x_n(vn) = \binom{n}{vn}(1 - \epsilon)^{(1-v)n}\epsilon^{vn}f'(x^*)^n d. \tag{3.11}$$

The logarithm of the average growth ratio of the perturbation per time step (for large n) represents the stability of a state against some moving frame. This "stability exponent", thus defined, is given for the homogeneous fixed-point state as

$$L(v) = \log|f'(x^*)| + \log\left|\frac{1 - \epsilon}{1 - v}\right| + \log\left|\frac{\epsilon(1 - v)}{v(1 - \epsilon)}\right|^v, \tag{3.12}$$

or

$$L(v) = L_0 + \log\left|\frac{1 - \epsilon}{1 - v}\right| + \log\left|\frac{\epsilon(1 - v)}{v(1 - \epsilon)}\right|^v. \tag{3.13}$$

The maximum occurs for $v = \epsilon$, which takes the value $L_{\max} = \log|f'(x^*)|$. From the above calculation, it is shown that (i) the disturbance is transmitted with the largest amplification at the speed $v = \epsilon$ and that (ii) the state is stable only for $|f'(x^*)| < 1$. If $|f'(x^*)| > 1$, the homogeneous state is unstable even for a very weak noise or by a round-off error.

The above computation is limited to the case for the stability of a spatially homogeneous state. It is possible to introduce the instability exponent to measure chaos in a convectively unstable system. In Fig. 3.22, we have plotted one snapshot of the one-way CML. (The time series at one lattice point also shows similar 'chaotic' behavior.) However, the conventional Lyapunov exponent is negative in this case. Again, the perturbation grows down-flow only in some inertial frame with nonzero velocity. Extending the above stability analysis for the homogeneous case, we can introduce the co-moving Lyapunov

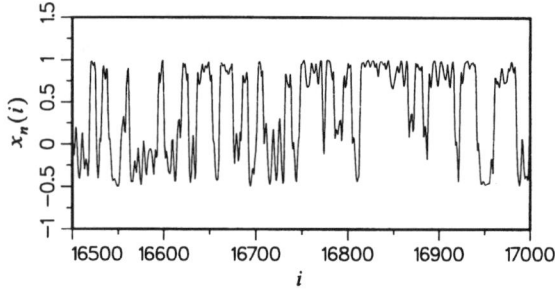

Fig. 3.22. Snapshot pattern of $x_n(i)$ of the CML (3.5) with a tiny noise added (random number over $[-10^{-9}, 10^{-9}]$). $a = 1.5$ and $\epsilon = 0.4$. The conventional Lyapunov exponent is negative, while $\lambda(v)$ is positive for some interval of velocities v, as will be seen in Fig. 3.23. (From R.J. Deissler and K. Kaneko, *Phys. Lett.* **119A** (1987) 397, with the permission of the publishers.)

exponent $\lambda(v)$, that is the Lyapunov exponent measured in the inertial frame with the velocity v [Deissler and Kaneko 1987; Kaneko 1986b]. It is the rate of orbital instability in the inertial frame with velocity v. For a CML, this exponent is computed as follows: first take a large enough domain $[i_L, i_R]$, and compute the Jacobi matrix for the CML mapping. Shift this domain with the velocity v as $[i_L + vn, i_R + vn]$ and compute the product of the Jacobi matrices $[\delta x_{n+1}(i)/\delta x_n(j)]$ along this shift. The long-term average of the logarithm of the eigenvalue of this product gives the co-moving Lyapunov exponent $\lambda(v)$.

In Fig. 3.23, we have plotted $\lambda(v)$ of the one-way CML, thus computed. Even though the conventional Lyapunov exponent $\lambda(0)$ is negative, $\lambda(v)$

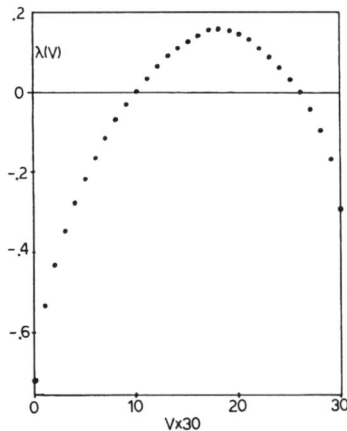

Fig. 3.23. The co-moving Lyapunov exponent $\lambda(v)$ for the CML (3.5). (From K. Kaneko, *Physica D* **23** (1986b) 436, with the permission of the publishers.)

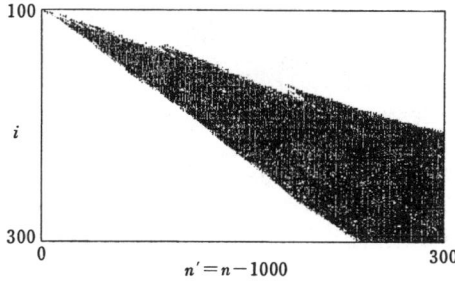

Fig. 3.24. The difference pattern of the CML corresponding to Fig. 3.22. At time step $n = 1000$, we perturb a lattice point as $y_n(100) = x_n(100)+0.001$, and compute the evolutions of $x_n(i)$ and $y_n(i)$ with the same CML (3.5). When $|x_n(i) - y_n(i)| > 0.001$, the corresponding space–time pixel is painted black. $a = 1.5$ and $\epsilon = 0.6$. The disturbance propagates within some range of velocities, and this range agrees with the range such that $\lambda(v) > 0$. (From K. Kaneko, *Physica D* **23** (1986b) 436, with the permission of the publishers.)

takes positive values for some range of the velocity, and the chaos in Fig. 3.22 is quantitatively characterized. Only within the range of $\lambda(v) > 0$, the perturbation is amplified. Hence the propagation speed (area) of perturbation is obtained from $\lambda(v) > 0$. For example, we have computed how the difference of two evolutions propagates by perturbing the value $x_n(j)$ of one lattice point as $x'_n(j) = x_n(j) + \delta$. In Fig. 3.24 the plotted space–time domain is such that $|x'_m(i) - x_m(i)|$ is larger than some threshold. The perturbation propagates within some range of velocity, which agrees with the region $\lambda(v) > 0$.

The relationship between the convective Lyapunov exponent and the propagation of the perturbation is not limited to the one-way CML. For the diffusive CML (3.1), we have plotted the propagation of the difference pattern in Fig. 3.25, when a single lattice point is perturbed. Figure 3.25a corresponds to the frozen random state, where the perturbation propagates smoothly within a domain, and then is trapped at the domain boundary, until 'tunneling' to the next domain occurs. Thus, the step-like propagation in Fig. 3.25a is generated. On the other hand, in fully developed spatiotemporal chaos, the propagation is rather smooth with a constant velocity. Indeed this velocity is given by the condition for $\lambda(v) > 0$. The corresponding co-moving Lyapunov exponent $\lambda(v)$ is plotted in Fig. 3.25c.

3.5.4 Phase Diagram

To what type of dynamics is our CML (3.5) attracted down-flow? The pattern dynamics of the CML is extremely rich, but it can be divided into two clearly distinct classes: spatiotemporal patterns and spatial patterns with perfect temporal periodicity.

A phase diagram displaying the regions in which the major classes dominate is given in Fig. 3.26 [Willeboordse and Kaneko 1994, 1995]. The class

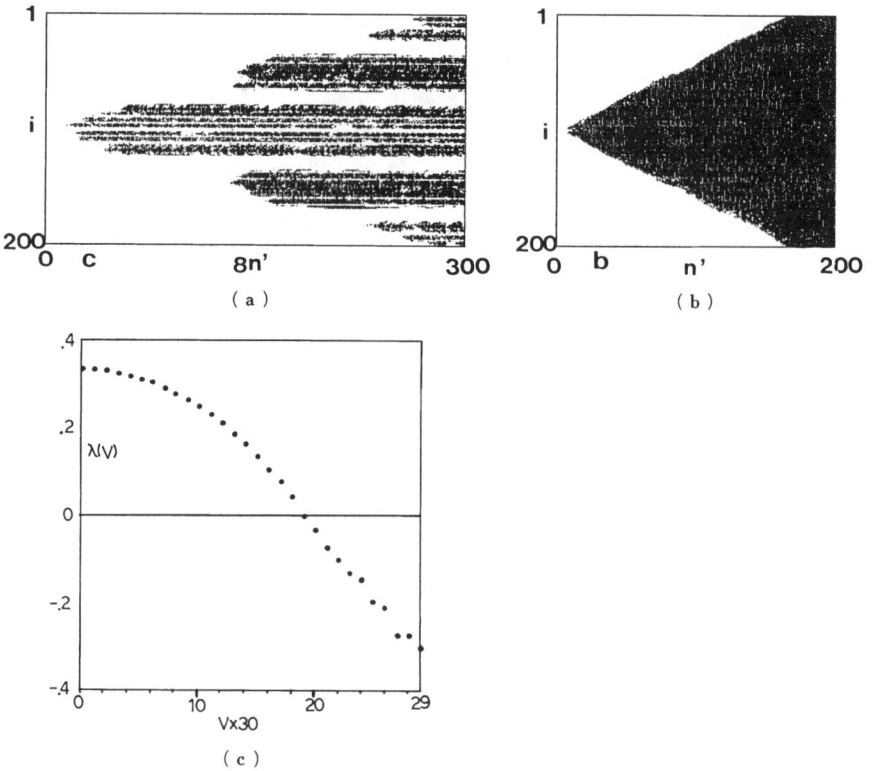

Fig. 3.25 a–c. The difference pattern of the CML (3.1). The computation and the plot were obtained in the same manner as Fig. 3.24, while the CML (3.1) was adopted instead of the one-way CML (3.5). $n' = n - 1000$. (a) $a = 1.41$, $\epsilon = 0.08$ (there are domains as in Fig. 3.4, and the disturbance propagates through tunneling among the domains). (b) $a = 1.9$, $\epsilon = 0.6$. The range of the velocity allowing for the propagation is given by $\lambda(v) > 0$. For comparison, plotted in (c) is $\lambda(v)$ (here $\lambda(v) = \lambda(-v)$). (From K. Kaneko, *Physica D* **23** (1986b) 436, with the permission of the publishers.)

of spatiotemporal patterns is indicated by STP, with the remaining regions belonging to the class of spatial patterns with temporal periodicity, being the three classes of spatial chaos (SC), spatial quasiperiodicity (SQP) and spatial periodicity (SP).

When the coupling is small, the motion is generally aperiodic, and complex spatiotemporal pattern dynamics is observed. As an exception, periodic motion with a spatially zigzag pattern is observed in the region around 'SP' at $\epsilon \approx 0.1$. It is a period-2 state both in space and in time. This zigzag pattern corresponds to that observed in diffusively CMLs in Sect. 3.3.4. Now two domains with π phase difference can exist as in the case of Sect. 3.3.4, which are separated by a defect. This defect again shows chaotic motion.

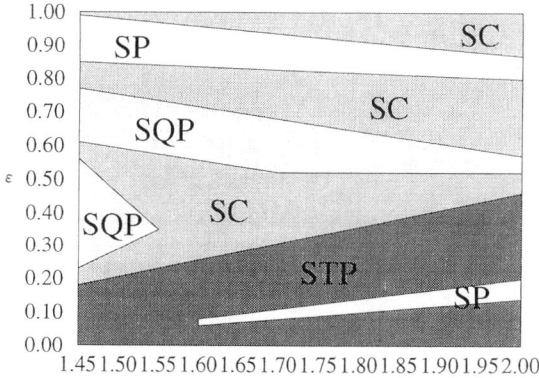

Fig. 3.26. Phase diagram of the CML (3.5). The system size is $N = 384$. Only the predominant patterns are indicated and labeled as SP (spatially and temporally periodic), SQP (spatially quasiperiodic but temporally periodic), SC (spatially chaotic but temporally periodic), and STP (spatially and temporally nonperiodic), respectively. Random initial conditions were used and the boundary was fixed to $x(0) = 1$ (although the exact value does not seem to matter). (From F.H. Willeboordse and K. Kaneko, *Physica D* **86** (1995) 428, with the permission of the publishers.)

Depending on the parameter values a and ϵ, these defects are arranged periodically in space, propagate with some fluctuation, or repeat creation and pair-annihilation intermittently (see Willeboordse and Kaneko [1995] for details). Except for this zigzag region, the down-flow attractors in our CML at the down-flow are chaotic in space and in time, when the coupling is small.

3.5.5 Spatial Chaos

For larger coupling, temporally periodic states are commonly observed, although the spatial pattern is not necessarily simple. Indeed the region termed as 'spatial chaos' is found in the phase diagram. Here by the term 'spatial chaos', we mean that the spatial sequence $x_n(i)$ with the increase of i (from left to right) is regarded as chaotic by reinterpreting this sequence as a time series (see Fig. 3.27).

Spatial chaos was first discussed by Aubry [1979], where he discussed whether the atomic configuration of a certain one-dimensional crystal at equilibrium can be chaotic in space. Shinjo and Sasada [1987] also discussed the possibility of a chaotic configuration in a ground state in solid-state physics. In our case, we are concerned with a spatial configuration of a temporally oscillating state, rather than the ground state. A similar problem is discussed by Otsuka and Ikeda [1989] for a chain of nonlinear optical elements.

Consider a periodic state with the temporal period p. In other words $x_{n+p}(i) = x_n(i)$ down-flow. From the condition $x_{n+p}(i) = x_n(i)$ and our CML dynamics (3.5), it is possible to obtain the equation that $x_n(i)$ should

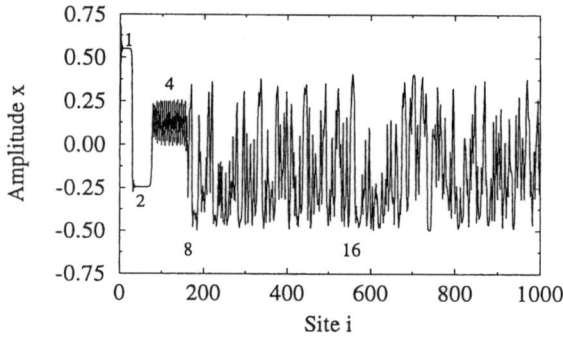

Fig. 3.27. Snapshot pattern of the CML (3.5) ($a = 1.5$, $\epsilon = 0.55$). The period increases as 2, 4, 8, and is kept at 16 at the down-flow. The pattern is a spatial chaos. (From F.H. Willeboordse and K. Kaneko, *Physica D* **86** (1995) 428, with the permission of the publishers.)

satisfy, given the set of values $\{x_n(i-p), x_n(i-p+1), \cdots x_n(i-1)\}$. Thus a p-dimensional map from $\{x_n(i-p), x_n(i-p+1), \cdots x_n(i-1)\}$ to $x_n(i)$ is obtained. (This mapping is not necessarily given in an explicit form but is given by an implicit form as a solution of some polynomial equation.) In the region with temporal periodicity (i.e., except the STP region in the phase diagram), this p-dimensional map has a unique solution. In the spatial-periodicity region, this p-dimensional spatial map gives a periodic attractor, while at spatial chaos the attractor of the p-dimensional map is chaotic. Since the state is periodic *in time*, this spatial pattern is repeated after p time steps.

If the spatial sequence is chaotic, an identical spatial sequence cannot appear from a different initial condition. Hence there can be infinitely many attractors with different spatial patterns, depending on the initial condition. When one considers the use of this CML as an information-processing system, one can assume that a memory can be put in each attractor. Then an infinite memory might be possible in spatial chaos.

Unfortunately, this astonishing capacity can not be realized, because the spatial chaos with temporal periodicity is believed to be convectively unstable. By representing the Lyapunov exponent for the above spatial map as λ_{spa} and the co-moving Lyapunov exponent of the original open flow system by $\lambda(v)$, the following conjectures are proposed that are expected to hold:

(a) A positive Lyapunov exponent in the spatial map ($\lambda_{\text{spa}} > 0$) implies a positive maximum co-moving Lyapunov exponent (maximum of $\lambda(v) > 0$) in an open-flow system.

(b) A spatial pattern with a positive maximum co-moving Lyapunov exponent will bifurcate temporally at some lattice site down-flow, as long as any tiny noise is applied to the system.

We therefore believe that spatially chaotic patterns do not last over large domains in the lattice, in the presence of noise (with amplitude δ). Still, it

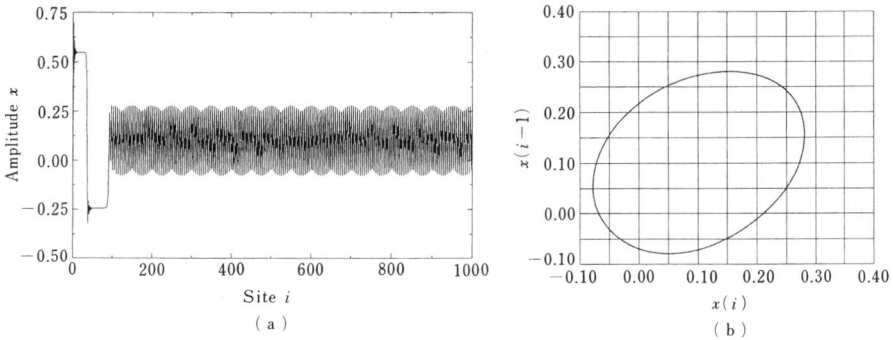

Fig. 3.28. (a) Snapshot pattern of the CML (3.5), $a = 1.7$, $\epsilon = 0.6$, (at the down-flow, the state is temporally period 4 which depicts a spatially quasiperiodic state), (b) the return map $(x_n(i), x_n(i+1))$ corresponding to (a). The closed curve represents the torus motion in space. (From F.H. Willeboordse and K. Kaneko, *Physica D* **86** (1995) 428, with the permission of the publishers.)

takes L lattice points, such that $\delta \exp((\lambda/v)_{\max} L) \approx O(1)$, for the growth of such a noise term to destroy the spatial chaos, with $(\lambda/v)_{\max}$ as the maximum of $\lambda(v)/v$. Within the lattice region smaller than L, spatial chaos can be kept and a large capacity of memory is possible.

After the collapse of spatial chaos (of a temporally periodic state) by noise, the state is replaced by a spatiotemporally irregular pattern down-flow. It is interesting to note that thus far we have not found any differences between the original spatiotemporal chaos and the irregular patterns which are the result of destroyed spatial chaos. In the Fourier spectrum, for example, there are no traces whatsoever of the original temporal periodicity.

In contrast with spatial chaos, the temporally periodic state with spatial periodicity or spatial quasiperiodicity can be stable. This corresponds to the region of SP and SQP in the phase diagram. An example of spatial quasiperiodicity is plotted in Fig. 3.28, where the maximum of the co-moving Lyapunov exponent is zero (there are also cases of SQP where the co-moving Lyapunov exponent is positive for some velocity and the state is convectively unstable).

3.5.6 Selective Amplification of Input

For an open-flow system, we have observed selective amplification of external input upflows. As an example one can apply a very tiny signal of some periodicity at $x_n(0)$. Depending on the input frequency, the input is amplified to alter the down-flow state drastically. In Table 3.2, the spatiotemporal periodicity down-flow is plotted, when a periodic input of amplitude 10^{-3} and a periodicity p is applied at $x_n(0)$. Even though the input amplitude is tiny, the amplification leads to the selection of periodic states as plotted in

Table 3.2. The attracted state down-flow when the boundary (the upstream) is periodically modulated. The periods in space (sp) and time (tp) are plotted corresponding to each value of a (horizontal axis) and input period (vertical axis).

| a | 1.65 | | 1.70 | | 1.75 | | 1.80 | | 1.85 | | 1.90 | |
input period	tp	sp	tp	sp	tp	sp	tp	sp	tp	sp	tp	sp
1:	8	4	32	–	32	–	32	–	8	4	8	4
2:	16	–	32	–	8	4	8	4	32	–	8	4
3:	12	6	24	–	24	–	48	–	48	–	24	12
4:	8	4	8	4	8	4	8	4	8	4	8	4
5:	5	5	5	5	10	10	5	10	10	10	40	–
6:	12	6	24	–	48	–	48	–	48	–	8	4
7:	28	–	28	14	28	–	28	–	28	14	56	–
8:	32	16	8	4	8	4	8	4	8	4	8	4
9:	18	9	36	–	18	27	18	18	18	18	36	–
10:	5	5	5	5	5	40	20	–	40	20	40	–

the Table (indeed we have checked the amplitude down to 10^{-12}). It should be noted that

(a) the input is transferred down-flow selectively based on the periodicity,
(b) the amplification rate of the input is almost infinite.

With the use of the p-dimensional spatial map mentioned above, one can check whether a stable cycle of period p is formed by the periodic input. If the maximum co-moving Lyapunov exponent for this p-dimensional map is not positive, a state with temporal period p and with spatial periodicity or quasiperiodicity is realized. In some cases, a state with $2^k \times p$ periodicity is realized, where the co-moving Lyapunov exponent for this $2^k \times p$-dimensional mapping is not positive. One can check which periodicity propagates, by the sign of the co-moving Lyapunov exponent. As an example, in Fig. 3.29, a spatially periodic state with period 9 is selected by applying an input of temporal period 3, on the spatial chaos.

An analysis on input-dependence in an open-flow system has recently been presented in relation with the (local) co-moving Lyapunov exponent [Fujimoto and Kaneko 1998] in connection with signal transduction in a cell. In Chap. 5, we will again discuss the information dynamics in a one-way coupled map lattice system.

3.6 Universality

How should we justify our CML approach? When we represent a local chaotic process as a logistic map, it is impossible to make a direct one-to-one correspondence between our result and some specific spatiotemporal chaos as occurs in e.g. Bénard convection or a chemical reaction. For example, the

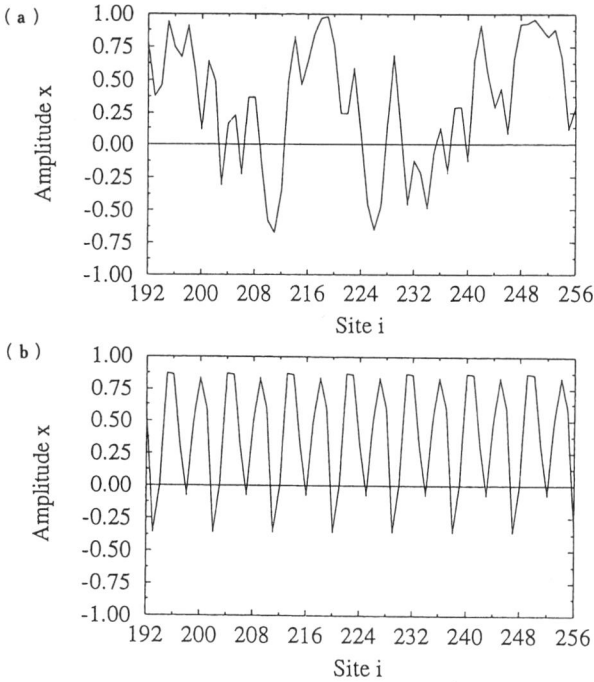

Fig. 3.29 a,b. Effect of modulating the boundary. The nonlinearity is $\alpha = 1.8$ and the coupling strength is $\epsilon = 0.7$. In (a), the boundary is fixed to 0.0, and the pattern is spatially chaotic while having a temporal periodicity of four. In (b), the boundary is modulated by a period-3 sawtooth-like wave with an amplitude of 0.01. The pattern is spatially periodic (period 9) and has a temporal periodicity of 3. In both cases, lattice sites 192–256 are shown. (From F.H. Willeboordse and K. Kaneko, *Physica D* **86** (1995) 428, with the permission of the publishers.)

local chaos in Bénard convection cannot be expressed by a logistic map. Hence, we cannot answer such question as "at what temperature between the top and bottom plates does spatiotemporal intermittency arise?". Then how is our approach relevant to the understanding of natural phenomena?

First, note that the present classes of phenomena are observed just by combining representative processes. The phenomena are not dependent on the details of each process. For example, instead of the logistic map, we can choose some other mappings or differential equations that show chaotic dynamics, and still observe similar phenomena. Furthermore, exactly what physical mechanism creates this chaos is not relevant to see such phenomena. In other words, we can conclude that the classes of phenomena are observed independently of the physical mechanism which creates the local chaotic dynamics. Each mechanism for thermal convection, chemical reaction, or electric convection and so forth, is irrelevant to the origin of the class of phenomena.

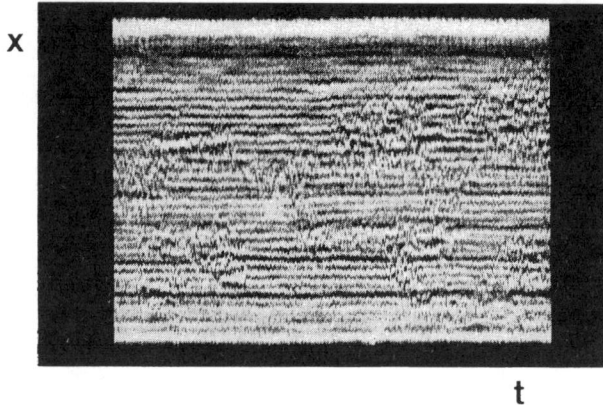

X

t

Fig. 3.30. Space–time evolution of the Bénard convection pattern in an annular cell. The light intensity of shadowgraphic images is plotted using a gray scale, over 1000 s, along the horizontal axis from left to right. (From F. Daviaud et al. *Phys. Rev.* **A 42** (1990) 3388, with the permission of the authors and the publishers.)

Thus the observed phenomena are expected to form some universality class. In other words, our approach has the power to predict universality classes.

In physics, quantitative universality is justified with the use of a renormalization group (RG), as can be seen in critical phenomena, onset of chaos, and field theory. For example critical exponents are determined by the symmetry of the system and independent of the details of the model system. A similar universality may exist in a CML, and may be justified by the RG method. Still, the universality in a CML may not necessarily hold at a quantitative level (for exponents). Rather, universality at a qualitative level may exist. It is hoped that the qualitative universality put forward by a CML may be justified in a wider context.

Once a universality class is predicted, we can expect that phenomena consisting of the same processes lead to the same behavior. Indeed, some classes of phenomena are observed in a rather wide range of fields. As mentioned, spatiotemporal intermittency is experimentally confirmed in Bénard convection [Ciliberto and Bigazzi 1988; Daviaud *et al.* 1990], electric convection in liquid crystals [Nasuno *et al.* 1989; Nasuno and Kai 1991], viscous fingers in fluid [Michalland *et al.* 1993], periodically pumped surface waves (Faraday waves) [Gollub and Ramshankar 1991], one-dimensional vortex lines [Willaime *et al.* 1993], and so forth. In Fig. 3.30, an example of STI in Bénard convection is presented. Now STI is believed to be one of the universal routes from spatial order to developed spatiotemporal chaos. Spatial bifurcation is also found in Bénard convection, while the traveling wave by the phase slip is demonstrated in the Bénard convection in an annulus [Flesselles *et al.* 1995].

In short, our CML, though it cannot make one-to-one quantitative predictions for parameters, provides us with predictions on qualitative universality classes, which help us understand the essence of the encountered phenomena.

Further, universality classes provide a way to reconstruct the observations of nature, rather than giving mere descriptions. In complex phenomena, without some viewpoint, we might end up with just a huge number of data. (Let us recall that the data regarded as 'dirty' and kept sleeping in the bottom of the desk, were revived as significant after the recognition of chaos.) The constructive approach provides a basis for the 'hypothesis generator' with which we can view nature.

Lastly, one can pose and answer a general question with this constructive approach. As for spatiotemporal chaos, the following questions are cast, and the answers are searched for in a general form:

1. Is spatial inhomogeneity or bifurcation formed through temporal chaos?
2. How is chaos suppressed to form spatial patterns? How general is this suppression of chaos? To what degree do temporal chaos and spatial order coexist?
3. Can collective order with spatial long-range order arise from microscopic (local) chaos [Chaté and Manneville 1992]? How are local chaos and spatial correlation related?
4. How is spatial order destroyed, triggered by chaos? Spatiotemporal intermittency is proposed as a general mechanism for the transition from a spatially ordered state to turbulence. Is it the only mechanism for the transition? How is it universal? What kind of types exist in the STI?
5. How are information propagation in real space and chaotic instability in phase space related? Is it possible to propagate chaos in real space?

3.7 Theory for Spatiotemporal Chaos

To analyse quantitatively the spatiotemporal complexity discussed so far, quantities in dynamical systems theory have been extended to space–time dynamics. They include Lyapunov exponents, dimension, information flow, correlation, and so forth. Here, we very briefly mention these studies. For details, see the references or some reviews (e.g., Kaneko [1993b]).

Lyapunov exponents characterize the logarithm of the eigenvalues of the expansion/reduction rate in the tangent space of phase space. For systems with N degrees of freedom, phase space is N-dimensional, and the dimension of the tangential space is N. Thus there are N Lyapunov exponents, which form a spectrum. The nature of the Lyapunov spectrum is studied in relation to disorder in spatiotemporal dynamics [Kaneko 1986b, 1993b; Giacomelli and Politi 1991]. The spectrum of fully developed spatiotemporal chaos is related to the random matrix. In spatiotemporal intermittency, separation of ordered and turbulent motions is represented by the Lyapunov spectrum.

Another characteristic is the Lyapunov vector, the eigenvector corresponding to the Lyapunov spectrum. The localization of the Lyapunov vector is noted in connection with Anderson localization, while the propagation of perturbations is studied as the overlap between Lyapunov vectors.

The propagation of perturbations is also related to the information flow discussed in Chap. 2. The domain allowing for this propagation is given by the region of positive co-moving Lyapunov exponents. Analytically, this co-moving Lyapunov exponent is calculated by summing up all amplification of perturbations over all paths in space–time, in a similar manner as the path integral method [Kaneko 1992c]. So far the application of this analytic method is limited to a CML with a piece-wise-linear map with a constant slope.

For the statistical mechanical theory of spatiotemporal chaos, unification of (low-dimensional) dynamical systems theory and statistical mechanics with many degrees of freedom is required. So far there are two approaches, although their application is limited to fully developed spatiotemporal chaos.

(1) Mapping to a higher-dimensional statistical mechanics.

The statistical mechanics of low-dimensional chaos theory has been fully developed since the studies by Ruelle [1978]. According to these studies, the chaotic dynamics can be mapped to the statistical mechanics of a one-dimensional spin system (see also Shimada [1979]).

Then, it is natural to expect that the thermodynamics of our k-dimensional CML can be mapped onto the statistical mechanics of $(k+1)$-dimensional spin systems. The interaction in the $(k+1)$-dimensional system is strongly anisotropic, since the interactions in "spatial" and "temporal" directions are quite different. Bunimovich and Sinai [1989] have shown the existence of a unique Gibbs measure for a CML with a hyperbolic mapping. One of the goals in this statistical mechanical approach is to understand the phase transition in the pattern dynamics in earlier sections, but it has not yet been accomplished (see for example Chaté and Courbage [1997] for recent progress).

(2) Self-consistent Perron–Frobenius operator.

Another approach to the statistical mechanics is the self-consistent approximation to the Perron–Frobenius operator [Kaneko 1989d; Houlrik *et al.* 1990]. The Perron–Frobenius (PF) operator has been a powerful aid in the study of the statistical mechanics of low-dimensional chaos [Ruelle 1978; Oono and Takahashi 1980].

As mentioned in Sect. 2.7, the probability distribution of x_n for the one-dimensional map $x \to f(x)$ is given by the solution of

$$H^{\mathrm{PF}}\rho(x) = \sum_{y=f^{-1}(x)} \rho(y)/|f'(y)|. \qquad (3.14)$$

This can be extended to N-dimensional dynamical systems to obtain the probability distribution $\rho(x(1), \cdots, x(N))$. The operator is written as the operator for the entire dynamical system and acts on the measure of N-dimensional space $\rho(x(1), \cdots, x(N))$. The operator is written as

$$H^{\mathrm{PF}}\rho(x(1), \cdots, x(N)) = \sum_{y(i) = \text{preimages}} \frac{\rho(y(1), \cdots, y(N))}{J(y(0), \cdots, y(N-1))}, \qquad (3.15)$$

where the sum is taken over all possible sets of $\{y(i)\}$, with $\{y(i)\}$ preimages of $x(i)$ (i.e., $y(i) \to x(i)$ by the map (3.1)), and $J(y(0), \cdots, y(N-1))$ is the Jacobian of the CML transformation (3.1).

In our model, the preimages are easily calculated, since they consist of two separated procedures, i.e., $y(i) \to x'(i) = f(y(i))$ and the spatial average by $x(i) = (1 - \epsilon)x'(i) + (\epsilon/2)(x'(i+1) + x'(i-1))$, and can be obtained explicitly.

Since the treatment of the N-dimensional distribution is practically impossible, we use a projection into a k-dimensional space by integrating out other variables. For example with the aid of a self-consistent approximation, we can obtain the equation for the one-body probability distribution $\rho(x)$, projected from the above N-dimensional probability distribution function as:

$$H^{\mathrm{SPF}}\rho(x) = (1 - \epsilon)^{-1} \int \int \sum_{y = f^{-1}(\frac{x - \epsilon/2(y_0 + y_2)}{1 - \epsilon})} \frac{\rho(y)\rho(y_0)\rho(y_2)}{|f'(y)|} \mathrm{d}y_0 \mathrm{d}y_2,$$

$$(3.16)$$

where the sums over preimages y are given by the solutions of $y = f^{-1}[(x - \epsilon(y_0 + y_2)/2)/(1 - \epsilon)]$. It is also straightforward to obtain the form for the two-body (and many-body) distribution function. For example the transition by STI is estimated by the one-body or two-body approximation.

On the other hand, it is important to judge if some experimental data are spatiotemporal chaos or not. In low-dimensional chaos, methods of reconstruction of dynamical systems from data of time series are proposed. Dimension, Lyapunov exponents, and dynamics itself are estimated from the data. In spatiotemporal chaos the dimension (the number of degrees of freedom) increases proportionally with the system size. (For a one-dimensional system, proportional to the length, for a k-dimensional system proportional to the k-th power of the length.) The dimension of a dynamical system is an 'extensive' quantity. Thus the estimate of dimension density is more important [Mayer-Kress and Kaneko 1989; Torcini et al. 1991; Bauer et al. 1993]. Also, an algorithm for the reconstruction of a CML is searched for [Kaneko 1989c]. With this algorithm, it will be possible to distinguish spatiotemporal chaos from random data.

Still, the quantities and the theory so far are extensions of conventional dynamical systems theory to space–time. Also, the power of the theory is rather limited, and may often remain at the level of mere reinterpretation of observed phenomena. Theory on spatiotemporal chaos is at a premature stage.

3.8 Applications of Coupled Map Lattices

Our strategy of studying dynamical phenomena in spatially extended systems by a CML is based on the separation of parallel procedures and their successive operations.

Most dynamical phenomena in spatially extended systems are described by a combination of some elementary local dynamics. Thus it is possible to construct some procedures with local dynamics and local spatial coupling. Here we list some examples.

3.8.1 Pattern Formation (Spinodal Decomposition)

When a system is quenched from a disordered state (at high temperature) to an ordered state (at low temperature), spatial-pattern formation commonly occurs, as is typically known as spinodal decomposition. Traditionally this problem has been studied by the time-dependent Ginzburg–Landau equation or by the kinetic Ising model using Monte Carlo methods.

Oono and Puri [1986, 1988] have proposed a cell dynamical system (or CML) for this problem. Their model is based on the coarse graining in cells with a large number of sites from the kinetic Ising model. The local dynamics of a cell leads to a map with two stable fixed points, for example

$$f(x) = \tanh(\beta x) \tag{3.17}$$

(see Fig. 3.31). The same diffusive coupling form as in the previous sections can be used since the phase transition dynamics includes a term which tends to make two neighboring regions align ("ferro-coupling").

When the order parameter is conserved, the above dynamics is not relevant since the constraint $\sum x(i,j) = $ constant has to be imposed. For this case, Oono and Puri constructed a model written as

$$x_{n+1}(i,j) = f(x_n(i,j)) - \ll f(x_n(i,j)) - x_n(i,j) \gg \tag{3.18}$$

where $\ll ... \gg$ denotes the spatial average of the appropriate neighbors. This method provides a powerful simulator to study the phase-separation process and has been widely adopted.

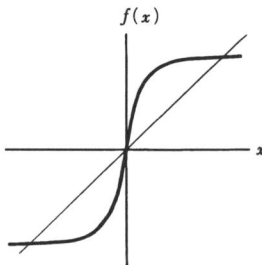

Fig. 3.31. A local sigmoid map to simulate the pattern formation.

3.8.2 Crystal Growth and Boiling

We can study crystal growth by introducing a temperature field, besides the dynamics of the order parameter represented by the above method. Here, another procedure is implemented corresponding to the latent heat that leads to the coupling between the thermal field and the order parameter [Kessler *et al.* 1990; Kaneko 1990a].

On the other hand, Yanagita [1992] has recently constructed a CML model for boiling, which displays the transition to bubbles, their floating by convection, and diffusion. The model clearly reproduces the nucleus–film transition, observed in experiments of boiling phenomena.

3.8.3 Convection

Rayleigh–Bénard convection has always been a typical experiment for chaos, spatiotemporal chaos, pattern formation, and turbulence. Here we briefly discuss a CML model for convection [Yanagita and Kaneko 1993, 1995].

First we choose a two-dimensional lattice (x, y), with y labeling the vertical direction, and assign velocity field $v^t(x, y)$ and internal energy $E^t(x, y)$ as field variables at time t. The field dynamics consists of Lagrangian and Eulerian parts. The Lagrangian part expresses the motion along the flow of the fluid. Here, we take a virtual fluid particle at the lattice point, and the field variable there is transported by the fluid velocity $v^t(x, y)$ given on that lattice point.

The Eulerian part is further decomposed into the buoyancy force, heat diffusion and viscosity. These processes are numerically simulated by the conventional method used to construct a CML model. In constructing procedures, we assume that $E^t(x, y)$ is associated with the temperature. For the Eulerian part we notice the following prescriptions: (1) a site with higher temperature receives a force in the upward direction; (2) heat diffusion leads to the diffusion for $E^t(x, y)$; (3) the velocity field $v^t(x, y)$ is also subjected to diffusive dynamics, due to the viscosity; (4) in an incompressible fluid, the pressure term keeps $\mathrm{div}\,v = 0$. We do not use this condition here, since the inclusion of pressure requires considerable computational resources and complicated modeling. Instead, we borrow a term from compressible fluid dynamics, which intuitively brings about this pressure effect, and refrains the $\mathrm{div}\,v$ term from growing too large. This term is given by a discrete version of $\mathrm{grad}(\mathrm{div}\,v)$. An "expanded" region with larger $\mathrm{div}\,v$ imposes a force to neighboring lattice points through this term.

Simulations of the model reproduce a wide range of phenomena in Bénard convection experiments: indeed the roll pattern of Fig. 3.1 is obtained from this CML, where the roll shows weakly chaotic oscillations as can be seen in Fig. 3.1b. For small aspect ratios, the oscillation of convective rolls and many routes to chaos are observed when increasing the Rayleigh number, which is proportional to the temperature difference between the top and bottom

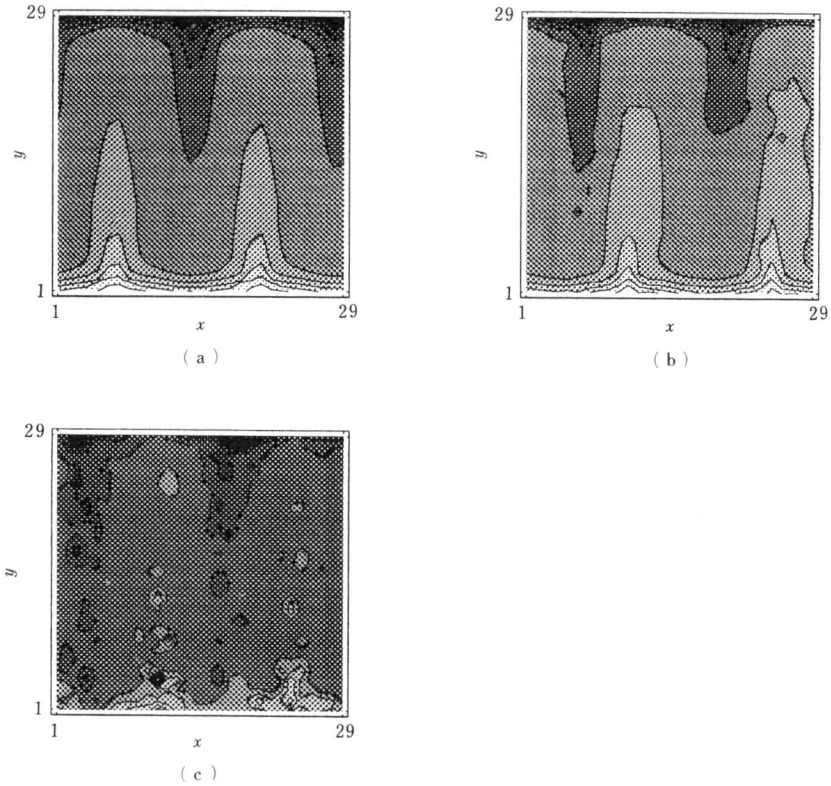

Fig. 3.32 a–c. The isotherms of Bénard convection. The temperature difference between the top and bottom plates increases for (**a**), (**b**), and (**c**). See for detailed accounts Yanagita and Kaneko [1995]. (From T. Yanagita and and K. Kaneko, *Physica D* **82** (1995) 288, with the permission of the authors and the publishers.)

plates. For large aspect ratios, spatiotemporal intermittency is observed, while roll formation in three-dimensional convection is also simulated. These results reproduce experiments well.

At high Raleigh numbers, we have plotted the isotherms when the convection is turbulent in Fig. 3.32. In particular, in Fig. 3.32c, one can see a localized region much hotter or cooler than the surrounding area. In fluid experiments, such localized structures are called 'plumes' and they are believed to be related to the transition between soft and hard turbulence. Our model shows this type of transition due to the plume structures where the velocity-distribution profile changes from Gaussian to exponential, as is consistent with experimental data.

3.8.4 Spiral and Traveling Waves in Excitable Media

Another possible application of a CML lies in the pattern dynamics of a system with an excitable state and a relaxation from it. We have seen such examples in reaction-diffusion processes within excitable media, and in some biological phenomena such as the heart rhythm and electrical activities in neural tissues.

A simple map with an excitable state was introduced by Nagumo and Sato [1972] (see Sect. 2.4). They used the map

$$f(x) = b \times (x - H(x)) + c \,, \tag{3.19}$$

where $H(x)$ is Heaviside's step function ($H(x) = 1$ for $x > 0$ and $H(x) = 0$ for $x < 0$). Here, $x > 0$ corresponds to the "fired state". The constant term "c" comes from an external stimulus applied to a single neuron.

We can easily construct a CML corresponding to the above model. Following the interpretation of the term "c", we can replace this constant term by stimuli from other lattice points. As a simple model, we assume that a constant pulse is emitted if $x > 0$ ("fired"). Then we obtain the following CML [Kaneko 1990a]:

$$\begin{aligned} x_{n+1}(i,j) = b \times (x_n(i,j) - H(x_n(i,j))) + d \times (H(x_n(i+1,j)) \\ + H(x_n(i-1,j)) + H(x_n(i,j+1)) + H(x_n(i,j-1))) \end{aligned} \tag{3.20}$$

(for a two-dimensional lattice). One can also choose a suitable neighbor instead of the nearest-neighbor coupling. As shown in Fig. 3.33, spiral formation and traveling waves are observed. See also Kapral [1993]. An important application of excitable media to biology is the heart rhythm, where the turbulent spiral formation is seen as fibrillation.

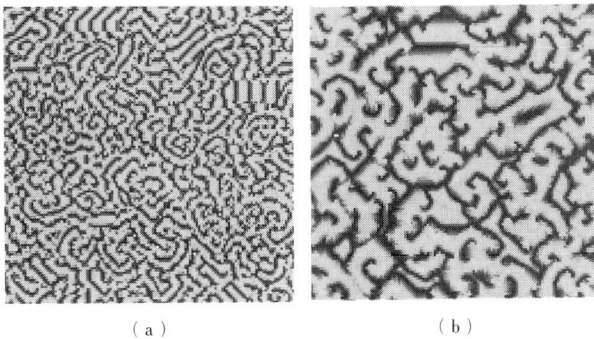

(a)   (b)

Fig. 3.33 a,b. Two-dimensional snapshot pattern of the CML (3.20). A gray-scale plot is adopted for each pixel according to the value of $x_n(i,j)$. (With the courtesy of Noriyuki Ouchi.)

3.8.5 Cloud Dynamics and Geophysics

Cloud dynamics plays an important role in the climate system, weather forecast, geophysics and so on. However this elementary process in meteorology is extremely complicated because it consists of different time and space scales and the phase transition from liquid to gas is coupled with the motion of the atmosphere. In order to investigate such a system, the construction of a phenomenological model is essential.

A CML model for cloud dynamics has recently been proposed by extending the method used for convection (Sect. 3.8.3). The model consists of the successive operations of the physical processes: buoyancy, diffusion, viscosity, adiabatic expansion, fall of a droplet by gravity, descent flow dragged by the falling droplet, and advection. Through extensive simulations, the phases corresponding to stratus, cumulus, stratocumulus and cumulonimbus are found with the change of the ground temperature and the moisture of the air [Yanagita and Kaneko 1997].

In geophysics, there are several applications of CMLs. In particular, they were used to study the formation of sand ripples [Nishimori and Ouchi 1993], while some progress has also been made in the study of sand-dune formation.

3.8.6 Ecological Systems

Let us recall that the logistic map was originally studied in the context of population dynamics. In general, population dynamics (of some species) is represented by nonlinear equations, and sometimes by a map, like e.g. the population of one year, when reproduction is limited to one season. In an ecological system, animals or insects usually move in space. Thus the population dynamics includes both the reproduction process represented by a map, and the spatial diffusion process. Hence the problem can be well modeled by CMLs [Bascompte and Solé 1998]. When there are two species relating to each other as predator/prey or host/parasite, the dynamics of the population can be represented by a two-dimensional map. When these organisms move in space, the model is reduced to a coupled map of two variables with diffusion coupling. Indeed extensive simulations have recently been carried out [Bascompte and Solé 1998] which show spiral patterns, chaotic Turing patterns [Solé *et al.* 1992], spatiotemporal chaos, supertransients, and so forth.

The immune response consists of the population dynamics of many antibodies linked in a network. Oscillations of some antibodies are observed, which may be chaotic in time. CML modeling appears to be promising there.

3.8.7 Evolution

In the above case, the number of possible species is fixed. In evolution, the number of possible types of organisms changes with time. One simplification of this type of problem is the use of a genetic algorithm where each type of organism is represented by a different genotype, coded by a gene as a bit sequence.

Let us discuss some (population) dynamics of many individuals, coded by genes. If genes are represented by a bit sequence, the mutation process in gene space is given by the diffusion in the bit sequence. As a simple abstraction, assume that the gene space is represented by a bit space (as is often adopted in genetic algorithms [Holland 1986]) like $i = 0101010$, and that a single point mutation is given by a flip-flop $0 \leftrightarrow 1$ of one bit. When some nonlinear population dynamics is included to take into account saturation, competition, or predator–prey (host–parasite) interaction, the total population dynamics is given by the local nonlinear dynamics and the diffusion process on a hypercubic lattice of length 2, corresponding to the bit sequences. The minimal model for this process is given by the following coupled map on a hypercubic lattice [Ikegami and Kaneko 1992; Kaneko 1994a]:

$$x_{n+1}(i) = (1 - \epsilon)f(x_n(i)) + \frac{\epsilon}{K} \sum_{j=1}^{K} f(x_n(\sigma_j(i))), \qquad (3.21)$$

where $\sigma_j(i)$ is a "type" whose j-th bit is different from species i (with only one bit difference), and K is the total bit length of the species (the total number of "types" is 2^K). We often use a decimal representation of the bit sequence; for example 42 stands for the sequence 101010, and $\sigma_2(42) = 40$.

The model is also of theoretical interest, since it lies between the CMLs discussed in this chapter and the globally coupled map to be discussed in the next chapter. In a globally coupled chaotic system, we have N connections per element, while a d-dimensional CML (with nearest-neighbor coupling) has $2d = o(N)$ connections per element. In our hypercubic system with $N = 2^K$ elements, we have $K = \log_2 N$ connections per element.

In the model, we have often observed a state with a few synchronized clusters when the nonlinearity parameter a is not large. Here a synchronized cluster means that two elements in the cluster oscillate in complete synchronization (see Sect. 5.6.4). In the present case, the split into two clusters is organized according to the hypercubic structure [Kaneko 1994a].

In a system with interacting population dynamics, it is interesting to study how the diversity of genes is maintained. In Kaneko and Ikegami [1992], the concept "homeochaos" is proposed as a mechanism for sustaining dynamic stability with diversity (see Sect. 5.6.5).

3.8.8 Closing Remarks

There are many other possible applications of CMLs. Indeed, one can expect a wide range of applications, since the model is constructed by only choosing a suitable set of local order parameters and by introducing a small number of procedures characterizing the essence of the dynamics of the phenomenon under consideration. Also there are several proposals for applying a CML to information processing. This problem will be discussed in Chap. 5, together with the applications of globally coupled maps which are discussed in the next chapter.

4. Networks of Chaotic Elements

4.1 GCM Model

As an example of the high-dimensional dynamics discussed in Chap. 1, let us consider a network of chaotic elements. In a network system many elements that can display chaotic dynamics interact with each other and evolve in time. Here we introduce the globally coupled map (GCM) as the simplest example of such a network of chaotic elements. We discuss the observed phenomena and the universal concepts revealed therein in some detail, since the model provides us with a 'dynamic many-to-many relationship', 'constructive model', and 'dynamics between the whole and its parts'. We believe that through the study of the GCM, we can work towards a methodology studying complex systems.

The present GCM shares with the CML in Chap. 3 the fact that many chaotic elements interact with each other. The only difference is that, in contrast with the local interaction in the CML, the interaction in the GCM is global. Here 'local' interaction of a given element means that the interaction is limited to the elements near it. From the viewpoint of standard statistical physics, this means that the 'local' interaction decays fast enough (exponentially) over the distance between the two elements. On the other hand, an interaction decaying as some power of the distance ($r^{-\alpha}$) is regarded as long-ranged. Examples of long-ranged interactions are seen in vortices in fluids, ensembles of stars with gravitational interaction, and so forth. Furthermore, there are cases where elements interact with all other elements with the same or similar strength. For example, when many elements evolve with some constraint, such as a conservation law, a global interaction through a mean field often appears. As an example consider the circuit in Fig. 4.1, where many nonlinear elements couple through the electric current. In this case, the change of the electric current in each element is governed by the current itself and the voltage applied to the element. The latter is given by $V - R\sum_j I(j)$, and thus the sum of all currents governs the dynamics, and completely global interaction appears. This kind of problem is studied in solid-state physics, in e.g. Josephson-junction arrays, charge-density waves and so forth. A similar phenomenon occurs in nonlinear optics such as the multi-mode laser, where many modes compete for the supplied energy.

R   resistivity
$F(I_i, V_i)$ function characterizing
                element i
I   electric current
I_i current through
                the element i
$V_0(t)$ voltage
V_i voltage for
                the element i

$$\dot{I}_i = F(I_i, V_i)$$
$$V_i = V_0(t) - R_i I = V_0(t) - R \sum_{k=1}^{N} I_k$$

X   concentration of
                nutrition in soup
$F(x_i, X)$ function characterizing
                the change of the state
                of cell i
x_i internal state of cell i
$f(x_i)$ function determining
                the activity
                to get the nutrition

$$\dot{x}_i = F(x_i, X)$$
$$\dot{X} = -\sum_{k=1}^{N} f(x_k)$$

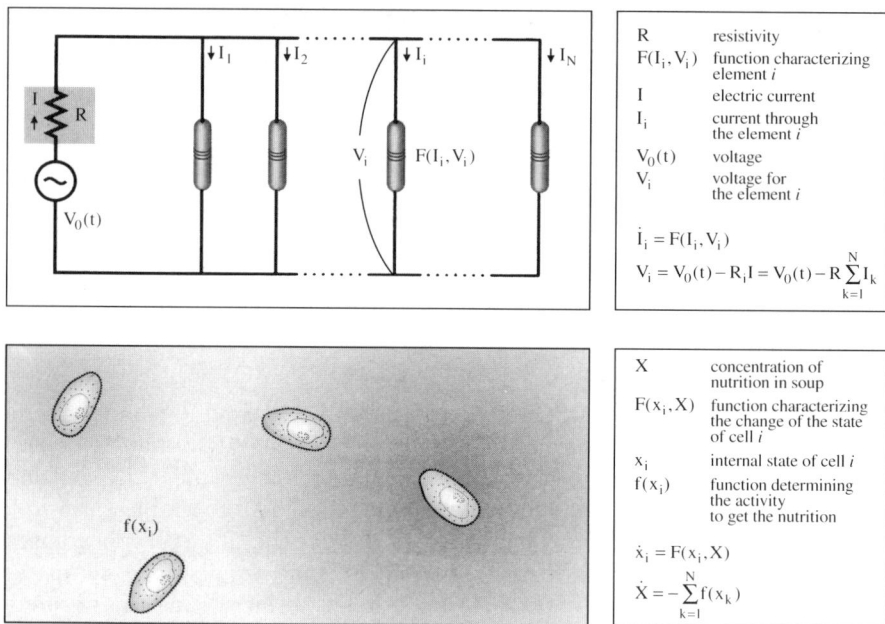

Fig. 4.1 a,b. Schematic representation of two examples of globally coupled systems. (a) An electric circuit, (b) cells competing for resources. (From K. Kaneko, Nikkei-Science (May, 1994d) 34 (in Japanese), with the permission of the publishers).

The importance of globally coupled chaotic systems is not restricted to dynamical systems theory or physical phenomena. They are relevant to biological phenomena as well.

Recent extensive investigations on neural networks can provide us with a simple example of applying statistical mechanics to simple information processing. The real neural dynamics in the brain consists of an ensemble of complex elements with complex coupling. Most current neural-network studies adopt oversimplified elements (0-1 or a sigmoidal function) with moderately simplified coupling. On the other hand, it is known that even a single neuron or a small ensemble of neurons can exhibit complex dynamical behavior like chaos, as will be studied in Chap. 5 in detail. Then, it is natural and important to ask the following question: what happens if we use moderately simplified elements (with chaotic response) with oversimplified couplings instead as a different limit of simplification from the neural dynamics?

In cell biology, metabolic reactions often show nonlinear oscillations through autocatalytic processes, while global interaction among cells is important in developmental processes and cell differentiation. In an example, schematically shown in Fig. 4.1b, many cells compete globally for nutrition.

In an immune network, oscillations of antibodies, coupled to each other, are studied theoretically, while globally coupled population dynamics is common in ecological systems. In population dynamics, global constraint from the "environment" is important. The dynamics of each species is described by the inherent population dynamics of the species and the interaction with the environment. The dynamics of the environment, on the other hand, depends on all species.

In a similar context, Eigen and Schuster [1979] have proposed an equation for the evolutionary dynamics of the population of RNA. It consists of the population dynamics for each RNA species and the constraint from the food source which is globally coupled to each of the RNA species.

Furthermore, this global coupling is not limited to biological science. In economics and sociology (e.g., the stock market), again, local complex dynamics and global feedback are important. For example, in the stock market, the dynamics of the stock price depends on each firm's economic state, and on some weighted average of the stock prices such as the Dow Jones Industrial Average. In general, global coupling commonly appears if many agents interact through a market.

As a prototype of such global-coupling dynamics, we introduce the "globally coupled map" (GCM).[1] It is a minimal model that has local chaotic dynamics and global interaction. Although simple, it provides universal classes of phenomena, such as the formation of synchronized clusters, bifurcation of clustering, hierarchical clustering, chaotic itinerancy over ordered states, information cascades, collective dynamics with complete desynchronization, and so forth. Specifically, we study the mean-field version of the CML

$$x_{n+1}(i) = (1 - \epsilon)f(x_n(i)) + \frac{\epsilon}{N}\sum_{j=1}^{N} f(x_n(j)), \qquad (4.1)$$

where n is the discrete time and i indexes the elements ($i = 1, 2, \cdots, N$ = system size) [Kaneko 1989e]. Since this is a general model, we could in principle choose any function $f(x)$. Here, however, we are interested in coupled chaotic systems, and assume that the function $x_n \rightarrow x_{n+1} = f(x_n)$ exhibits chaotic dynamics (for some parameter values). In most examples, we choose the logistic map $f(x) = 1 - ax^2$ as the prototype for chaotic dynamics. This GCM is an abstract model. By regarding the variable $x_n(i)$ as an electric current, neural activity (local-field potential), energy, population, food resources, chemical concentrations, the price of goods, and so on, one can make this model a conceptual prototype. Indeed, from the results obtained with our GCM, we can make new predictions and provide a novel theoretical framework for the real-world phenomena mentioned above. For example, hierarchical clustering will be relevant to neural dynamics, ecology,

---

[1] Globally coupled systems without chaos have been studied for a long time. In particular, the entrainment of oscillators with different frequencies has been studied by Winfree [1980] and Kuramoto [1984]. A degenerate torus in a coupled oscillator was also studied by Hadley and Wiesenfeld [1989] in connection with Josephson-junction arrays.

and cell differentiation. Chaotic itinerancy is observed in a multi-mode laser system, and its relevance to neural-information processing will be discussed in Chap. 6. The clustering bifurcation is observed in simulations in Josephson-junction arrays, while collective dynamics is discussed for an optical system. The relevance of the results to biology is presented in Chap. 5 (see also Kaneko and Ikegami [1998]).

The model is a mean-field-theory-type extension of the coupled map lattice (CML) studied in Chap. 3. Our dynamics (4.1) consists of a parallel nonlinear transformation and a feedback from the "mean field". It can schematically be shown in the diagram of Fig. 4.2. Here, after the nonlinear transformation is applied to $x_n(i)$, each element interacts globally with the mean field to yield the value $x_{n+1}(i)$ at the next time step. Since this process is repeated, the dynamics is equivalent to a model in which we apply the global interaction first and the nonlinear transformation second. In other words, by the transformation $y_n(i) = f(x_n(i))$, our model is equivalent to

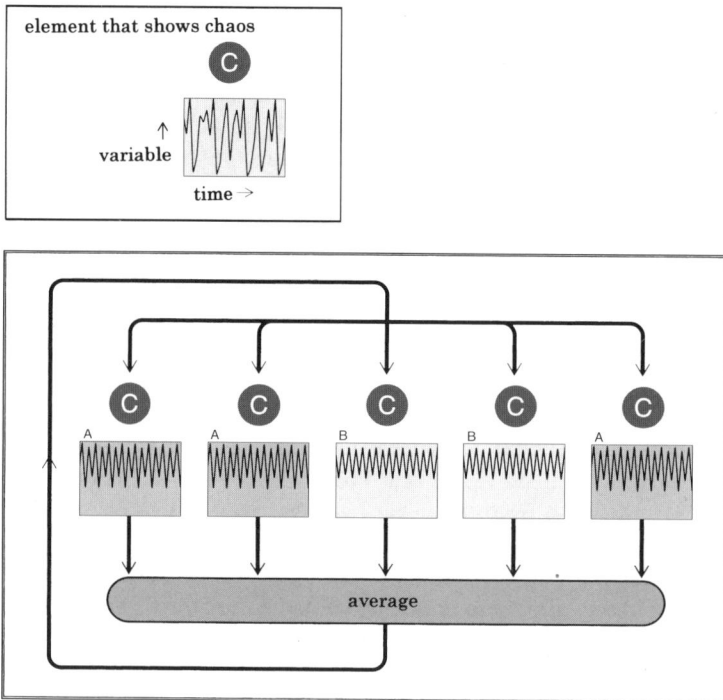

Fig. 4.2. Schematic representation of the globally coupled map. (From K. Kaneko, Nikkei-Science (May, 1994d) p. 34 (in Japanese), with the permission of the publishers).

$$y_{n+1}(i) = f\left((1-\epsilon)y_n(i) + \frac{\epsilon}{N}\sum_{j=1}^{N}y_n(j)\right). \tag{4.2}$$

This form may be more familiar to researchers in neural nets, especially if one chooses a sigmoidal function (e.g., $\tanh(\beta x)$) as $f(x)$ and if the coupling term ϵ depends on the elements. Indeed, the dynamics adopted in a standard neural network is represented by

$$y_{n+1}(i) = g\left[\sum_{j=1}\epsilon_{i,j}y_n(j)\right], \tag{4.3}$$

with $g(x) = \tanh(\beta x)$ a transformation with a bistable fixed point (see Fig. 3.31). In neural networks, a state with $x > 0$ corresponds to a fired neuron (or spin-up state), and a state with $x < 0$ to a nonfired neuron. On the other hand, the coupling $\sum_{j=1}\epsilon_{i,j}y_n(j)$ represents input from other neurons or spins. The coupling $\epsilon_{i,j}$ takes a positive or negative value, depending on whether the neuron is excitatory or inhibitory (ferromagnetic or anti-ferromagnetic for spin systems). By choosing $\epsilon_{i,j}$ random or by choosing specially coded values, many (quasi)stable states are formed in neural networks and spin glasses.

Our equation (4.1) has the following features that contrast with spin-glass-type dynamics: (1) instead of the stable dynamics of $g(x)$, we adopt a dynamics $f(x)$ with orbital instability; (2) instead of random or complex coupling, we adopt a homogeneous coupling. In our system, stochastic dynamics is generated by chaos intrinsically, without external randomness. Due to the random behavior generated, our GCM shares some common behavior with spin glasses, but much richer and more interesting dynamics will be observed.

4.2 Clustering

The fundamental feature of our network dynamics is the competition between local chaos and global averaging. Recall the sensitive dependence on initial conditions in chaos. Due to this sensitive dependence, the difference between two initially close elements is amplified, and hence the phases of the oscillations have a tendency to diverge. On the other hand, two elements also have the opposite tendency to be synchronized due to the global coupling. Hence it is expected that elements are desynchronized when local chaos is strong, and synchronized when the global coupling is large. Between these two extremes, other interesting states are found, depending on the strength of chaos and the coupling. The following attractors are possible.

(a) Coherent state with complete synchronization:

All elements $x_n(i)$ take the same value, and the oscillation is completely synchronized (and can be chaotic). Once the elements are synchronized, the motion is governed just by the single logistic map $x_{n+1} = f(x_n)$. The stability of this attractor is calculated by computing the linear stability of the synchronized state. Note that

$$x_{n+1}(i) - x_{n+1}(j) \approx (1 - \epsilon)f'(x_n(i))(x_n(i) - x_n(j)) \tag{4.4}$$

with a linear approximation. Then the condition for synchronization is given by

$$\lambda_{\mathrm{spl}}(i) \equiv \lim_{T \to \infty} (1/T) \log \left| \prod_n^T (1 - \epsilon)f'(x_n(i)) \right| \tag{4.5}$$

$$= \log(1 - \epsilon) + \lim_{T \to \infty} (1/T) \sum_n^T \log |f'(x_n(i))| \tag{4.6}$$

$$= \log(1 - \epsilon) + \lambda_0 < 0. \tag{4.7}$$

Here $\lambda_{\mathrm{spl}}(i)$ is the split exponent as will be discussed later in detail (p. 114).

Since we consider the case with complete synchronization, the dependence of $x_n(i)$ on the element i is not necessary, and the dynamics is represented just by a single logistic map. Hence λ_0 is nothing but the Lyapunov exponent of the single logistic map. From condition (4.7) it follows directly that the completely synchronized state for the stable cycle of the logistic map is stable in the GCM. Furthermore, even if the dynamics of the single logistic map shows chaotic behavior, synchronized chaos is stable if $-\log(1 - \epsilon) > \lambda_0$.

(b) Completely desynchronized state:

All elements have different values at all times and oscillate without any synchronization. As an extreme limit, consider the case with $\epsilon \to 0$. If the elements oscillate chaotically, any small difference between two elements is amplified, and thus the phase of the oscillations becomes different. Even if such a limit is not taken, the oscillation is totally desynchronized if the coupling is small or local chaos is strong.

Besides these two states that were expected, we have found the following nontrivial states.

(c) Creation of several clusters in which elements show synchronized oscillation:

A cluster is defined as the set of elements in which $x(i) = x(j)$, that is, $x(i) = x(j)$ for $i, j \in$ the same cluster. Some examples of the dynamical behavior of these attractors are shown in Fig. 4.3.

We can classify the state of our GCM by the number of clusters k and the number of elements for each cluster N_k (see Fig. 4.4). Unless otherwise mentioned, we label the clusters such that $N_1 \geq N_2 \geq \cdots N_k$.

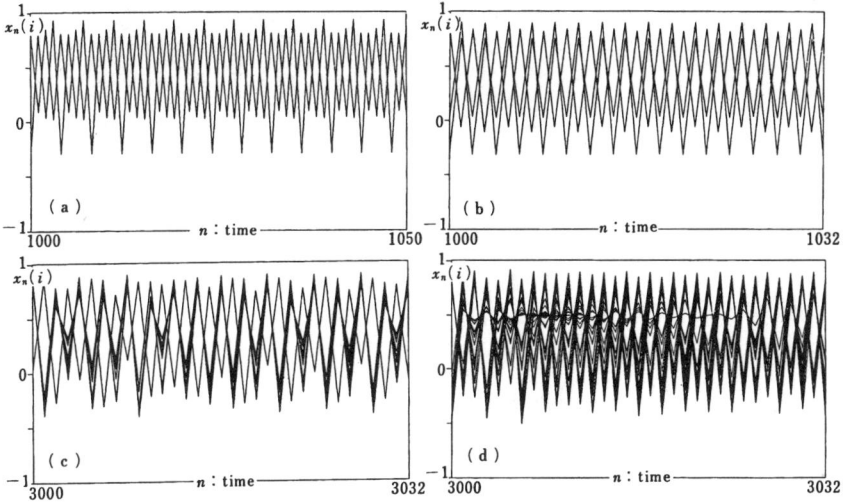

Fig. 4.3 a–d. Examples of time series of attractors: $x_n(i)$ for all i are plotted as a function of time n. If there are only k lines ($k < N$), the system has fallen onto a k-cluster state. $N = 50$. (a) Two-cluster attractor: $a = 1.65$, $\epsilon = 0.2$; $N_1 = 35, N_2 = 15$, (b) 3-cluster attractor: $a = 1.8$, $\epsilon = 0.2$; $N_1 = 18, N_2 = 18, N_3 = 14$. ($1000 \leq n \leq 1032$), (c) attractor with $k = 27 : a = 1.95, \epsilon = 0.2$; $N_1 = 24$ and $N_j = 1$ ($27 \geq j \geq 2$). ($1000 \leq n \leq 1032$), (d) N-cluster attractor ($N_j = 1$, $N \geq j \geq 1$): $a = 2.0$, $\epsilon = 0.1$ ($1000 \leq n \leq 1032$). (From K. Kaneko, *Physica D* **41** (1990c) 137, with the permission of the publishers.)

If a system is attracted to exactly a k-cluster solution with $(N_1, N_2, \cdots, N_k)$, it never leaves this state, since the motion for $x_n(i)$ and $x_n(j)$ at time $n > m$ is governed by exactly the same dynamics if $x_m(i) = x_m(j)$ holds ($x_n(i) = x_n(j)$ if $x_m(i) = x_m(j)$).

The dynamics of a k-cluster attractor with $(N_1, N_2, \cdots, N_k)$, can be written as the following k-dimensional map:

$$X^\nu_{n+1} = (1 - \epsilon)f(X^\nu_n) + \sum_{\mu=1}^{k} \epsilon_\mu f(X^\mu_n), \qquad (4.8)$$

where X^ν_n denotes the value of x_n in the ν-th cluster, and the "effective coupling" ϵ_μ is given by $\epsilon_\mu = \epsilon \times (N_\mu/N)$.

The coherent state in (a) corresponds to the case with $k = 1$, while $k = N$ corresponds to the desynchronized state in (b). For $k = 2$, elements split into two coherently oscillating groups. Indeed there are several states ranging from small k to $k \approx O(N)$. Examples of the time series of $x_n(i)$ for all the elements i are overlaid in Fig. 4.3. Since the time series are overlaid, there should be N lines if the elements are desynchronized. In Fig. 4.3a, $N = 50$ elements split into two clusters with 35 and 15 elements. In each cluster elements show synchronized oscillation. Similarly in Fig. 4.3b, elements split

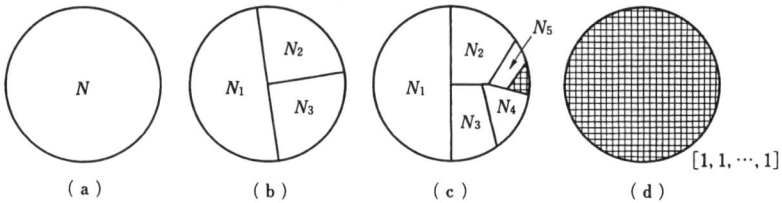

Fig. 4.4a–d. Schematic figure for clusterings: (a) coherent attractor (b) few clusters ($k = 3$) (c) many-cluster attractor with unequal partition sizes (d) many-cluster attractor with $k = N$. (From K. Kaneko, *Physica D* **41** (1990c) 137, with the permission of the publishers.)

into three clusters with 18, 18, and 14 elements. An example with many clusters is shown in Fig. 4.3c, where the dynamics consists of a cluster with 24 synchronized elements, and a further 26 desynchronized elements (the cluster number is $k = 27$). On the other hand, Fig. 4.3d shows a completely desynchronized case, and 50 lines are depicted.

Our attractor is characterized by the clustering condition $k, (N_1, N_2, \cdots, N_k)$. States with complete synchronization, few clusters, many clusters, and complete desynchronization are schematically plotted in Fig. 4.4.

As one can see from (4.8), the effective coupling depends on the clustering distribution. For a given cluster number k, the effective coupling ϵ_m changes with the distribution N_j, and bifurcations in the k-dimensional map are observed when changing the distribution of the elements. Indeed, later we will often see period-doubling bifurcations from the fixed point to period 2, 4, 8, $\cdots$, to chaos, and from quasiperiodicity to chaos. A variety of attractors with different types of dynamics coexists for the same parameters.

Note, however, that not all of the solutions given by (4.8) are stable, although the evolution given by (4.8) is always a solution of the GCM dynamics. To be an attractor, the state should be linearly stable, and the stability of the k-cluster state is not solely determined by the above k-dimensional map. To destroy the coherence within a cluster, a small disturbance $x(j) - x(i)$, with elements i, j belonging to the same cluster, must be amplified. For this, the condition is positivity of (4.5) or (4.6) (now not all elements are synchronized, and condition (4.7) cannot be adopted). From now on, we call $\lambda_{\mathrm{spl}}(i)$ the split exponent of element i. The condition for the stability is stated as "the split exponent for an element belonging to a cluster with $N_j > 1$ should be negative". The examples in Fig. 4.3 are stable in this sense, and are thus attractors.

The split exponent $\lambda_{\mathrm{spl}}(i)$ appears to be similar to the Lyapunov exponent but is actually different. The Lyapunov exponents give the amplification (reduction) rate of two close orbits in the N-dimensional phase space, while the split exponent gives such a rate for two elements. Often the orbit is chaotic (with some positive Lyapunov exponents), even though the split exponent is negative.[2]

As for the spectrum of Lyapunov exponents, it has been shown that only up to k positive Lyapunov exponents can exist for an attractor with k clusters [Kaneko 1990c]. This is confirmed by checking the linear stability of the synchronization between two elements. Let us recall that we have N coupled chaotic elements. Without coupling there are N positive Lyapunov exponents. In a clustered attractor up to k positive exponents exist. Thus there is a radical reduction of chaotic degrees from N to k. Hence there is strong suppression of chaos. This suppression is related to the pattern selection in the CML discussed in Sect. 3.3.1. In the GCM, the global coupling leads to such suppression of chaos.

4.3 Phase Transitions Between Clustering States

We have found that many states with different clusterings coexist as attractors. Then how do types of attractors change with parameters a and ϵ? According to the numerical results, the following phases appear with the increase of a (or the decrease of ϵ).

(i)   Coherent phase: coherent attractors occupy (almost) all basin volumes.
(ii)   Ordered phase: few-cluster attractors occupy (almost) all basin volumes.
(iii) Partially ordered phase: coexistence of many-cluster and few-cluster attractors.
(iv)   Desynchronized phase: all attractors have many ($\approx N$) clusters.

Typical attractors of each phase correspond to those in the schematic Fig. 4.4a–d, respectively.

We take an ensemble of initial conditions to investigate a variety of attractors and the basin volume of each attractor. If we take the total number of initial conditions which lead to a k-cluster attractor, and divide it by the number of initial conditions checked, we obtain a basin-volume distribution $Q(k)$ of k-cluster attractors. This gives the probability that an arbitrary chosen initial condition falls on a k-cluster attractor.

A simple quantification of the attractors is obtained by using the average cluster number $R \equiv \sum_{k=1}^{N} kQ(k)$. In the coherent phase $R = 1$, while $R = b \ll N$ for the ordered phase, $R = rN$ for the partially ordered ($r < 1$) phase, and $R = N$ for the desynchronized phase. The change of R with a is plotted in Fig. 4.5.

According to the behavior of R, the phase diagram of our GCM is obtained in Fig. 4.6, determined from $Q(k)$ which is calculated from $M = 500$ samples,

---

[2] Such synchronization of chaotic elements has recently been discussed as 'chaotic synchronization' [Pecora and Caroll 1990]. This is essentially an application of the present synchronization condition. For synchronization of two chaotic elements, see Fujisaka and Yamada [1983].

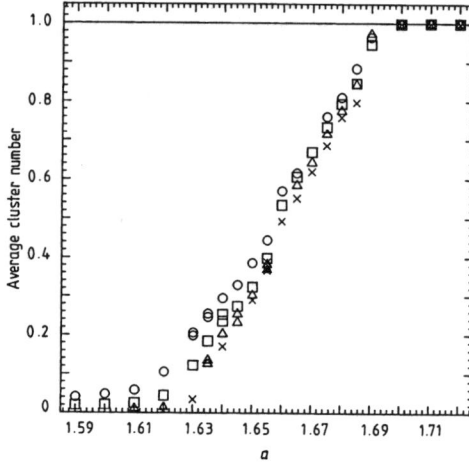

Fig. 4.5. The scaled cluster number c, defined by the average cluster number divided by N, is plotted versus of a. $\epsilon = 0.1$. $N = 100$ (○), $N = 200$ (□), $N = 400$ (△), and $N = 800$ (×). Obtained from 5000 randomly chosen initial configurations, after discarding 40 000 steps as transients. To check the accuracy, we have sometimes carried out a few runs for different sets of initial conditions, where more than one mark overlap. (From K. Kaneko, *J. Phys.* **A24** (1991a) 2107, with the permission of the publishers.)

after a transient of 2000 steps, with a system size of $N = 200$. In the diagram, "(1,2)" means that the dominant cluster sizes are 1 or 2, and "(1,2,3)" that the dominant cluster sizes are 1,2,3 and so on.

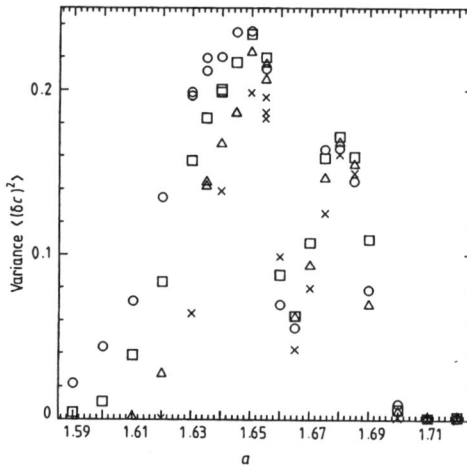

Fig. 4.6. Variance of the scaled cluster number $\langle (\delta c)^2 \rangle$ as a function of a, for $\epsilon = 0.1$. $N = 100$ (○), $N = 200$ (□), $N = 400$ (△), and $N = 800$ (×). (From K. Kaneko, *J. Phys.* **A24** (1991a) 2107, with the permission of the publishers.)

The fluctuation of the cluster number for initial conditions is measured as the variance $\sum_k k^2 Q(k) - R^2$. The variance is plotted in Fig. 4.6. The variance is larger in the partially ordered (PO) phase, and it does not decrease with the increase of size N. This means that there is a variety of attractors with a variety of clusters in the PO phase, while the most typical ones have $k = O(N) < N$, and clusters ranging from $O(N)$ to 1.

The existence of these four phases is rather common in globally coupled nonlinear systems. They include the coupled circle map with global coupling [Kaneko 1991b; Sinha *et al.* 1992a] given by

$$\theta_{n+1}(i) = \theta_n(i) + (K/N) \sum_j \sin(2\pi(\theta_n(j) - \theta_n(i))),$$

or

$$\theta_{n+1}(i) = \theta_n(i) + \Omega + (K/N) \sum_j \sin(2\pi\theta_n(j)),$$

the globally coupled Ikeda map [Perez *et al.* 1992], and globally coupled ordinary differential equations that correspond to Josephson-junction arrays [Dominguez and Cerdeira 1993]. Some common phases are also observed in globally coupled Ginzburg–Landau equations [Nakagawa and Kuramoto 1993, 1994a, b; Hakim and Rappel 1992].

Now we will study the behavior in each phase in more detail.

4.4 Ordered Phase and Cluster Bifurcation

In the ordered phase, the possible numbers of clusters k are limited, and for most initial conditions the attractors consist of only a few possible clusters, which are written as the numerals in the ordered phase in Fig. 4.7. In Fig. 4.8, we have plotted the ratio of initial points ('basin volume') that fall onto the attractors of each cluster number. In the ordered phase, the typical number of clusters in the attractors increases with a, as 1, 2, 3, $\cdots$. Although one might expect a 'doubling increase' (as 1, 2, 4, 8, $\cdots$) like in the 'logistic map', this is not the case.

Why does the cluster number increase one by one instead of by a doubling cascade? Note that in a two-cluster attractor, elements belonging to a different cluster oscillate out-of-phase. When elements oscillate with the same phase of oscillation, the dynamics of the mean field and of each element are synchronous. If the oscillation is chaotic, the orbital instability is enhanced by this synchronization. Then the synchronization is destabilized (unless condition (4.7) is satisfied), and the synchronized cluster splits into two. If the two clusters oscillate out-of-phase, the mean-field motion does not enhance the instability. Hence the two clusters can exist, smearing out the orbital instability. If the number ratio of the two clusters is 1:1, the contributions of the two clusters to the mean field cancel each other out. If the number

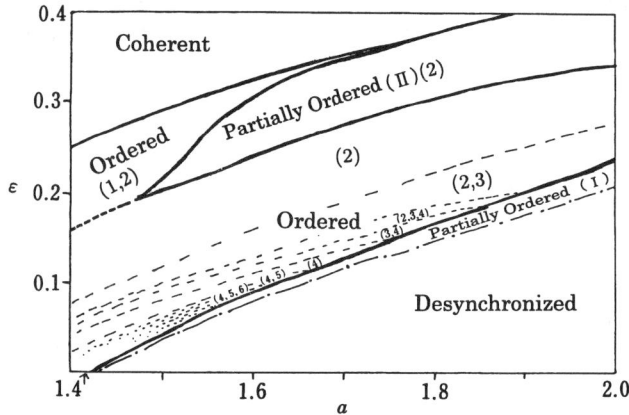

Fig. 4.7. Rough phase diagram: phases are determined by $Q(k)$, calculated from 500 randomly chosen initial conditions and $N = 200$. The parameters range from $a = 1.4$ to 2.0 by 0.01 and $\epsilon = 0.02$ to 0.4 by 0.02. The numbers such as (1,2,3) represent the dominant cluster numbers (with the basin-volume ratio more than 10%). The arrow at the horizontal axis shows an accumulation of period-doubling bifurcations of the logistic map. (From K. Kaneko, *Physica D* **41** (1990c) 137, with the permission of the publishers.)

ratio is not balanced, the dynamics of the elements belonging to the larger cluster and the mean field are in-phase, and orbital instability remains for the motion of this cluster. Now the synchronization among the elements of the larger cluster has a tendency to be destroyed. As a result the cluster may split. If one of the split clusters is absorbed into the other, original cluster (with a smaller number of elements), the system returns to the two-cluster state, by a balanced number ratio. Thus there is a threshold for the number ratio of two clusters to exist as an attractor.

When the instability is much larger, the two clusters split from the larger cluster. These two new clusters ('sub'-clusters) oscillate out-of-phase with respect to period-4. With this mutual out-of-phase oscillation, the instability is suppressed. Now a state with three clusters exists as an attractor. In this manner the number of clusters increases as 3, 4, 5, $\cdots$ as the nonlinearity a is increased.

Let us study the dynamics of the two-cluster state in more detail. When the elements fall into two clusters, the dynamics (4.1) is written as the two-dimensional coupled map

$$X_{n+1}^j = (1 - \epsilon)f(X_n^j) + \sum_{m=1}^{2} \epsilon_m f(X_n^m), \qquad (4.9)$$

with X_n^1, X_n^2 equal to $x_n(i)$ for each cluster, and $\epsilon_{1,2} = \epsilon \times (N_{1,2}/N)$.[3]

---

[3] Here we adopt a slightly different labeling of the clusters (N_1, N_2). We call the cluster "1" if site 1 belongs to it. Thus $N_1 \geq N_2$ is not necessarily satisfied.

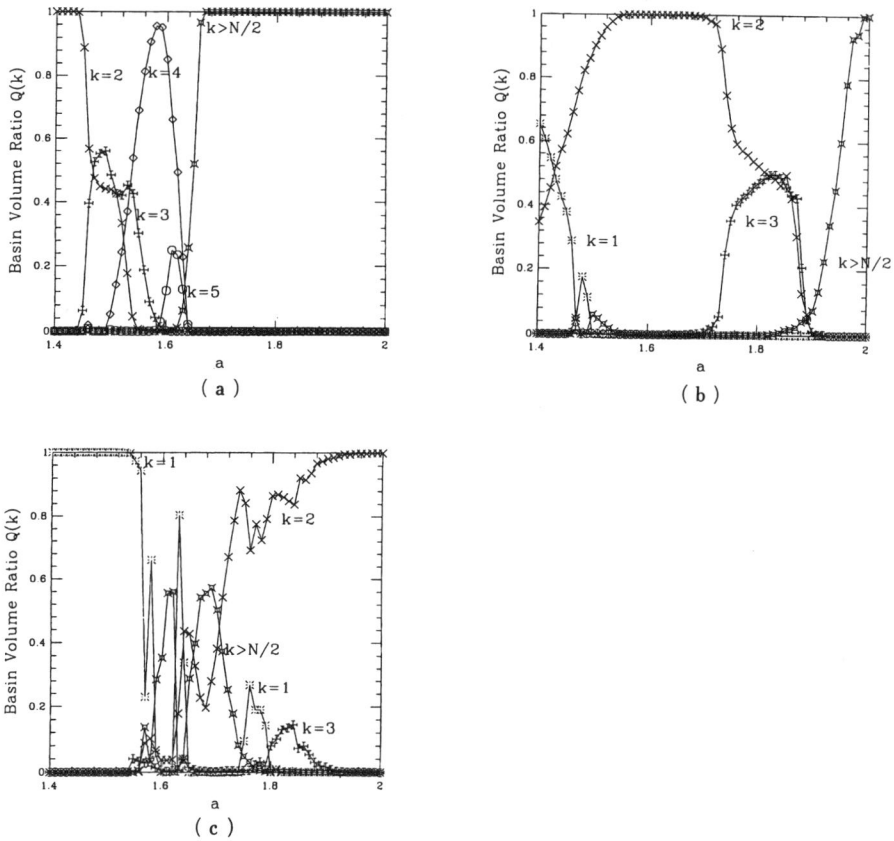

Fig. 4.8a–c. Cluster distribution $Q(k)$ for $\epsilon = 0.1$ ($1 < k \leq 5$) and Q_L plotted as a function of a for $1.4 < a < 2.0$. $\epsilon = 0.1, N = 200$. Calculated from 1000 randomly chosen initial conditions, after 3000 transients. (a) $\epsilon = 0.1$, (b) $\epsilon = 0.2$, (c) $\epsilon = 0.3$. (From K. Kaneko, *Physica D* **41** (1990c) 137, with the permission of the publishers.)

Thus the change of N_1 corresponds to the change of a bifurcation parameter in the two-dimensional map. We have to note, however, that the above reduction is possible only after the system has fallen onto two-clusters.

Fig. 4.9a gives a time series of a two-cluster attractor with a 50:50 partitioning, while the time series of an attractor with a 59:41 partitioning is given in Fig. 4.9b. In (a) the two clusters show the same oscillation except for the phase difference, while in (b) the cluster with 59 elements shows a motion with a larger amplitude.

In Fig. 4.10, two examples of the cluster bifurcation of two-cluster attractors are shown, where $X_n^1 = x_n(1)$ ($n = 2000, 2001, \cdots, 2260$) are plotted as a function of N_1. This bifurcation is nothing but that of the two-dimensional map (4.10). Once the state falls onto the two-cluster attractor, the dynamics

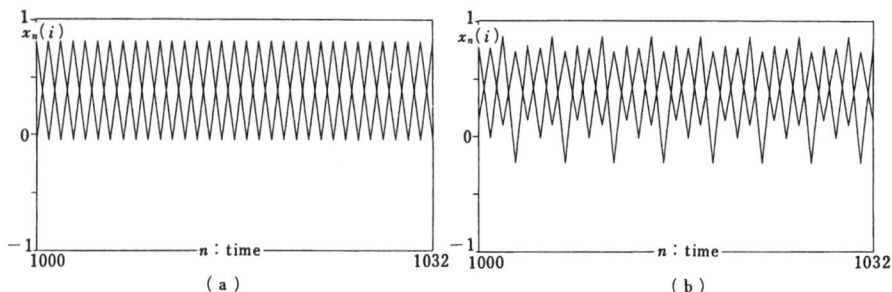

Fig. 4.9a,b. Time series of two-cluster attractors for $a = 1.88$, $\epsilon = 0.3$, and $N = 100$. $x_n(i)$'s are plotted as a function of time n ($n = 1000, 1001, \cdots, 1032$). (a) ($k = 2, 50, 50$), (b) ($k = 2, 59, 41$). (From K. Kaneko, *Physica D* **41** (1990c) 137, with the permission of the publishers.)

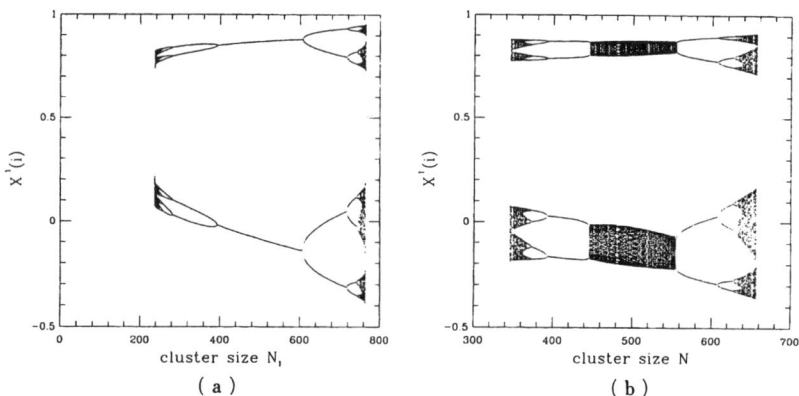

Fig. 4.10a,b. Cluster bifurcation for two-cluster attractors: $X_n^1 = x_n(1)$ ($n = 2000, 2001, \cdots, 2260$) are plotted as a function of N_1. $N = 1000$. $\epsilon = 0.2$. To change the attractors, the switching method in Sect. 5.2.1 is used. (a) $a = 1.6$; $N_{\mathrm{thr}} = 764$, (b) $a = 1.75$; $N_{\mathrm{thr}} = 656$. (From K. Kaneko, *Physica D* **41** (1990c) 137, with the permission of the publishers.)

is just a two-dimensional map, and typical bifurcations are observed there. In Fig. 4.10a, X_n^1 clearly exhibits the period-doubling "bifurcation" route to chaos as N_1 is changed. The other route is a quasiperiodicity transition with a period-doubling bifurcation, which is observed for smaller ϵ (e.g., 0.2). As an example, see Fig. 4.10b. The motion is quasiperiodic with some lockings if $N_1 \approx N_2$. As $N_1 - N_2$ is increased further, the attractors exhibit period-doubling to chaos from period-2.

Still, we have to note that as a problem for the GCM, this is not a bifurcation in the usual sense. The fact is that the parameters are fixed here and all that we have done is to arrange the attractors in the order of N_1. In other words, we have found a simple way to organize many attractors, such that the various attractors can be seen just as in a bifurcation cascade.

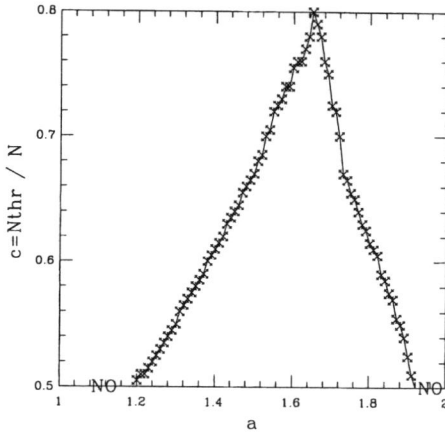

Fig. 4.11. $c = N_{thr}/N$ as a function of a (for $\epsilon = 0.2$). Calculated from a system with $N = 200$, and with the use of the switching method in Sect. 5.2.1. c goes to 0.5 at $a = 1.2$ and $a = 1.92$. Below 1.2 and beyond 1.92, any two-cluster state can no longer be an attractor. (From K. Kaneko, *Physica D* **41** (1990c) 137, with the permission of the publishers.)

By switching from one attractor to another in order to increase N_1, one can observe this bifurcation-like phenomenon. Within one GCM system, there is a variety of dynamics between which switches are possible. This behavior will be relevant for the application to information processing and electric activity in neural systems, as will be discussed in Chap. 5.

Although $\epsilon_1 = \epsilon N_1/N$ can take any value when viewed in the context of a two-dimensional mapping, as for an attractor in the GCM, however, ϵ_1 is not necessarily allowed to take any value between 0 and 1. As mentioned, the split exponent λ_{spl} should be negative so that the synchronization of the elements in a cluster is maintained. If the chaotic instability is too strong then λ_{spl} becomes positive, and the two clusters cannot remain as attractors. The state splits into three clusters, or returns to a two-cluster state with a more balanced number ratio. The above two-cluster attractors exist for $N - N_{thr} \leq N_1 \leq N_{thr}$ ($N_{thr} \geq N/2$). This is why the two-cluster attractor in Fig. 4.10 is plotted only within some range of N_1/N. Indeed the motion is quasiperiodic or periodic (with a short period) for $N_1/N \approx 1/2$, and with the increase of N_1/N, the motion becomes chaotic.

The threshold N_{thr} was numerically obtained and depends on the value of a and is proportional to N ($N_{thr} = cN$). The coefficient c depends on a as in Fig. 4.11. It increases with a and then decreases to 0.5, where the two-cluster solution loses its stability.

In 3-cluster attractors, we can see a co-dimension-2 bifurcation with the change of N_1 and N_2. Again, we have observed quasiperiodicity, lockings, and period-doubling to chaos. There is a threshold of N_1 and N_2, beyond which 3-cluster attractors are no longer stable. Near the edge of the threshold, the motion is chaotic. Through the change of cluster sizes, we can tune the dynamical state of our system.

In a similar manner, we can see the bifurcation in a k-dimensional space for k-cluster attractors. If we take all the possible attractors of different cluster numbers, we can have bifurcations *not only in the parameter space with a fixed number of parameters, but also with regard to a variable number of parameters and dimensions.*

4.5 Hierarchical Clustering and Chaotic Itinerancy

4.5.1 Partition Complexity

In the partially ordered (PO) phase, there is a variety of attractors with different numbers of clusters, and different ways of partitioning $[N_1, N_2, \cdots, N_k]$. The clustering here is typically inhomogeneous: partitions $[N_1, N_2, \cdots, N_k]$ are far from equally large. An idealized partitioning is schematically formed as follows (see Fig. 4.4c or Fig. 4.12): first, split the system into two equal clusters. Take one of them and split it again into two equal clusters while leaving the other without a split. By repeating this process, the partitioning is given by $[N/2, N/4, N/8, \cdots]$. In this case, the differences of the phases of the oscillations are also hierarchical. The difference is roughly π for the first split, and then $\pi/2$ for the second, and so forth. Although this partitioning is a drastic simplification, such a hierarchical structure in the partitioning and in phase space is typically observed in the PO phase.

Now let us discuss the coexistence of attractors with different partitionings. As mentioned, each attractor is coded by its number of clusters, and by the partition sizes $[N_1, N_2, \cdots, N_k]$. Depending on the attractors (and accordingly on initial conditions), there is a variety of partitionings.

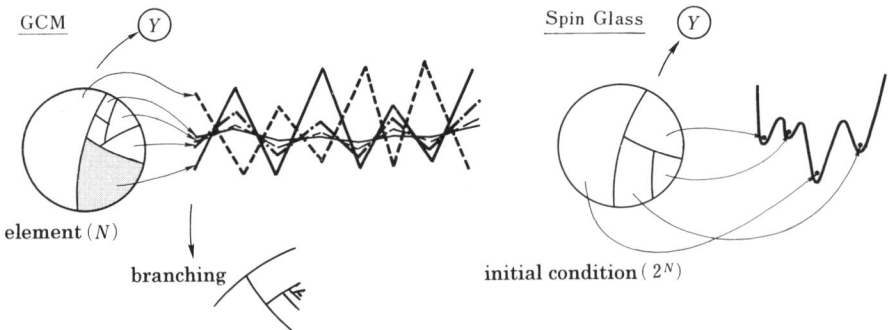

Fig. 4.12. Schematic representation of the partitioning in the GCM and in spin glasses. For the GCM, inhomogeneous partitionings and inhomogeneous tree structures are shown schematically in relationship with the time series of the elements in each cluster.

To study the nonuniformity of partitionings we introduce the probability that two elements fall on the same cluster. This probability is calculated by

$$Y = \sum_{j=1}^{k} (N_j/N)^2. \tag{4.10}$$

For an attractor with two clusters of equal partition size $Y = 1/2$; for a three-cluster attractor with equal partition $Y = 1/3$; and for a completely desynchronized attractor $Y = 1/N$. Since Y depends on the attractor, we have measured the distribution $\pi(Y)$ as the probability that a randomly chosen initial condition leads to an attractor with Y. From an ensemble of randomly chosen initial conditions, we obtain the histogram of Y, in order to obtain $\pi(Y)$. In Fig. 4.13 $\pi(Y)$ is plotted for $a = 1.6$, 1.64, 1.66, 1.68,

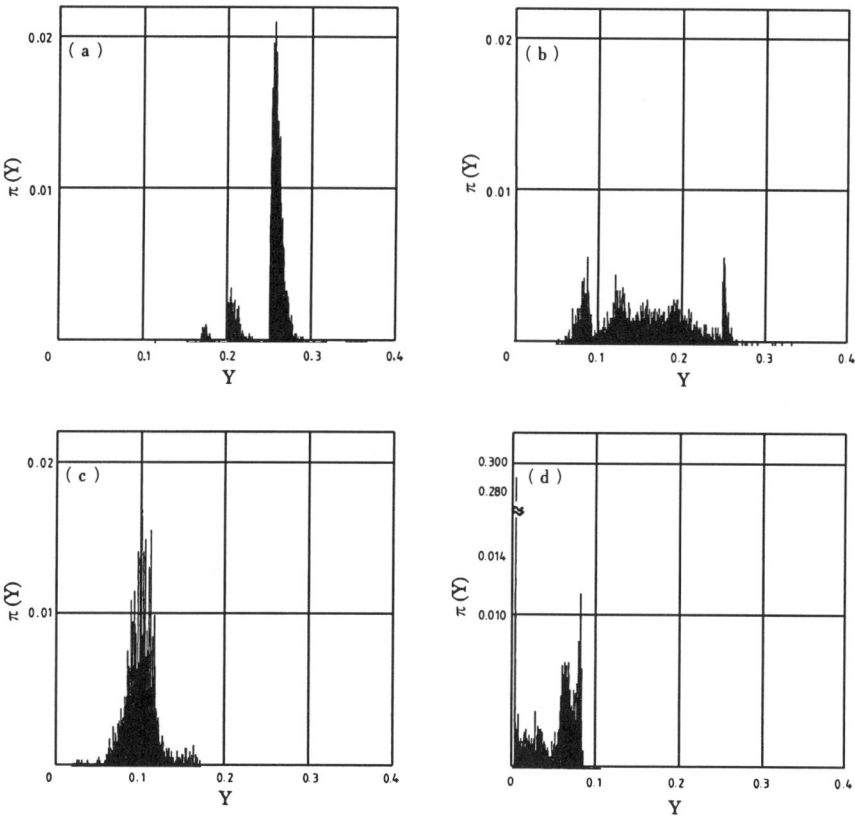

Fig. 4.13 a–d. Probability distribution of partitions $\pi(Y)$, calculated from 5000 randomly chosen initial conditions. $\epsilon = 0.1$, and $N = 200$. (a) $a = 1.6$, (b) $a = 1.64$, (c) $a = 1.66$, (d) $a = 1.68$. (From K. Kaneko, *J. Phys.* **A24** (1991a) 2107, with the permission of the publishers.)

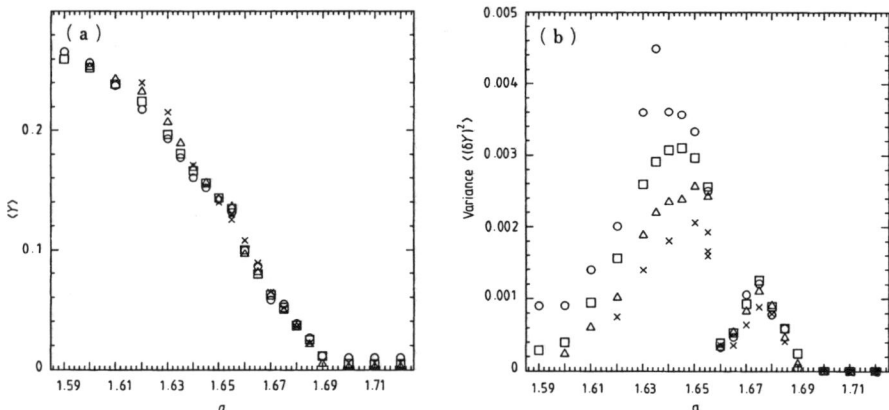

Fig. 4.14. (a) $\langle Y \rangle$ is plotted versus a, for $\epsilon = 0.1$. $N = 100$ ($\circ$), $N = 200$ ($\square$), $N = 400$ ($\triangle$), and $N = 800$ ($\times$). (b) $\langle (\delta Y)^2 \rangle$ is plotted versus a, for $\epsilon = 0.1$. $N = 100$ ($\circ$), $N = 200$ ($\square$), $N = 400$ ($\triangle$), and $N = 800$ ($\times$). (Semi-log plot.) (From K. Kaneko, *J. Phys.* **A24** (1991a) 2107, with the permission of the publishers.)

and $\epsilon = 0.1$. As a is increased, peaks located slightly above $1/M$ appear successively with increasing $M(= 4, 5, 6, \cdots)$. These peaks have left endpoints at $1/M$ and correspond to M-cluster attractors.

The average $\langle Y \rangle$ is plotted in Fig. 4.14a. In the ordered phase, peaks with the left endpoints at $1/M$ appear successively with increasing $M(= 4, 5, 6, \cdots)$, which correspond to the M-cluster attractors that are dominant at the given parameters. As a is increased, $\langle Y \rangle$ decreases further in the PO phase until it reaches $1/N$ in the desynchronized phase.

To investigate the variation depending on the initial conditions, the variance $\langle (\delta Y)^2 \rangle = \langle Y^2 \rangle - \langle Y \rangle^2$ is plotted in Fig. 4.14b. In the ordered phase the variance is large, but it decreases with the increase of N, and is expected to approach 0 for $N \to \infty$. In the PO phase, the variance seems to approach N independent values. In other words, a variety of partitionings coexists as attractors, even in the large-N limit, and there is no single 'typical' partitioning.

This diversity in partitioning was first discussed in the spin-glass system described in Sect. 1.5.6. Given a random interaction $J_{i,j}$ there are many quasi-stable states. Starting from a given initial spin configuration $S(i)$, the basin of attraction of each metastable state is measured and thus the probability Y that the two randomly chosen initial configurations lead to the same metastable state. This probability Y depends on the random interaction $J_{i,j}$. The distribution $\pi(Y)$ is introduced as the distribution of Y over different samples, with different sets of random couplings. According to Derrida and Flyvbjerg [1988], $\pi(Y)$ has peaks at $Y = 1/2, 1/3$ and so forth. The variance of the distribution is enhanced in the spin-glass phase, and remains finite even in the limit of $N \to \infty$. It is proposed that there is a universal relationship

Table 4.1. Comparison of our model (GCM) with the SK model for spin glasses.

System	Spin glass	Network of chaotic elements
model	SK model	GCM model
phase	spin glass	partially ordered
randomness	given	created by chaos
	fixed	dynamic
tree	for a thermodynamic state	for an attractor
	inhomogeneous	inhomogeneous
	fixed	dynamic
Y	probability that two initial conditions fall on the same metastable state	probability that two elements fall on the same cluster
distribution of Y	over random coupling	over initial conditions
variance of Y	remains finite at $N \to \infty$	remains finite at $N \to \infty$
relation between $\langle Y \rangle$ and $\langle (\delta Y)^2 \rangle$	universal	nonuniversal

between the variance and the average as

$$\langle (\delta Y)^2 \rangle = \frac{1}{3} \langle Y \rangle (1 - \langle Y \rangle). \tag{4.11}$$

This relationship is satisfied by the Sherrington–Kirkpatrick model for spin glasses, random maps, random energy models, and so on.

Let us compare the PO phase in the GCM with spin glasses. Instead of "two initial conditions" in spin glasses, we take two elements $x(i)$. In our case, the sampling is taken over all initial conditions rather than the random couplings. Instead of the basin volume of initial configurations attracted to a metastable thermodynamic state, we choose the cluster size of an attractor. Although the remnant of the variance of Y is common to both cases, the universal form (4.12) is not satisfied in our case. For clarity the various aspects are shown in Table 4.1. The dynamic tree structure in the table will be discussed in Sect. 4.5.3.

4.5.2 Hierarchical Clustering

So far we have characterized an attractor by the number and sizes of the clusters. Still, the clustering is not only characterized by these. In both the examples of Fig. 4.15a, we have four clusters. However, the way of partitioning is clearly different. We have the following situation: in (b), first, elements split into two 'mega'-clusters and each of them splits into two again to form the four-cluster state, while in (a) after the first split into two, only one of the clusters splits into three and the other remains unchanged.

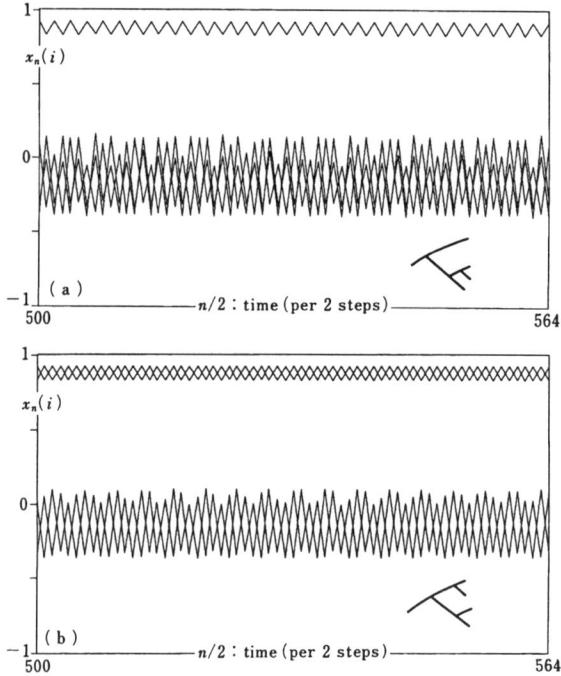

Fig. 4.15 a,b. Time series of 4-cluster attractors with different tree structures. $a = 1.58$, $\epsilon = 0.1$, and $N = 50$. $x_{2n}(i)'s$ are plotted for all elements i. With the partitions **(a)** $[19, 15, 15, 1]$ and **(b)** $[16, 16, 10, 8]$. Corresponding tree structures are also plotted. (From K. Kaneko, *Physica D* **41** (1990c) 137, with the permission of the publishers.)

In order to characterize such differences, we introduce a coarse-grained measurement of $x_n(i)$ and then the notion of precision-dependent clustering. Here the coarse-grained measurement is defined as

$$\bar{x}_n^P(i) \equiv [P \times x_n(i)]/P, \qquad (4.12)$$

where $[\cdots]$ is the integer part of $\cdots$ and P is an integer which gives the precision. The above coarse-graining gives digital values with m/P ($m = 0, \pm 1, \pm 2, \cdots$). The precision-dependent clustering is defined by the clustering for $\bar{x}_n^P(i)$. If $\bar{x}_n^P(i) = \bar{x}_n^P(j)$ holds, the elements i and j are regarded as belonging to the same cluster within that precision P. The precision-dependent cluster number k_n^P is defined as the number of clusters at time n with the precision P.

As the precision is increased, the clusters spilt and their number k_n^P increases until it reaches the exact cluster number k. To illustrate this process, we introduce the *precision-dependent tree*. In Fig. 4.16, $\bar{x}_n^P(i)$ is plotted versus the precision P, for $i = 1, \cdots, N$, and a fixed n. The number of lines for a given P gives the number of clusters k_n^P at that precision.

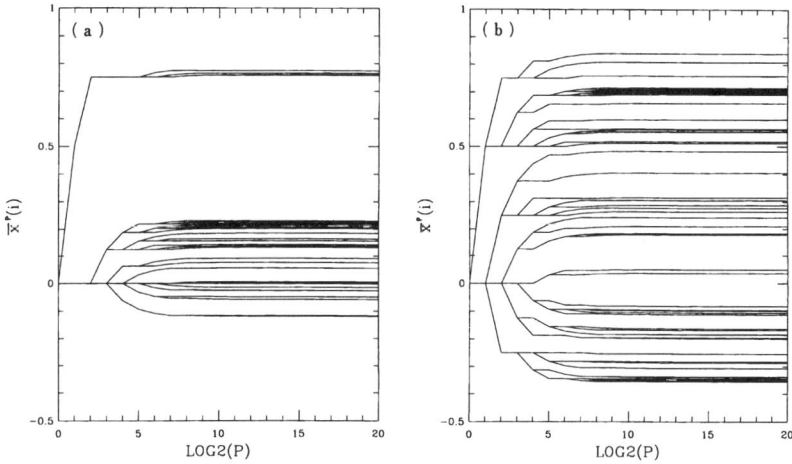

Fig. 4.16 a,b. Precision-dependent tree: $\overline{x}_n^P(i)$ is plotted for all i, at $n = 50\,000$, with the change of the precision P as $P = 2^m (m = 1, 2, \cdots, 30)$. $N = 1000$ $a = 1.92$, $\epsilon = 0.2$. Two examples from different initial conditions are shown in (a) and (b). (From K. Kaneko, *Physica D* **41** (1990c) 137, with the permission of the publishers.)

The tree structure is rather homogeneous in the desynchronized phase (see Fig. 4.17), while it is strongly inhomogeneous in the PO phase. Some branches bifurcate a lot, while others do not do so at all in the PO phase. A typical tree consists of one big branch without sub-branches, and many

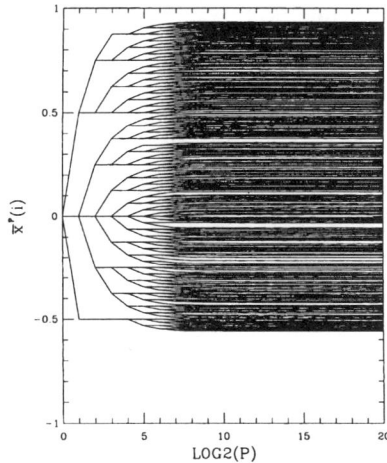

Fig. 4.17. Precision-dependent tree for a desynchronized state $\overline{x}_n^P(i)$ is plotted for all i, at $n = 50\,000$, with the change of the precision P as $P = 2^m (m = 1, 2, \cdots, 30)$. $N = 1000$ $a = 1.9$, $\epsilon = 0.1$. (From K. Kaneko, *Physica D* **41** (1990c) 137, with the permission of the publishers.)

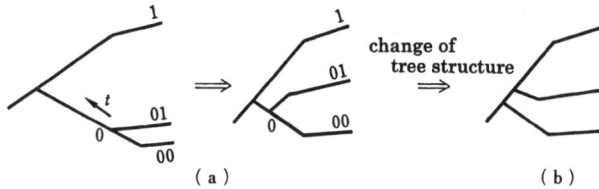

Fig. 4.18 a,b. Schematic representation of the change of the tree structure with three clusters.

smaller branches with bifurcations (see Fig. 4.18b). In most attractors of this type, the number of elements N_1 in the single big branch is comparable to the sum of all the other branches' elements. Let us now discuss the tree structure, taking the idealized example mentioned in Sect. 5.5.1, with the partition $[N/2, N/4, N/8, \cdots]$. The phases of the first and the remaining clusters differ by π, the phases of the second and later clusters differ by $\pi/2$, and so forth (see Fig. 4.12). This leads to the inhomogeneous tree structure shown in Fig. 4.12.

Summing up, an attractor in the PO phase has a large number of clusters, the number of elements in each cluster ranges from small to large, and the distances between the variables $x_n(i)$ in each cluster differ substantially. The attractor often has an inhomogeneous structure.[4]

Note that this hierarchical clustering structure in the PO phase is spontaneously formed. A hierarchical structure is rather common in biological and social phenomena, and also underlies our categorical recognition. The present model provides a simple example of the formation of a hierarchical structure, starting from identical elements. The significance of this observation is discussed in Chap. 5.

4.5.3 Hierarchical Dynamics

Although the exact clustering is fixed once the system has fallen onto an attractor, the precision-dependent clustering can be time-dependent. Since our dynamics involves chaos, a small difference $|x(i) - x(j)|$ can be amplified during the temporal evolution. Thus the elements for which $\overline{x}_n^P(i) = \overline{x}_n^P(j)$ holds can change in time. Then the above tree structure is no longer fixed, and changes with time. In this case, our attractor is represented as a dynamically-changing tree as is shown below. The dynamics of our GCM may be better represented by a hierarchical code, which leads to the description of hierarchical dynamics.

Here the meaning of *hierarchical dynamics* is as follows [Kaneko 1990c]: in a hierarchical dynamical system, there are many units of different levels,

---

[4] In the partially ordered phase, there are a lot of attractors with different tree structures.

organized in a tree structure. A unit interacts strongly with the other units of the same level. A unit at a lower level is strongly governed by the unit at a higher level, while there is a feedback from the lower level to the higher level, which can lead to a change of the tree structure.

Haken has proposed the slaving principle [Haken 1979], which focuses on the constraint from a higher level to a lower level. In the slaving principle, the feedback from the lower level to the higher is weak and the time scale of the lower level is much faster. Then the lower-level dynamics just follows the higher-level dynamics. The lower-level variable can be adiabatically eliminated and we can write down the evolution of the system only by the higher-level variables.

Our hierarchical dynamics does not follow the slaving principle. Due to the feedback from the lower level to the higher, the tree structure, i.e., the construction of levels themselves, changes in time. For some time span, the adiabatic elimination of fast variables may be valid. However, after some duration, the feedback from the lower to the higher level becomes stronger and will eventually destroy the level structure itself. Later a new tree structure is formed, which lasts over some time steps, but again collapses to form the next structure. The change of the hierarchical structure repeats forever.

We believe that hierarchical dynamics is essential to changes in society and biological organization, dynamic functional change in the brain, and dynamic processes in categorization. Here we show explicitly that our system belongs to the hierarchical dynamical system in the above sense.

As the simplest case we consider a 3-cluster attractor with the tree structure shown in Fig. 4.18. The three clusters can be coded by 1, 01, and 00. From (4.8) with $k = 3$, the dynamics of each cluster X^1, X^{01}, X^{00} is written as

$$X^1_{m+1} = (1 - n^0\epsilon)f(X^1_m) + n^{01}\epsilon f(X^{01}_m) + n^{00}\epsilon f(X^{00}_m) \tag{4.13}$$

$$X^{01}_{m+1} = (1 - (n^1 + n^{00})\epsilon)f(X^{01}_m) + n^1\epsilon f(X^1_m) + n^{00}\epsilon f(X^{00}_m) \tag{4.14}$$

$$X^{00}_{m+1} = (1 - (n^1 + n^{01})\epsilon)f(X^{00}_m) + n^1\epsilon f(X^1_m) + n^{01}\epsilon f(X^{01}_m) \tag{4.15}$$

where $n^1 = N^1/N$, $n^{01} = N^{01}/N$, $n^{00} = N^{00}/N$ and $n^0 = n^{01} + n^{00}$. Let us introduce the dynamics of a mega-cluster which has a higher level than 11 and 10. The simplest way for this to be accomplished is the introduction of the weighted average of the two clusters X^{01} and X^{00}, $X^0 = (n^{01}X^{01} + n^{00}X^{00})/n^0$. This gives the dynamics of the node at the higher level. The motions of the lower-level clusters X^{01} and X^{00} are represented by X^0 and a small deviation from it (by the notion of "precision-dependent clustering" we can assume that the deviation is smaller than $X^0 - X^1$). Introducing the notation of $X^{01} = X^0 + \delta^{01}$ and $X^{00} = X^0 + \delta^{00}$ (note that $n^{01}\delta^{01} + n^{00}\delta^{00} = 0$), and taking the order up to $(\delta)^2$, we obtain:[5]

$$X^1_{m+1} = (1 - n^0\epsilon)f(X^1_m) + n^0\epsilon f(X^0_m) + \frac{\epsilon n^0 n^{00}}{2n^{01}}(\delta^{00}_m)^2 f''(X^0_m) \qquad (4.16)$$

$$X^0_{m+1} = (1 - n^1\epsilon)f(X^0_m) + n^0\epsilon f(X^0_m) + \frac{(1 - n^0\epsilon)n^{00}}{2n^{01}}(\delta^{00}_m)^2 f''(X^0_m) \quad (4.17)$$

$$\delta^{00}_{m+1} = (1 - \epsilon)f'(X^0_m)\delta^{00}_m + \frac{(1 - \epsilon)(1 - n^{00}/n^{01})}{2}(\delta^{00}_m)^2 f''(X^0_m). \qquad (4.18)$$

First we note that the dynamics of X^0 and X^1 are the same as (4.9) for the two-cluster, up to the order of δ. Thus our system behaves as if two-clusters were interacting. In the second order of δ there is a correction to this dynamics. Thus we can view our 3-cluster dynamics as that of two-clusters with some additional small corrections by the motion of the sub-clusters (X^{00} and X^{01}). The motion of sub-clusters is governed by the higher-level cluster ("slaved"). However, there is a first-order correction which separates the sub-clusters from the higher-order cluster. This correction dynamics (the first term in (4.18)) again depends on the higher-level cluster. We note that this separation dynamics is, up to the first order, a linear map with the slope $(1 - \epsilon)f'(X^0_m)$, given by the derivative of the motion of the higher-order cluster.

If the condition

$$\left| \prod_m (1 - \epsilon)f'(X^0_m) \right| > 1 \qquad (4.19)$$

holds, the difference δ is amplified. Then the difference between X^{00} and X^{01} increases until it is of the order of the difference between X^0 and X^1. Now the tree structure changes its type from Fig. 4.18a to Fig. 4.18b. Hence the tree structure changes dynamically.

Note that condition (4.19) has the same form as the term which determines the positivity of the split exponent (4.5). Since the elements in each of the three clusters do not separate, the split exponent itself $\lambda_{\text{spl}}(i) = \langle|\log(1 - \epsilon)f'(x_n(i))|\rangle$ is negative. However, the 'split exponent' of a mega-cluster given by the logarithm of (4.19) is not negative (otherwise, the distance between the clusters 01 and 00 would decrease, and they would form a single cluster). Thus condition (4.20) is satisfied as long as we have three (rather than two) clusters. The condition means that the separation of sub-clusters occurs when chaos is strong enough to overcome the stabilizing effect of the global coupling. Thus we may call the above collapse of the original tree structure by chaos a "chaotic revolt against the slaving principle".

The above procedure can straightforwardly be extended to cases with more clusters and branchings of more than two. By applying the above

---

[5] For the logistic map $f(y) = 1 - ay^2$, the above equations are rigorous even at higher orders, since $f''(y) = -2a$ (constant), and the higher-order derivatives of $f(y)$ vanish.

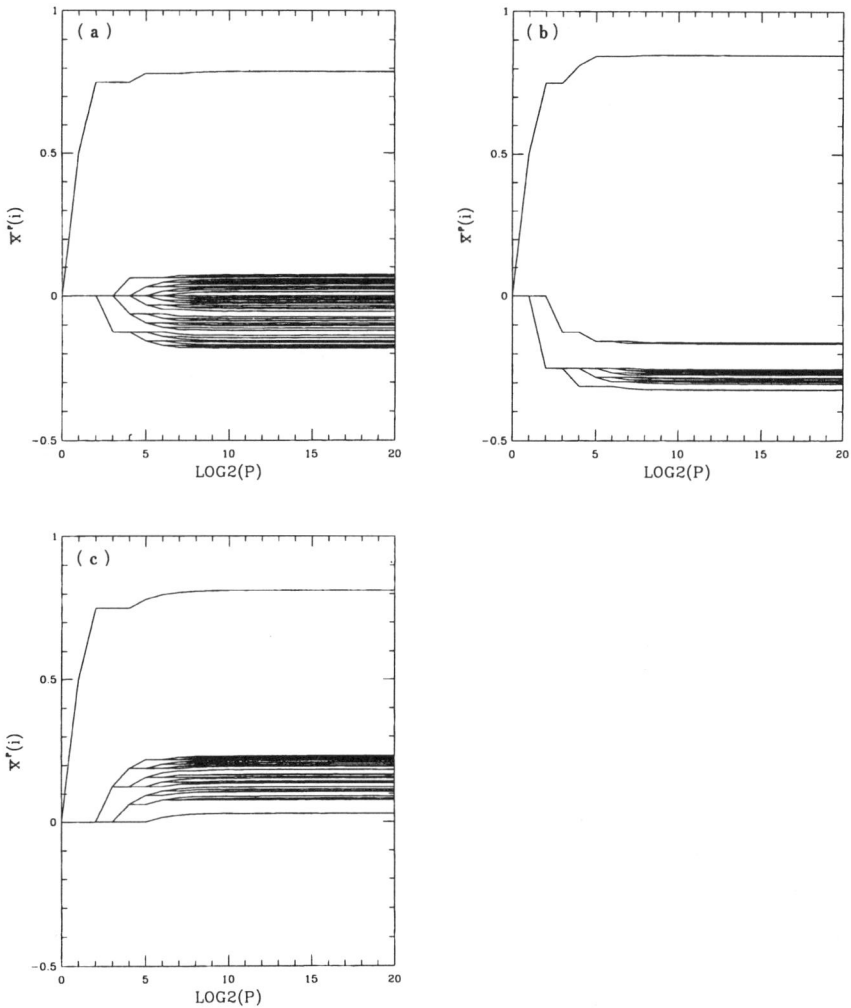

Fig. 4.19 a–c. Temporal evolution of precision-dependent clustering. $\bar{x}_n^P(i)$ is plotted for all i, at **(a)** $n = 50\,032$, **(b)** $50\,064$, and **(c)** $50\,096$ steps. $a = 1.91$, $\epsilon = 0.2$, and $N = 1000$. Corresponding to the attractor of Fig. 4.16a. (From K. Kaneko, *Physica D* **41** (1990c) 137, with the permission of the publishers.)

procedure at every level of branching, we can obtain the equations of the hierarchical dynamics. In other words, the motion of a sub-cluster at a given level is slaved by the dynamics of the higher-level cluster, and is also governed by the interaction among the sub-clusters at the same level. The interaction from a lower to a higher level appears as a second-order perturbation.

In Fig. 4.19, we have plotted an example of the dynamical change of the tree structure. Here, clusters split into two first, and then split into smaller clusters. One of the branches in the tree structure does not show any further branching, while the other branching structures change in time. The dynamic structure itself is inhomogeneous by the branching.

Now, it is clear that the tree dynamics in our GCM leads to hierarchical dynamics. As already discussed in Sect. 1.5, the hierarchical structures in biological and social systems are not fixed, but change in time. Also, it has been proposed that human memory should be recognized as a dynamic categorization process [Edelman 1987]. Our model provides the simplest example for such hierarchical dynamics.

4.5.4 Chaotic Itinerancy

As a simple way to see the temporal change of precision-dependent clusterings, we have plotted k_n^P in Fig. 4.20. Here k_n^P represents the degrees of freedom which determine the next step within the precision P. From the dynamics of k_n^P, we can see how the effective degrees of freedom change in time.

In Fig. 4.20, examples of the change of k_n^P are plotted. In the desynchronized phase, the temporal variation of k_n^P is very small, and shows a small Gaussian fluctuation around a constant value for large P.

In the partially ordered phase, there is a characteristic behavior. Much larger variations are observed over much longer time scales. In Fig. 4.20, k_n^P varies from 2 to 200, even for $P = 2^{18}$. In particular, in Fig. 4.20b, k_n^P stays at 1 for ~ 600 steps and then grows to 100.

What causes the change in the number of effective degrees of freedom? As schematically shown in Fig. 4.21, two elements almost synchronize for a while. During this time interval, almost all elements are attracted to a few synchronized clusters and stay there. The dynamics is essentially governed by a few effective degrees of freedom. After a while, this state with a few effective degrees of freedom is destabilized, and elements are separated again.

The dynamics is represented as an itinerancy over several ordered states through disorganized, desynchronized states. It consists of a quasistationary high-dimensional state, exits to "attractor-ruins" with low effective degrees of freedom, residence therein, and chaotic exits from them. The motion at "attractor-ruins" is quasistationary. For example, if the number of effective degrees of freedom is two, the system splits into two groups, in each of which elements oscillate almost coherently. The system is in the vicinity of a two-cluster state, which is not a stable attractor, but whose stable manifolds are rather high-dimensional. After staying at an attractor-ruin, the orbit exits from it due to the chaotic instability. The unstable manifold for this exit is expected to be low-dimensional. The low-dimensionality of this unstable manifold gives a constraint for the itinerancy. For a motion along this unstable manifold, the difference between two almost synchronized elements grows exponentially in time, till the motion goes to a rather high-dimensional chaotic state without clear coherence. This high-dimensional state is again

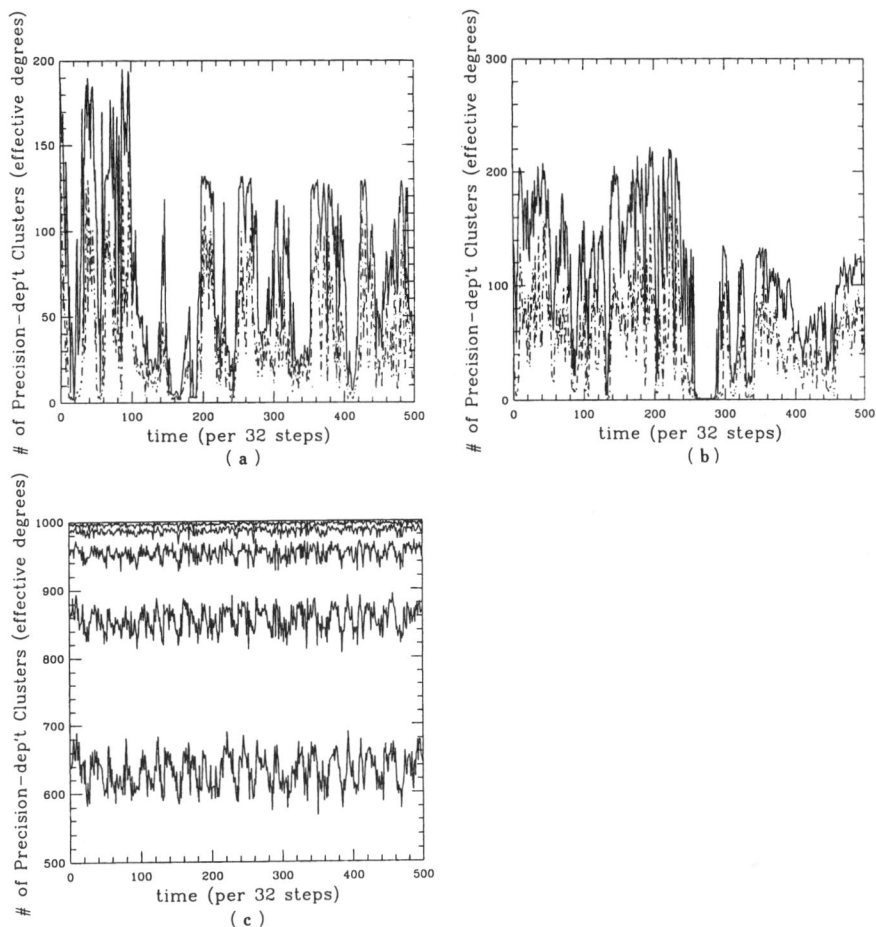

Fig. 4.20a–c. Temporal evolution of the number of precision-dependent clusters. k_{32n}^P is plotted as a function of the time step n (per 32 steps). Calculated after 50 000 transients. $N = 1000$, and starting with random initial conditions. $P = 2^{12}$ (dashed line) and $P = 2^{18}$ (solid line). (a) $a = 1.92$, $\epsilon = 0.2$, (b) $a = 1.62$, $\epsilon = 0.3$, (c) $a = 1.9$, $\epsilon = 0.1$. $P = 2^{10}$, $P = 2^{12}$, $P = 2^{14}$, $P = 2^{16}$, $P = 2^{18}$ from bottom to top (for the desynchronized phase). (From K. Kaneko, *Physica D* **41** (1990c) 137, with the permission of the publishers.)

quasistationary, although there are some holes that connect to the attractor-ruins. Once the orbit is trapped in a hole, a sudden collapse of the high-dimensional chaos appears, and ordered motion reoccurs.

This dynamics, called chaotic itinerancy (CI), is a novel universal class in high-dimensional dynamical systems.[6] This CI dynamics has been found in

---

[6] In the partially ordered (PO) phase, there are a variety of attractors with different numbers of clusters, and different ways of partitionings $[N_1, N_2, \cdots, N_k]$.

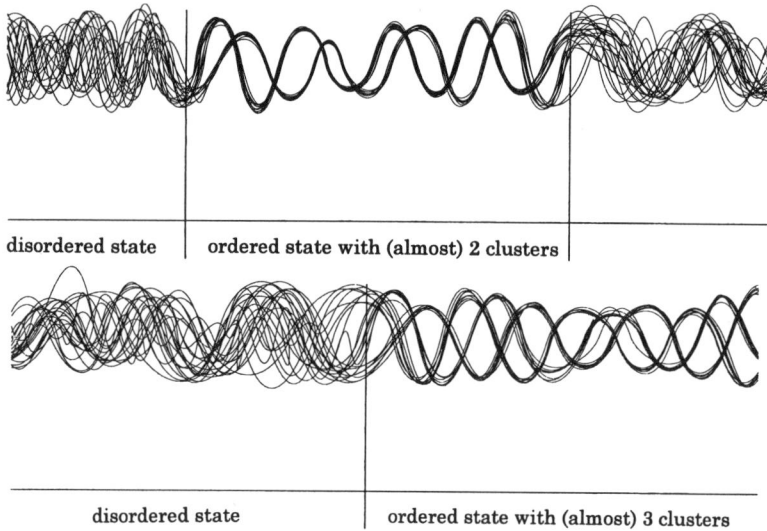

Fig. 4.21. Schematic representation of chaotic itinerancy. The time series of all elements are overlaid schematically to show the change of synchronization. (From K. Kaneko, Nikkei-Science (May, 1994d) p.34 (in Japanese), with the permission of the publishers.)

a model of neural dynamics by Tsuda [1990a, b, 1991a, b], to be discussed in Chap. 6, in optical turbulence by Ikeda *et al.* [1989], and in the present GCM [Kaneko 1990c, 1991b] independently at the same time. The term 'Chaotic Itinerancy' was suggested by Peter Davis. We have unanimously cooperated to propose this notion as a novel universal class. Experimentally it has later been observed by Arecchi [1991], in an experiment on a laser system.

CI is commonly observed in a variety of high-dimensional dynamical systems. It should be noted that CI supports the notion of meta-dynamics, in the sense that it gives successive transitions over states which have a faster scale of dynamics within. CI is also observed as a long-transient regime before the system finds a final attractor [Kaneko 1991b]. In the ecological model by Hastings and Higgins [1994], long-term change in the collective dynamics is observed, which may be regarded as a transient CI.

The CI in our GCM reminds us of coherent structures in fluid turbulence. In turbulence, large vortex structures often emerge intermittently, and collapse again. If a few large vortices govern the whole fluid motion, then the motion is described by a low number of degrees of freedom up to a high

Around the transition region, it is found that most attracted states are weak. By any small perturbation, the orbit can be kicked out of the state which is termed a Milnor attractor [Kaneko 1997, 1998c]. The global dynamics in this case is formed as successive itinerancy over such Milnor attractors. This meta-dynamics over Milnor attractors gives one interpretation of chaotic itinerancy.

precision. This state corresponds to our ordered state governed by a few number of clusters up to high precision. As observed in our CI, the long-term motion in a fluid makes an itinerancy over such coherent structures via high-dimensional fully turbulent states.

It should be noted that CI provides an example of successive changes of relationships among elements. In the CI of our GCM, the degree of synchronization between two elements changes successively with time. Such fast-scale change of mutual correlation is studied as dynamics of functional coupling in the cerebral cortex by Aertsen and Vaadia [Aertsen *et al.* 1994; Aertsen and Vaadia 1992]. They have found that the functional coupling among neurons varies temporally over a rather short time scale, by showing that the degree of synchronization among pairs of neurons changes both temporally and by the choice of pairs.

By associating each ordered state with a memorized state, CI in the GCM provides a model for a spontaneous recall in a hierarchical memory. The significance of CI to dynamic brain functions is discussed in detail in Chap. 5. In future, it is hoped that dynamical categorization processes will be studied in terms of CI.

4.6 Marginal Stability and Information Cascade

4.6.1 Marginal Stability

In both the PO phases, I and II, of the phase diagram shown in Fig. 4.7, chaotic itinerancy and dynamics of tree structures are observed. Elements first split and then tend to synchronize. Here we study the degree with which two elements tend to synchronize.

As mentioned, the tendency to split is measured by the split exponent given by

$$\lambda_{\mathrm{spl}}(i) = \lim_{T \to \infty} \frac{1}{T} \sum_{n}^{T} \log |(1 - \epsilon) f'(x_n(i))|. \tag{4.20}$$

When an element forms a synchronized cluster with some other elements, $\lambda_{\mathrm{spl}}(i)$ is negative, while it is positive when elements are desynchronized with the other elements. While the above exponent is a temporal average (over infinite time steps), it is useful to introduce a "local" split exponent, averaged over only a given finite number of time steps T, for the characterization of the temporal change of clustering:

$$\lambda_{\mathrm{spl}}^{T}(i, n) = \frac{1}{T} \sum_{m=n}^{n+T} \log |(1 - \epsilon) f'(x_m(i))|. \tag{4.21}$$

In the partially ordered phase, the exponents often fluctuate around 0, but they can also remain negative over long periods of time, while the corresponding elements tend to be synchronized. Such long-term maintenance of negative values often leads to a significant reduction in the difference of the two elements. Indeed, in the PO phase, we have found that this reduction lasts over so many time steps that the difference between the two elements goes down to the minimum bit precision given by the computation in our digital computer. For example assume that $\lambda_{spl}^T(i,n) = -\beta$ over T steps. Then the distance between two elements may be reduced to $\exp(-\beta T)$. For $\beta = 0.2$ and $T = 200$ steps, for example, this would be $\exp(-40) = 4.2 \times 10^{-18}$. For a double-precision computation, a difference smaller than 10^{-15} is neglected. Hence the two elements, for a double-precision computation, are regarded as identical, and the evolution afterwards should remain the same. Even though $\lambda_{spl}^T(i,n)$ later changes its sign to a positive value, the two elements cannot separate any more in a computer (note that the round-off error in a computer is not noise. The round-off obeys a deterministic algorithm. Thus a temporal evolution from the 'same' value and with the same equation leads to the same result).

Indeed such a digitization-induced pseudo-attractor often appears in the PO phase [Kaneko 1994b]. In Fig. 4.22, we have plotted the time series of $\lambda_{spl}^T(i,n)$ and $x_n(i)$. Here the system is attracted to a 4-cluster state with $N_1 = 21, N_2 = 16, N_3 = 7, N_4 = 6$. However, the split exponent is slightly positive for the third and fourth clusters. Although the synchronization in the third and fourth clusters is unstable, the elements cannot split due to the limit of the precision in our computation. Even if we adopt a higher-precision computation, there remains a finite probability that $\exp(-T\lambda_{spl}^T(i,n))$ reaches the lowest bit precision, when the computation is carried out over sufficiently many time steps.

Hence it is hard to simulate model (4.1) 'correctly' in our digital computer. In order to remove this kind of numerical artifact, we introduce a tiny noise term to (4.1). We have added a term $\eta_n(i)$ as a random number homogeneously distributed over $[-\sigma, \sigma]$. Here σ is chosen to be very small, ranging from 10^{-15} to 10^{-8}. In Fig. 4.22b, the noise is applied from the time step shown by the arrow. After some time steps, the system falls onto an attractor with 4 clusters with $N_1 = 21, N_2 = 18, N_3 = 8, N_4 = 3$, where all split exponents are negative.

Throughout this section, we add this tiny noise to the system. With this noise, all elements are desynchronized, and no attractor with $k < N$ is observed. In the second regime of the PO phase (PO II), the distribution of the exponent $\lambda_{spl}^T(i,n)$ (for $T = 128$) is given in Fig. 4.23, sampled over all elements and many time steps. The distribution has a sharp peak around zero, implying that the stability for synchronization is marginal. Hence the splitting of clusters and synchronization of elements are balanced. The two processes are temporally separated as distinct regimes. Over long periods of time, the elements are almost synchronized (up to high precision), till amplification of tiny differences dominates later.

Due to the balance between synchronization of elements and splitting of clusters, the average value of $\lambda_{spl}(i)$ is close to zero. In Fig. 4.24, the exponent

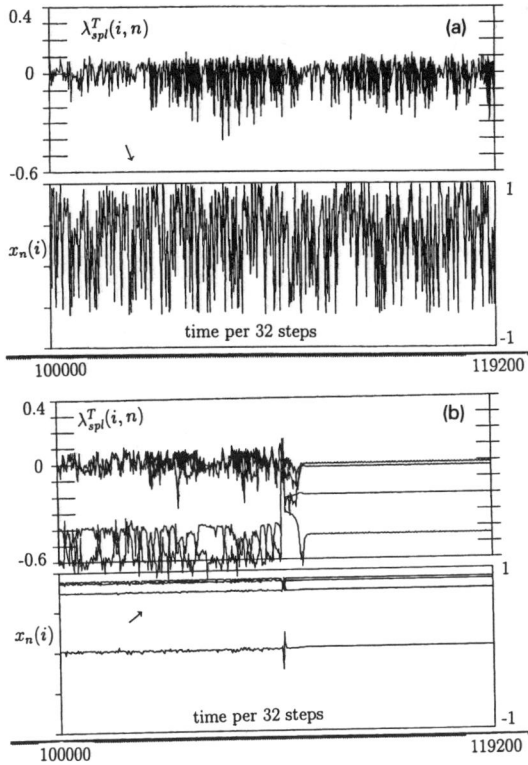

Fig. 4.22 a,b. Overlaid time series of $x_n(i)$ with the corresponding $\lambda_{\text{spl}}^T(i,n)$. Up to the time steps indicated by the arrow, the simulation is carried out without a noise term, while the noise term with $\sigma = 10^{-8}$ is introduced later. Plotted per 32 time steps, which are chosen as the averaging time of $\lambda_{\text{spl}}^T(i,n)$ (i.e., $T = 32$). (a) $a = 1.59$, $\epsilon = 0.3$, and $N = 100$. The system had fallen on a "pseudo-attractor" with two clusters ($N_1 = 69$ and $N_2 = 31$), before the noise was added. (b) $a = 1.90$, $\epsilon = 0.2$, and $N = 50$. The system had fallen on a "pseudo-attractor" with four clusters ($N_1 = 21, N_2 = 16, N_3 = 7, N_4 = 6$), and switched to a real attractor with $N_1 = 21, N_2 = 18, N_3 = 8, N_4 = 3$ later. (From K. Kaneko, *Physica D* **77** (1994b) 456, with the permission of the publishers.)

is plotted versus a. We can clearly see that the split exponent remains close to zero in the (type-II) PO phase around $1.57 < a < 1.61$, and $\epsilon \approx 0.3$.[7] Thus the marginal stability is sustained not at a critical point, but over some interval of parameters. This is a characteristic feature, in contrast with

---

[7] In the figure, drops are detected for some parameters. We note that the coherent attractor is stable when $\lambda + \log(1 - \epsilon) < 0$ is satisfied, where λ is the Lyapunov exponent for the single logistic map. Due to window structures of the logistic map, λ has many drops when plotted versus a. Hence there must be some (exceptional) parameters in which the above condition is satisfied (e.g., where a logistic map shows a stable cycle). Thus $\lambda_{\text{spl}} < 0$ holds for some parameters.

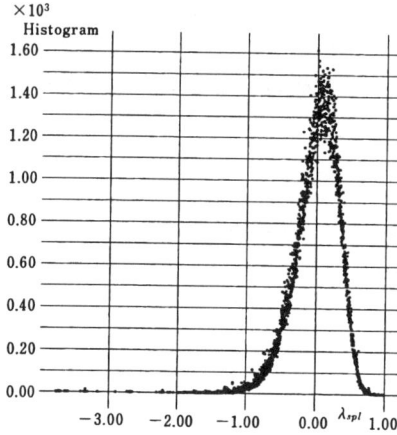

Fig. 4.23. Distribution of $\lambda_{spl}(i)$ for the GCM with a tiny noise, $a = 1.60$, $\epsilon = 0.3$, and noise amplitude $\delta = 10^{-6}$. The exponents are measured from the average over 128 time steps. The histogram is obtained by sampling over all elements ($N = 100$), and over 10^5 times the above 128 steps (totally 128×10^5 steps are iterated after the transient of 10^5 steps). (From K. Kaneko, *Physica D* **77** (1994b) 456, with the permission of the publishers.)

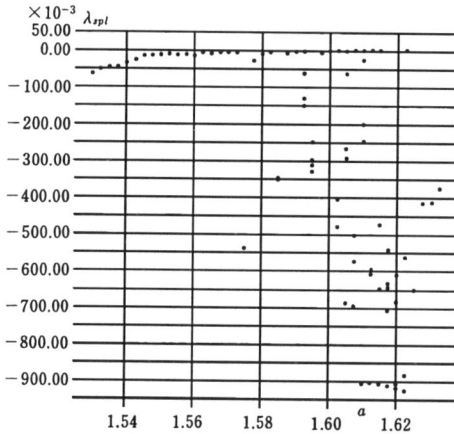

Fig. 4.24. The average $\lambda_{spl}(i)$ is plotted versus a, with $\delta = 2^{-20}$, and $\epsilon = 0.3$. The average is calculated over all elements and over 10^6 time steps. To see the coexistence of attractors, five runs from different initial conditions are carried out for $1.59 < a < 1.62$. (From K. Kaneko, *Physica D* **77** (1994b) 456, with the permission of the publishers.)

the usual critical phenomena. Temporally the regimes for synchronization and splitting are separated. After a long-term synchronization trend with $\lambda^T_{spl}(i, n) < 0$, the elements start to split. This switching is repeated forever.

In the PO phase shown as PO phase I, the split exponent depends on elements which belong to a different size of cluster. The marginal instability as above is not observed therein. Some elements take positive while other elements take negative values. Note that there the average is rather close to zero again, even though the balance as above is not satisfied.

4.6.2 Information Cascade

It is interesting to view our dynamics as a cascade process of clustering. As is studied, an attractor in the partially ordered phase can be described by successive splitting into smaller clusters. Let us again use the precision-dependent cluster already defined as $\bar{x}_n^P(i) \equiv [P \times x_n(i)]/P$, with P an integer giving the precision.

Here we study this splitting of clusters in an analogy with the cascade process of infinitely many vortices in fluid turbulence. In fluid turbulence, the picture of a cascade process of vortices is well established. Is it then possible to detect the cascade process visually in our model? To see the cascade process of clustering, we use the precision-dependent clustering to define the information flow in bit space. Let us first define the entropy measuring the variety of partitioning elements into clusters as

$$S_n(P) = -\sum_{i=1}^{k^P} (N_i^P/N) \log(N_i^P/N), \qquad (4.22)$$

at a given precision $1/P$ and at time n. With the increase of P, $S(P)$ increases. The increase from $S(P = 2^j)$ to $S(P = 2^{j+1})$ gives the growth of partition information from the j-th bit to the $(j+1)$-th bit. Thus we define the partition information creation (PIC) at the j-th bit [Kaneko 1994b] by

$$I_n^j = S_n(2^{j+1}) - S_n(2^j). \qquad (4.23)$$

This partition information creation tells us how much the partition variety increases at each bit. If a splitting process occurs at a bit j, the partition entropy S increases at that bit. Here, as a direct extension of the conditional probability, it may be useful to introduce the conditional cluster $N_{i,i_m}^{j+1,j}$, the splitting rate of each cluster N_i when the precision is increased from $1/P = 2^{-j}$ to $1/P = 2^{-(j+1)}$, where i_m indicates a "sub"-cluster generated from the i-th cluster with the above increase of the precision. Thus the PIC can be written as

$$I_n^j = -\sum_{\ell=1}^{k^{P+1}} (N_\ell^{P+1}/N) \log(N_{i,i_m}^{j+1,j}/N_i), \qquad (4.24)$$

where the summation over ℓ is equivalent with that over i and i_m (i.e., $\sum_\ell = \sum_{i=1}^{k^P} \sum_{i_m}$).[8] Thus the PIC gives the mutual information between the clusterings at the j-th and $(j+1)$-th levels.

In Fig. 4.25, the PIC is displayed in bit and time space. Over long periods of time, no creation occurs up to the 30-th bit, implying that the elements are coherent down to a scale of 2^{-30}. After such a quiescent region, a small difference is amplified to a larger scale, as can be seen in successive plots of the PIC. If we wait longer, the system returns to the almost-coherent state again. The dynamics here can be seen as an information cascade (i.e., avalanche). Successive amplification of small differences can be seen in the propagation to higher bits (smaller j). The propagation roughly has a constant speed, i.e. the differences are amplified exponentially. Indeed, the split exponent is positive in this temporal region. This avalanche in bit space stops at some scale. Then the propagation to smaller scales takes over, where negative splitting exponents determine the propagation speed in bit space.

As can be seen in the figure, the PIC flows intermittently to higher and lower bits in the partially ordered phase. Long-term persistent propagation of the PIC is observed, leading to large fluctuations of the partition variety Y in Sect. 4.5.1. The PIC is localized in bit space. We also note that two PIC avalanches can coexist in bit space (see Fig. 4.25c).

In the ordered state, with a small number of clusters, no propagation of the PIC to higher bits is observed even when some noise is introduced. In the desynchronized phase, on the other hand, we have observed a weak stationary PIC, localized at a low bit (see Fig. 4.26). The fluctuation of PIC is very small there.

The avalanche of partition information reminds us of the cascade process of eddies in turbulence. A large vortex splits into smaller vortices, which again feed much smaller ones, and so forth [Kolmogorov 1941; Mandelbrot 1974; Frisch et al. 1978]. At some time steps, a large vortex is formed with ordered motion, leading to a "coherent structure" where the motion is described by a few degrees of freedom up to some precision. If we replace the term "vortex" by "cluster", this process in fluid turbulence is exactly what occurs in our system.

In turbulence in fluid dynamics, the cascade process leads to the celebrated theory of the power law of energy spectra by Kolmogorov [1941], where the energy at the wavenumber K decreases with $K^{-5/3}$. In our problem, although we do not have energy, we can detect some anomalous behavior in the spectrum $P(K) = \langle (1/N)| \sum_j \exp(2\pi i K x_n(j))|^2 \rangle$ (see for detail Kaneko [1994b]).

A remarkable correspondence with fluid turbulence is also discernible by Lyapunov spectra, which give how a small disturbance is amplified or

---

[8] Note that the summation $\sum_{i_m}$ runs over the number of split "sub"-clusters with the increase of the precision from 2^{-j} to $2^{-(j+1)}$. If no splitting occurs it is not necessary.

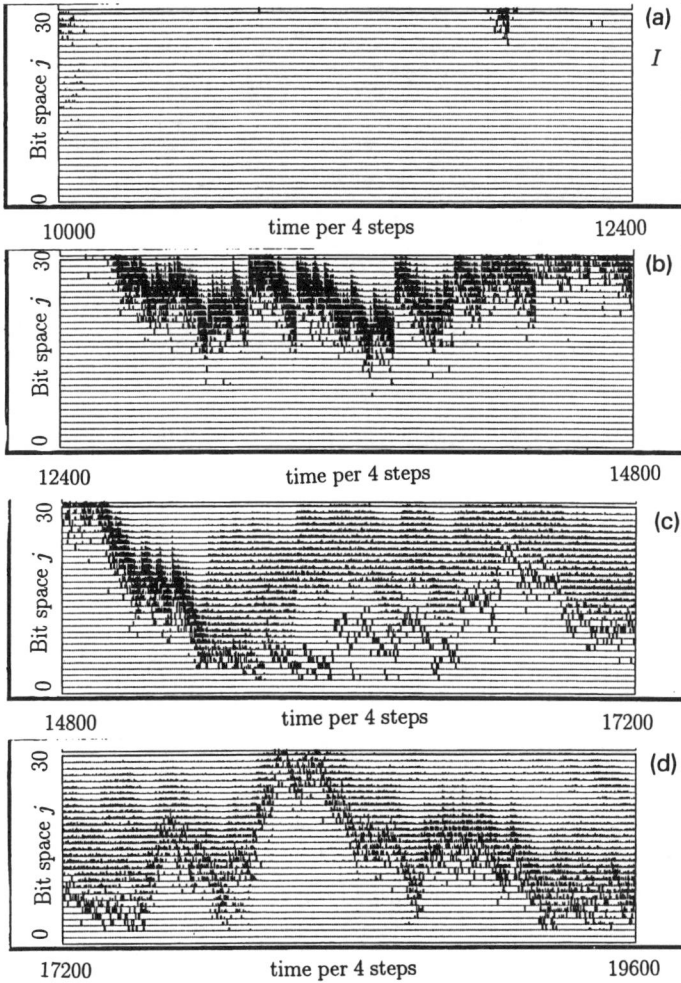

Fig. 4.25 a–d. Diagram of the PIC: the PIC is displayed as a diagram in bit space and time. At each pixel of the corresponding bit (j) and time (n), a box is painted whose length is proportional to I_n^j. (The scale is chosen such that the maximal size up to the next pixel gives $I_n^j = 1$.) $a = 1.59$, $\epsilon = 0.3$, $N = 100$, and $\delta = 2^{-50}$. Plotted per four steps, over (**a**) 10 000–12 400, (**b**) 12 400–14 800, (**c**) 14 800–17 200, and (**d**) 17 200–19 600 time steps. (From K. Kaneko, *Physica D* **77** (1994b) 456, with the permission of the publishers.)

reduced. They are computed as the averaged logarithms of the eigenvalues of a long-time product of Jacobi matrices $J_n(i,j) = (1 - \epsilon)f'(x_n(i))\delta_{i,j} + (\epsilon/N)f'(x_n(j))$. In Fig. 4.27, Lyapunov spectra are plotted. (For reference, the spectrum of the stable coherent state at nearby parameters is also plotted, which has one positive and $(N - 1)$-fold degenerate negative exponents.) For

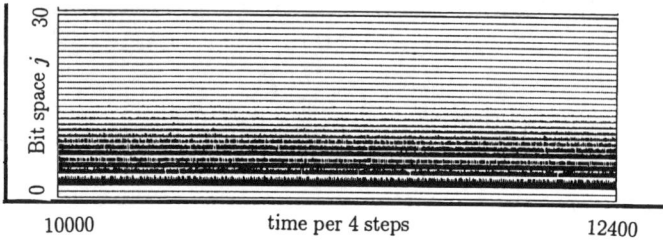

Fig. 4.26. Diagram of the PIC: the PIC is displayed in the same way as in Fig. 4.25. $a = 1.70$, $\epsilon = 0.1$, $N = 100$, and $\delta = 2^{-30}$. Plotted per four steps, over 10 000–12 400 steps. (From K. Kaneko, *Physica D* **77** (1994b) 456, with the permission of the publishers.)

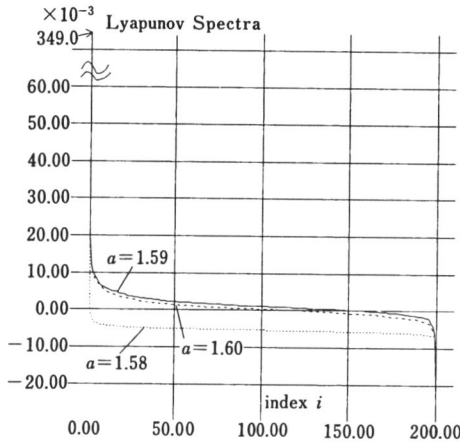

Fig. 4.27. Lyapunov spectra: calculated from the product of Jacobi matrices over 2×10^4 steps, for $N = 200$, $\epsilon = 0.3$, and $\delta = 10^{-9}$. Spectra for $a = 1.58$, 1.59, and 1.60 are overlaid. For $a = 1.58$, the system falls onto a coherent cluster state ($k = 1$), which is a stable attractor. Here the first Lyapunov exponent is positive (0.349), while all others are negative (-0.04). For $a = 1.59$ and 1.6, the maximal exponent is again close to 0.349, while the second largest is approximately 0.02. (From K. Kaneko, *Physica D* **77** (1994b) 456, with the permission of the publishers.)

the (type-II) partially ordered phase ($a = 1.59, 1.61$), we note that many exponents are accumulated around $\lambda = 0$, while a few distinct positive exponents also exist. Such accumulation at the null exponent is also found in the shell model for fluid turbulence [Yamada and Ohkitani 1988]. Our form of the spectra may suggest that (i) chaos is created by very few degrees of freedom, (ii) the chaotic instability is transferred to other modes, and that (iii) this transfer is marginal. This picture seems to be consistent with the bit avalanche in the previous section.

Table 4.2. Comparison with GCM and Fluid turbulence.

	Fluid turbulence	GCM
unit	vortex	cluster
interaction	long-ranged	global
ordering	attractive force between vortices	mean-field coupling
separation	chaotic advection field	chaos in each element
cascade	splitting into small vortices	splitting into small clusters
spatial power spectra	Kolmogorov's law	anomalous behavior of $P(K)$
Lyapunov spectra	accumulation at null exponents	accumulation at null exponents

The correspondence between our dynamics and fluid turbulence is summarized in Table 4.2. Besides the correspondence with fluid turbulence, the concept of an information cascade may be relevant to the computational process in the human brain or to a possible new information-processing mechanism.

4.7 Collective Dynamics

4.7.1 Remnant Mean-Field Fluctuation

Here we study the desynchronized phase, where all elements are desynchronized, all Lyapunov exponents are positive, and the number of finite-precision clusters fluctuates around N. One might expect that there is neither order nor coherence in this phase. If $x(i)$ takes random values almost independently in the desynchronized phase, one expects that the aggregate

$$h_n \equiv (1/N) \sum_j f(x_n(j)) \qquad (4.25)$$

might obey the law of large numbers and the central limit theorem. To study this, we consider the distribution function $P(h)$ of the mean field sampled over long time steps, and the mean field (MSD), $(\delta h)^2 \equiv \langle h^2 \rangle - \langle h \rangle^2$, where $\langle \ldots \rangle$ is the average by the distribution $P(h)$. If the above expectation were true, the MSD of mean-field h would decrease with N as N^{-1}, and the distribution function $P(h)$ would be Gaussian. The interaction term in (1) could be replaced by a noise term whose root-mean-square is $O(1/\sqrt{N})$. Then, in the limit of $N \to \infty$, the dynamics (1) would reduce to N independent logistic maps given by $x_{n+1}(i) = (1 - \epsilon)f(x_n(i)) + \epsilon h^*$, with a constant mean field $h^* = (1/N) \sum_i f(x(i))$.

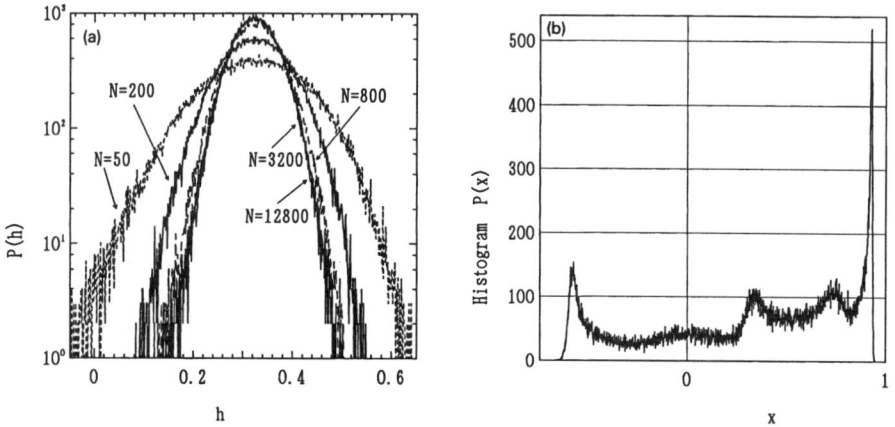

Fig. 4.28. (a) Semi-log plot of the histogram of the mean field h_n. Sampled over 10^5 time steps, after discarding the initial 10^4 steps, using the bin width 10^{-3}. $a = 1.95$ and $\epsilon = 0.1$. Sizes are $N = 50$ (broken line), 200, 800 (broken line), 3200, and 12 800. Histograms for the last two sizes agree within a statistical error, (b) for comparison, the distribution of a single point $x_n(i)$ is plotted for the same parameters with $N = 3200$. (From K. Kaneko, *Physica D* **55** (1992b) 368, with the permission of the publishers.)

To examine the above expectation, we have numerically measured the distribution of the mean field h_n [Kaneko 1990d, 1992b]. Numerically, the distribution function $P(h)$ is rather close to the Gaussian form (see Fig. 4.28). The distribution sharpens with an increase in size up to some size, after which the sharpening stops. The distribution approaches N independent degrees for large N. In Fig. 4.28b, the distribution of a single point $x_n(i)$ is also plotted, which is far from Gaussian, and keeps some singular feature of an invariant measure of a single logistic map.

To see the convergence of the mean-field fluctuation, we plot the MSD. If each element $x(i)$ were approximated by an uncorrelated random number, it is expected that $(\delta h)^2 \propto 1/N$. In Fig. 4.29, the MSD is plotted versus the change of N, for several parameters a. The decreasing of the MSD with N stops around the crossover size $N = N_c$. The change of the MSD versus ϵ is plotted in Fig. 4.30. Here, N_c depends on a, and approaches infinity with $\epsilon \to 0$.

The distribution form is close to the Gaussian form in contrast with the single-point distribution (Fig. 4.28b), but it deviates at the tail. To check the form of the distribution itself, we have measured the flatness of the system given by $(\langle (h - \langle h \rangle)^4 \rangle / (\langle (\delta h)^2 \rangle))^2$, which should be 3 for a Gaussian distribution. The flatness is plotted as a function of the size in Fig. 4.31. As can be seen, the flatness deviates from 3 in general for $N > N_c$, although the deviation is rather small for some parameter values of a. The distribution

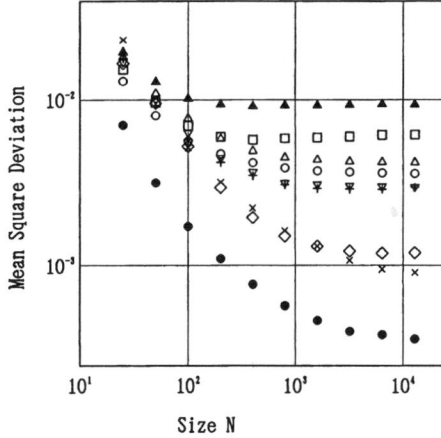

Fig. 4.29. Mean-square deviation (MSD) of the distribution of the mean field h, plotted as a function of the system size. The MSD is calculated over 10^5 time steps after 10^4 transients. $\epsilon = 0.1$. The parameter a is 1.81 (•), 1.84 (○), 1.86 (□), 1.88 (◇), 1.91 (△), 1.94 (▲), 1.96 (▽), and 1.98 (×). (From K. Kaneko, *Physica D* **55** (1992b) 368, with the permission of the publishers.)

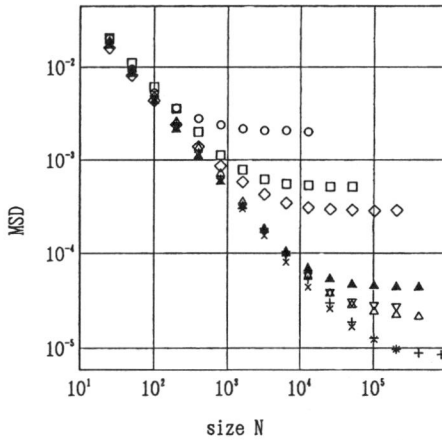

Fig. 4.30. Mean-square deviation (MSD) of the distribution of the mean field h, plotted as a function of the system size. The MSD is calculated over 10^5 time steps after 10^4 transients. $a = 1.99$. The coupling ϵ is 0.1 (○), 0.05 (□), 0.03 (◇), 0.02 (▲), 0.01 (△), 0.009 (▽), 0.008 (+), and 0.005 (×). (From K. Kaneko, *Physica D* **55** (1992b) 368, with the permission of the publishers.)

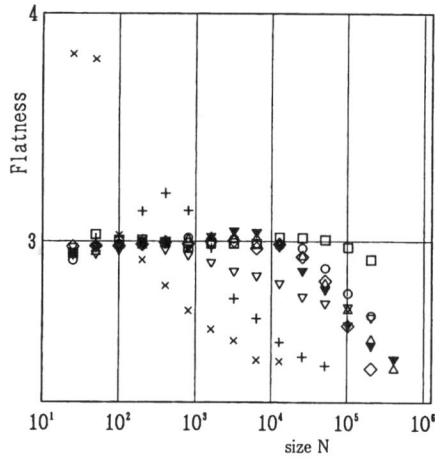

Fig. 4.31. The flatness $(\langle(\delta h)^4\rangle/\langle(\delta h)^2\rangle)$ corresponding to Fig. 4.30 is plotted as a function of the size. Notations are same as in Fig. 4.30. (From K. Kaneko, *Physica D* **55** (1992b) 368, with the permission of the publishers.)

approaches a Gaussian form up to N_c and then deviates a little at the tail of its distribution.[9]

The MSD for $N \to \infty$, and the crossover size N_c, depend on a and ϵ. Indeed the dependence on a is not simple. As shown in Fig. 4.32, there are many tongue structures corresponding to windows. This complex dependence on the parameter originates in the sensitive dependence of the behavior (the Lyapunov exponent) on the parameter a of the single logistic map, as mentioned in Sect. 3.3.1. As for the scaling with ϵ, the MSD decreases as ϵ^2, and $N_c \approx \epsilon^{-2}$. See Ershov and Potapov [1997] and Shibata and Kaneko [1998a] for detailed accounts.[10]

4.7.2 Hidden Coherence

Since the law of large numbers is a general property in a bounded uncorrelated random process, the existence of the size-independent fluctuation we have discovered suggests a remaining correlation between the elements.

To study the correlation between elements, we have measured the mutual information among elements. The mutual information is calculated through a single-point probability $P_i(y)$ that $x(i)$ takes the value y, and a two-point joint probability $p_{i,j}(y,z)$ that $x(i)$ takes the value y and $x(j)$ takes the

---

[9] Since the distribution here is bounded, the distribution cannot be exactly Gaussian at its tail. When the law of large numbers and the central limit theorem are satisfied, the Gaussian distribution is attained after the variance is scaled by $1/N$, where the distribution can be unbounded in the limit of $N \to \infty$.

[10] Corresponding to windows in the logistic map, the clustered state can appear at $\epsilon \to 0$. Hence we need to check carefully the scaling behavior.

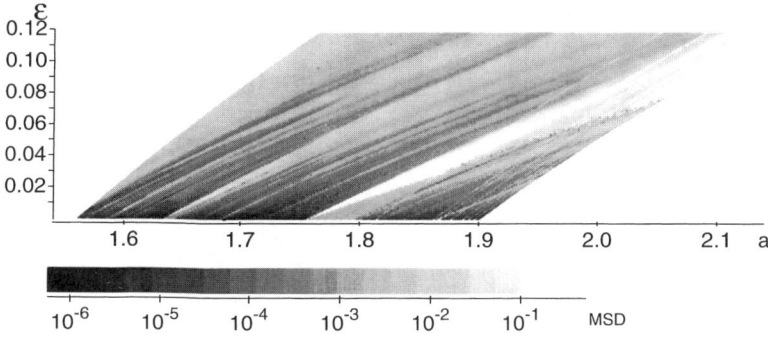

Fig. 4.32. Mean-square deviation (MSD) of the distribution of the mean field h, plotted in (a, ϵ) by using the gray scale. The gray scale shows the MSD value of the GCM for the corresponding parameters (a, ϵ), where the darkest gives MSD $\approx 10^{-6}$ and the lightest gives MSD $\approx 10^{-1}$. (With the courtesy of Tatsuo Shibata.)

value z $(i \neq j)$. These probability distributions are calculated from a long time sample. Mutual information $\mu_{i,j}$ is given by

$$\mu = \mu_{i,j} = -\int\int p_{i,j}(x,y) \log \frac{p_{i,j}(x,y)}{P_i(x)P_j(y)} dxdy. \qquad (4.26)$$

Since the distributions are independent of sites i, in the desynchronized phase, the mutual information is also site-independent and the suffices in $\mu_{i,j}$ can be removed.

In Fig. 4.33, we have plotted μ versus the change of the size N. Although the correlation is very small, there remains a finite mutual correlation even if N gets large. In the figure, we have also plotted the mutual information, for comparison, in the presence of noise. The remnant correlation in the GCM is clear, in contrast with the case of random noise, where the correlation goes to zero with the increase of N.

Hence the 'complete' desynchronization phase in our GCM keeps some coherence among the elements. Then what kind of collective order is kept there?

(a) It is not a low-dimensional collective dynamics.

One possibility for the origin of the hidden coherence is that the mean-field dynamics in the limit of $N \to \infty$ shows a simple low-dimensional dynamics. In Fig. 4.34a,b return maps (h_n, h_{n+1}) are plotted. They suggest some structure, but they are not tori, and they do not look like low-dimensional chaotic attractors. For example the width around the one-dimensional curve in Fig. 4.34b does not decrease with a further increase of N.

To check whether the time series of h_n can be represented by a low-dimensional attractor, we have examined the correlation dimension following the method by Grassberger and Procaccia [Shibata and Kaneko 1998a]. No examples showing low-dimensionality are observed.

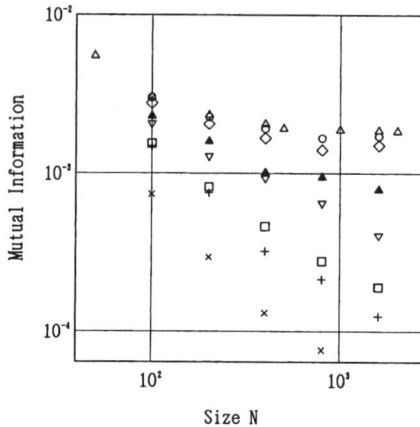

Fig. 4.33. Two-point mutual information. Calculated from $2 \times 10^7/N$ time steps after discarding 10^4 transients. Noiseless case ($\triangle$), with random numbers homogeneously distributed over $[-\sigma, \sigma]$. $\sigma = 0.005$ ($\circ$), 0.0075 ($\diamond$), 0.002 ($\blacktriangle$), 0.004 ($\triangledown$), 0.005 ($\square$), 0.01 ($+$), and 0.02 ($\times$). $a = 1.99$ and $\epsilon = 0.1$. (From K. Kaneko, *Physica D* **55** (1992b) 368, with the permission of the publishers.)

As another direct characteristic, we have measured the power spectrum for the mean field h_n through $|\sum_n h_n e^{2\pi i n \omega}|^2$ (see Fig. 4.35). The spectrum is continuous since the dynamics of the mean field is aperiodic. However, broad peaks remain in the power spectrum. The peaks get sharper with further increase of size N up to N_c. For $N > N_c$, the sharpening stops and the spectrum is invariant against an increase of N.

The sharpening suggests the emergence of a partly coherent motion. Indeed at some parameters there often appears a peak at a very low frequency (which approaches zero with $\epsilon \to 0$), implying a component of quasiperiodic motion with a very long time scale [Shibata and Kaneko 1998a]. Note that the power spectrum of the time series $x_n(i)$ of a single element i shows neither the sharpening of peaks nor the peaks at low frequency.

In contrast with the present case, there are some examples where the low-dimensionality in the collective motion is confirmed in spite of the chaotic motion of each individual element. Chaté and Manneville [1992] found periodic and quasiperiodic motion for the average of $x_n(i_1, \cdots, i_d)$ for a coupled map lattice in a d-dimensional space with $d > 1$ (see also Kaneko [1989b]). For a globally coupled tent map with a 'multiplicative' coupling, torus motion was discovered by Pikovsky and Kurths [1994]. Even in the present GCM, such low-dimensional motion is found when the elements are heterogeneous (nonidentical) [Shibata and Kaneko 1997], as will be discussed later. For comparison we have also plotted the return map (h_n, h_{n+1}) of the heterogeneous case (see Sect. 4.7.5) in Fig. 4.34c, d. For Fig. 4.34c, a torus can clearly be detected, and the width of the scattered points around the curve decreases as $1/\sqrt{N}$ with N.

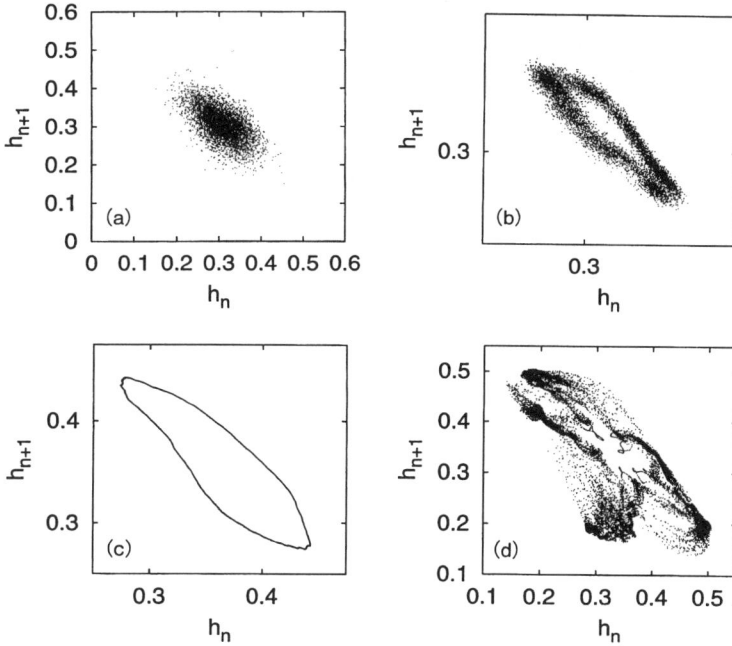

Fig. 4.34a–d. Return map of the mean field (h_n, h_{n+1}) of the GCM, plotted over 50 000 steps after transients have decayed away. $N = 10^7$. (a) $a = 1.69755$ and $\epsilon = 0.008$, (b) $a = 1.699$ and $\epsilon = 0.008$, (c) heterogeneous case, where the parameter a is distributed over $[1.8975, 1.9025]$, $\epsilon = 0.11$, (d) heterogeneous case, where the parameter a is distributed over $[1.9156, 1.9244]$, $\epsilon = 0.1$. (With the courtesy of Tatsuo Shibata.)

(b) It is not a frozen relationship.

One might also expect that the hidden coherence could be temporal in nature as in the frozen order observed in the spin glasses [Mezard *et al.* 1987]. To check this possibility, we have measured the probability that two closer elements stay closer over some time. In practice, we have defined the following relative closeness $S_n^{i,j}$ between two elements introducing a precision δ to judge the closeness:

$$2S_n^{i,j} = 1 \quad \text{if} \quad |x_n(i) - x_n(j)| < \delta,$$
$$= 0 \quad \text{otherwise.}$$

The temporal correlation function, given by $C(t) = \sum_{i,j} (\langle S_{m+t}^{i,j} S_m^{i,j} \rangle - \langle S_m^{i,j} \rangle^2)$, always decays exponentially to zero with time for our turbulent state, unless the precision δ is too large (≈ 0.5). Thus our coherence is not a frozen relationship between two elements.

Nevertheless there are some examples in GCM which show a frozen correlation among the elements. When elements oscillate with a different 'band'

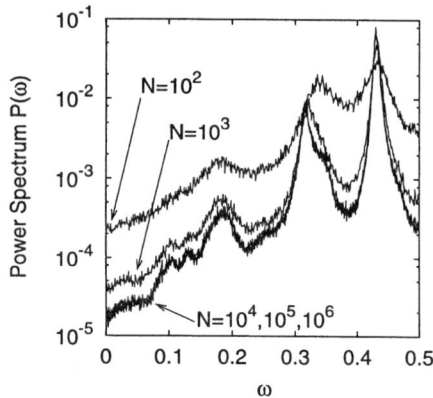

Fig. 4.35. Power spectra of the time series of the mean field h_n, for $N = 200$, 10^3, 2×10^4, and 2×10^5. The last two spectra agree within the given accuracy. They are calculated from 100 sets of 1024-step time series (totally 1024×100 steps after 10^4 transients). $a = 1.9$, $\epsilon = 0.1$. (With the courtesy of Tatsuo Shibata.)

motion as to the π phase difference between large-small-large and small-large-small, there are situations in which elements are desynchronized but keep the relationship of this π phase difference. This example will be discussed in Sect. 4.8.2.

4.7.3 Instability of the Fixed Point of the Perron–Frobenius Operator

Then what is the origin of the hidden coherence and the collective dynamics of the mean field? Here we note that the mean field changes according to the change of the distribution of elements. Let us analyse the collective dynamics in the limit of $N \to \infty$, in relationship with the dynamics of the distribution of $x_n(i)$ [Kaneko 1992b, 1995; Pikovsky and Kurths 1994].

In the limit of $N \to \infty$, the snapshot distribution $\rho_n(y)$ is defined by the probability that $x(i)$ takes the value y ($\rho_n(y) \equiv (1/N) \sum_{j=1}^{N} \delta(y - x_n(j))$). Hence our dynamics can be written as

$$x_{n+1} = (1 - \epsilon)f(x_n) + \epsilon \int \rho_n(y)f(y)\mathrm{d}y. \tag{4.27}$$

If the law of large numbers were valid, the above distribution would be time-independent, and be given by the invariant measure $\rho^*(y)$ for the logistic map $x_{n+1} = (1 - \epsilon)f(x_n) + \epsilon h^*$, where h^* is determined by the self-consistent condition $h^* = \int \rho^*(y)f(y)\mathrm{d}y$. Our observation in the previous subsection suggests that the above self-consistent probability solution does not exist or is unstable even if it exists.

Recall that the single point distribution of a one-dimensional map is calculated with the use of the Perron–Frobenius operator [Ruelle 1978; Oono and Takahashi 1980] (see Sect. 2.7). Hence the evolution of the probability distribution is given by

$$\rho_n(x) = \frac{1}{1-\epsilon} \sum_{pre=+,-} \frac{\rho_{n-1}(y_{pre})}{|f'(y_{pre})|}, \tag{4.28}$$

where the two preimages are given by

$$y_{pre} = f^{-1}\left(\frac{x - \epsilon \int f(z)\rho_{n-1}(z)dz}{1-\epsilon}\right) \tag{4.29}$$

with the two roots of the inverse functions $f^{-1}(y) = \pm\sqrt{(1-y)/a}$ for our logistic map. Equation (4.28) can be derived from the conservation of probability by the one-dimensional map (4.27).

Let us consider the fixed-point solution of (4.28). Assuming the existence of the fixed-point solution $\rho^*(y)$, the problem is reduced to the single logistic map given by $x_{n+1} = (1-\epsilon)f(x_n) + \epsilon h^*$, with $h^* = \int f(x)\rho^*(x)dx$.

To study the stability of this 'self-consistent' solution, we note that the invariant measure is straightforwardly calculated with the Perron–Frobenius operator for the one-dimensional map, because the above equation is just a one-dimensional map. Our problem reduces to a one-dimensional mapping of the form $x_{n+1}(i) = (1-\epsilon)f(x_n(i)) + \epsilon h_{in}$, where h_{in} is an "input mean field" to be determined later [Kaneko 1992b,1995, Perez and Cerdeira 1992]. Here we first consider the one-dimensional map with h_{in} as a given parameter, and compute the invariant measure $\rho^*_{h_{in}}(x)$ from which we obtain the "output" mean field $h_{out} = \int f(x)\rho^*_{h_{in}}(x)dx$. The fixed-point solution of our original self-consistent equation, of course, should obey $h_{in} = h_{out} = h^*$. The obtained invariant distribution $\rho^*(x)$ from the one-dimensional map surely satisfies the self-consistent equation.

Instead of confirming the stability of $\rho^*(x)$ for the original PF equation, we check the stability of the self-consistent solution against a perturbation to change only the mean-field value. In Fig. 4.36, we have plotted h_{out} as a function of h_{in}. Note that the function is not smooth but rugged. Indeed this change is due to the structural instability of the logistic map. Since windows exist in any neighborhood of the logistic parameter leading to chaos, complicated behavior is expected. Also, the invariant measure $\rho^*(x)$ for the single logistic map often has singularities at (infinitely) many points x.

As is shown in Fig. 4.36, it seems there are solution(s) that satisfy $h_{in} = h_{out} = h^*$. However, the condition $|dh_{out}/dh_{in}| < 1$ is not satisfied there. Any small change in h_{in} results in a large variation of h_{out}. Thus, the fixed-point solution of our self-consistent Perron–Frobenius equation (4.28) is expected to be unstable.

The result of a numerical integration of (4.28) is given in Fig. 4.37. Although we carried out the integration choosing various initial conditions, an approach to a fixed-point function was not observed. This numerical result is consistent with the instability of the fixed-point solution.

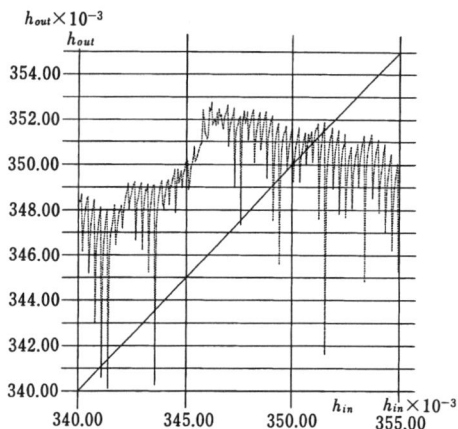

Fig. 4.36. Relation between h_{in} and h_{out}, obtained from the PF equation for a single tent map with $a = 1.9$ and $\epsilon = 0.1$. Only the region with $0.585 < h_{in} < 0.605$ is plotted, while the linear dependence (with a small slope) extends to the interval $[0,1]$. For reference the line $h_{in} = h_{out}$ is drawn. (From K. Kaneko, *Physica D* **86** (1995a) 158, with the permission of the publishers.)

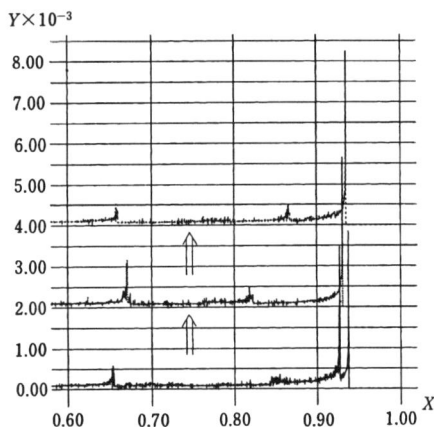

Fig. 4.37. Dynamics of the probability distributions $\rho_n(x)$ for the self-consistent PF equation. $a = 1.9$ and $\epsilon = 0.1$. Successive snapshot distributions for $n = 500$, 501, 502, 503 are overlaid. (From K. Kaneko, *Physica D* **86** (1995a) 158, with the permission of the publishers.)

Mechanism of Collective dynamics.

Instead of the self-consistent fixed-point relationship between the mean field h and the distribution $\rho(x)$, the dynamics of both h_n and $\rho^*(x)$ are mutually related. With the change of the mean field, the distribution is altered, since the one-dimensional map of each element changes its form according to the distribution. On the other hand, the mean field changes following the distribution. These changes occur with some delay, which leads to oscillatory dynamics. This oscillation has a longer time scale than the logistic map which each element follows. If the time scales of this mean-field dynamics and of the logistic map were clearly separated, the scales between the collective dynamics and the 'microscopic' chaos (of each logistic map) would also be separated. Then the low-dimensional collective dynamics could follow, as already mentioned. In the present case, the scales are not clearly separated. Shibata and Kaneko [1998b] measured the finite-size Lyapunov exponent [Paladin *et al.* 1995] for the collective dynamics. The scale of the collective chaos is much larger than that of the microscopic chaos, but the ratio between the two does not diverge in the limit of $N \to \infty$.

A noteworthy point in the hidden coherence is that both the high-dimensional motion and the low-dimensional part coexist in the collective chaos. Furthermore, these two parts seem to have a different scaling behavior with ϵ. The amplitude of the low-dimensional part (say the amplitude of the 'torus'-like motion in Fig. 4.34b scales with some power of ϵ (ϵ^α with $\alpha < 1$) [Shibata and Kaneko 1998a], while the width around this torus scales as ϵ [Ershov and Potapov 1997].

The existence of the high-dimensional part in the collective motion seems to be related to the singularities in the probability distribution of the logistic map [Shibata and Kaneko 1998a]. Indeed the measure given by the Perron–Frobenius equation has (infinitely) many singular points for the image of the point of $f'(x) = 0$. Then, is it possible to eliminate this high-dimensional part by smearing out such singularities? A straightforward method to smooth singularities is the addition of noise to the system. In the next section, we discuss the effect of noise on the collective dynamics.

4.7.4 Destruction of Hidden Coherence by Noise and Anomalous Fluctuations

Since the collective dynamics is formed by a mutual correlation among the elements, the above hidden coherence among elements may be destroyed with the addition of noise. To study the effect of noise, we have simulated the model

$$x_{n+1}(i) = (1 - \epsilon)f(x_n(i)) + \frac{\epsilon}{N} \sum_{j=1}^{N} f(x_n(j)) + \sigma \eta_n^i, \qquad (4.30)$$

where η_n^i is a white noise generated by an uncorrelated random number homogeneously distributed over $[-1, 1]$. In Fig. 4.38, the MSD $= \langle (\delta h)^2 \rangle$ is plotted versus the increase of the system size. If the noise strength σ is larger than a threshold σ_c (≈ 0.005 for the parameters in the figure), the MSD seems to decrease with the size N. The decrease, however, does not obey the expected $1/N$ behavior but has an anomalous dependence on size; $\langle (\delta h)^2 \rangle \propto N^{-\beta}$ with $\beta < 1$.

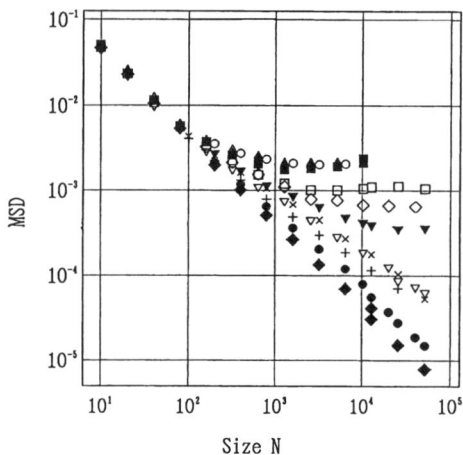

Fig. 4.38. Mean-square deviation of the distribution of the mean field h in our GCM with the addition of noise. Plotted as a function of the system size. $a = 1.99$ and $\epsilon = 0.1$. The noise strength σ is 0.02 ($\blacklozenge$), 0.01 ($\bullet$), 0.0085 ($+$), 0.007 ($\triangledown$), 0.006 ($\times$), 0.005 ($\blacktriangledown$), 0.0045 ($\diamond$), 0.004 ($\square$), 0.002 ($\blacksquare$), 0.0008 ($\blacktriangle$), 0.0005 ($\circ$), and 0.0001 ($\triangle$) from bottom to top. For $\sigma > \sigma_c \approx 0.005$, the MSD decays with $N^{-\beta}$. (From K. Kaneko, *Physica D* **55** (1992b) 368, with the permission of the publishers.)

This anomalous decay is thought to be due to a power-law decay of the correlation among the elements. Assume that the mutual correlation between two elements is given by $N^{-\beta}$. Then, the MSD is roughly estimated by $(1/N^2) \sum_{i,j} \langle \delta h_n(i) \delta h_n(j) \rangle = (1/N) \sum_j \langle \delta h_n(1) \delta h_n(j) \rangle \approx (1/N) N^{1-\beta} = N^{-\beta}$. In fact, the decay of mutual information plotted in Fig. 4.33 shows a decay as a power function. This power-law dependence still needs an explanation, and may originate in the hierarchical structure of our attractor.

This destruction of the hidden coherence leads to a strange conclusion. Take a globally coupled system with a desynchronized and highly chaotic state, and add noise to the system. Then the mean-field fluctuations decrease due to the noise. Such decrease of fluctuations by noise reminds us of the noise-induced order [Matsumoto and Tsuda 1983] studied in Sect. 2.5. Indeed, the peak in the power spectrum of the mean-field dynamics in the present GCM is sharpened by noise [Perez *et al.* 1993]. Quite recently Shibata *et al.* [1999] have clearly shown that the Lyapunov exponent of the mean field as well as the dimension of the system is reduced by the addition of noise, by solving the self-consistent Perron–Frobenius equation in Sect. 4.7.3. With the addition of noise, high-dimensional structures in the mean-field dynamics are destroyed successively, and the bifurcation from high-dimensional to low-dimensional chaos, and then to a torus proceed with the increase of the noise amplitude. With a further increase of noise to $\sigma > \sigma_c$, the mean field goes to a fixed point through a Hopf bifurcation. The appearance of low-dimensional 'order' through the destruction of small-scale structure in chaos is also found in noise-induced order, while a similar destruction is important in quantum chaos.

On the other hand, this noise-induced decrease of fluctuations can be used to distinguish high-dimensional chaos from random noise. If the irregular behavior originates from random noise, (further) addition of noise will result in an increase of the fluctuations. If the external application of noise leads to the decrease of fluctuations in some experiment, it is natural to assume that the irregular dynamics there is due to high-dimensional chaos with a global coupling of many nonlinear modes or elements.

4.7.5 Heterogeneous Systems

So far we have assumed that all elements are exactly identical. What happens if each element is slightly different? To address this question, we choose the following heterogeneous globally coupled map:

$$x_{n+1}(i) = (1 - \epsilon)f^i(x_n(i)) + \frac{\epsilon}{N}\sum_j f^j(x_n(j)), \qquad (4.31)$$

with

$$f^i(y) = 1 - ay^2 + s(i), \qquad (4.32)$$

where $s(i)$ is a homogeneous random number distributed over $[-\sigma, \sigma]$. Since the map $f^i(y)$ is equivalent to the logistic map with the parameter $a' = a(1 + s(i))$ by the transformation $y'' = y/(1 + s(i))$, a larger $s(i)$ leads to a logistic map with a higher peak. Thus the nonlinearity of each map varies for each element.

The above map may be regarded as a model with a static noise $(1 - \epsilon)s(i)$, since the map is written as $x_{n+1}(i) = (1 - \epsilon)f(x_n(i)) + (1 - \epsilon)s(i) + (\epsilon/N)\sum_j f(x_n(j)) + (1/\sqrt{(N)}) \times$ constant. The only difference between this problem and that in the previous section is that the "noise" $s(i)$ here is static, and fixed in time.

In Fig. 4.39, we have plotted the mean-square deviation as a function of the size. The results are rather intriguing. For a small variance of the static noise, there is no significant difference from the noiseless case. If the variance σ is larger, the mean-square deviation $\langle(\delta h)^2\rangle$ again starts to decrease with size as $N^{-\beta}$ with a variable exponent β. What is surprising here is that this decrease continues only up to a size N'_c, and that a further increase of size leads to the *increase* of the MSD, till it approaches a level close to the noiseless case. Metaphorically speaking, our system behaves as a noisy system up to N'_c and then "notices" that $s(i)$ is not a temporal noise and goes back to the noiseless case.

Indeed, it is shown that the heterogeneity enhances regularity in the collective dynamics. Low-dimensional quasiperiodic motion is often found

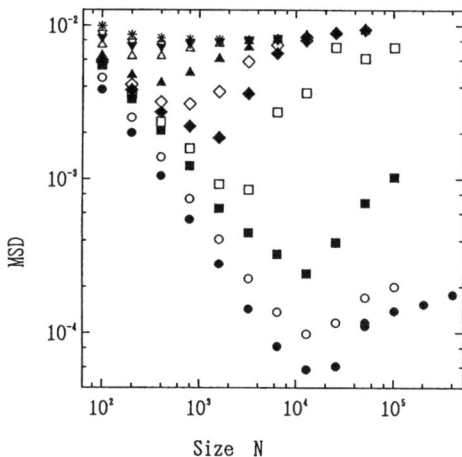

Fig. 4.39. Mean-square deviation of the distribution of the mean field h, for our GCM with inhomogeneous parameters. Plotted as a function of the system size. $\epsilon = 0.1$. The distributed parameter $s(i)$ is chosen randomly from the interval $[-\sigma, \sigma]$. $a = 1.92$. The width σ is 0.05 ($\bullet$), 0.035 ($\circ$), 0.025 ($\blacksquare$), 0.024 ($\square$), 0.023 ($\blacklozenge$), 0.022 ($\diamond$), 0.02 ($\blacktriangle$), 0.015 ($\triangle$), 0.01 ($\blacktriangledown$), 0.0075 ($\triangledown$), 0.0005 ($\times$), and 0.00025 ($+$). (From K. Kaneko, *Physica D* **55** (1992b) 368, with the permission of the publishers.)

for the mean field, even if each element shows chaotic dynamics (see Fig. 4.37). Recall that there are several windows in the logistic map. Through the mutual relation between the mean-field dynamics and the distribution of elements, the regularity of the periodic motions in the windows manifests itself. Then some coherence among the elements and regular oscillations in the mean field appear. The mechanism of this order in the collective dynamics is found to be due to the formation of an 'internal bifurcation structure' among the elements, and the self-consistent dynamics between the structures and the mean field [Shibata and Kaneko 1997].

The order in the collective dynamics is important when considering its significance for natural phenomena. In natural phenomena, it is rather hard to imagine an ensemble of completely identical units. Even in physical phenomena, units (such as in Josephson-junction arrays) cannot be exactly identical. For a biological system where heterogeneity is more important, the collective dynamics in the present case will be more significant.

4.7.6 Significance of Collective Dynamics

The present hidden coherence is characteristic of a globally coupled chaotic system, where the use of deterministic chaos, rather than random noise, is essential. With this deterministic origin of random behavior, coherence within highly disorganized dynamics is made possible. Recall here that the late Professor Kazuhisa Tomita [1985] coined the term 'coherent irregularity' instead of deterministic chaos.

Recognition that a nonlinear (chaotic) system with global coupling universally allows for the coexistence of weak coherence and random behavior will be relevant to a wide range of physical and biological phenomena. In physics, the observation of hidden coherence is suggested in a multi-mode laser [Bracikowski and Roy 1991] and in a Josephson-junction array.

Collective dynamics with hidden coherence will be important to the study of the brain. The firing of each neuron or the change of the local-field potential is not regular in general. Furthermore, a huge number of neurons (or ensembles of neurons) interact with each other. If there is no correlation at all among the activities in these neurons or ensembles, the variation of the average activity should be negligibly small, considering the number of elements involved. On the other hand, large temporal variations remain in EEGs, which measure the average electric activity in our brain (over a large domain that includes a huge number of neurons). Hence the neurons (or their ensembles) keep some kind of coherence, in spite of their irregular, desynchronized activity. Our hidden coherence provides a mechanism for such irregular collective dynamics.

In most statistical mechanical analysis of neural networks, it is assumed that the fluctuation decays as $1/\sqrt{N}$ with the increase of the number of neurons N. The present result in our globally coupled system presents us with a strong cause for caution with regard to such statistical treatment (recall that the connection among neurons is usually believed to be rather global (or long-ranged)). Of course, various inhomogeneities in couplings and elements exist in neural systems that may alter the nature of the mean-field dynamics. However, we have to note again that the collective dynamics appears even in the heterogeneous case.

4.8 Universality and Nonuniversality

4.8.1 Universality of Clustering and Other Transitions

The discovered phenomena such as clusterings, chaotic itinerancy, partially ordered phases, and hidden coherence are widely observed in globally coupled systems. As another class of globally coupled dynamical systems, we have investigated the globally coupled circle map

$$x_{n+1}(i) = x_n + \frac{K}{2\pi}\sin(2\pi x_n(i)) + \Omega + \frac{\epsilon}{2\pi N}\sum_j \sin(2\pi x_n(j)), \qquad (4.33)$$

which is thought to belong to the same class as globally coupled oscillators (see Hadley and Wiesenfeld [1989]). Again the map shows coherent, ordered, partially-ordered, and desynchronized phases as well as a quasiperiodic state [Kaneko 1991b]. Chaotic itinerancy and collective dynamics by hidden coherence are also observed. Of course, this universality is not limited to maps. Clusterings, phases with different degrees of synchronization, and hidden coherence are also observed when coupling ordinary differential equations showing chaotic dynamics. For example they are observed in a coupled Josephson-junction array system [Dominguez and Cerdeira 1993]. Similar phenomena are found in models for laser arrays or multi-mode lasers, for which the correspondence with experiments is discussed.

So far each element was chosen to be chaotic. Then the question arises whether the same behavior will be observed if each element does not show chaos by itself while the coupling induces some chaotic behavior. To address this, we have studied the following simple GCM [Kaneko 1991b]:

$$x_{n+1}(i) = x_n + \frac{K}{2\pi} \sin(2\pi(x_n(i) - x_n(j))). \tag{4.34}$$

In this model, all elements are synchronized for $-2 < K < 2$. There are ordered phases with two-cluster attractors for $2 < K < 3.9$, and partially ordered phases with a variety of clusters for $3.9 < K < 6.4$. For a further increase of K, there is a transition to the desynchronized phase. A partial system of the above model consisting of a few elements can show chaos due to the coupling term. The total system, then, is regarded as a globally coupled chaotic system. Universality of the phases as well as other characteristic features are confirmed in the model. Since the PO phase in this model is observed over a wide interval of parameters, the model is suitable for the study of CI. In fact, the number of effective degrees of freedom alternates between $2 \sim 5$ and N.

When K is negative, the two-cluster state appears for $-3.9 < K < -2$, which is replaced by the three-cluster state for $-4.7 < K < -3.9$. Here, a direct transition to a desynchronized phase is found for $K < -4.7$, without any PO phase. Indeed, no attractor with $k > 3$ clusters is observed.

As an example of coupled ordinary differential equations, Nakagawa and Kuramoto [1993, 1994a, b] studied the globally coupled complex Ginzburg–Landau equation

$$\dot{W}_j = W_j - (1 + \mathrm{i}c_2)|W_j|^2 W_j + K(1 + \mathrm{i}c_1)\left(\frac{1}{N}\sum_{k=1}^{N} W_k - W_j\right)$$

$$(j = 1, 2, \cdots, N) \tag{4.35}$$

(see also Hakim and Rappel [1992]), where W_j represents the complex amplitude of the element j. Here again the individual element does not show chaos, and no interesting behavior is observed unless coupling induces some chaotic behavior (for example, no chaos is observed for $c_1 = 0$). When $c_1 \neq 0$, the coupling introduces a phase rotation, and leads to chaos for some parameter values. In this model, clustering and collective dynamics are observed. Besides these common features with our GCM, collective dynamics maintaining clear coherence is also discovered in this model.

4.8.2 Globally Coupled Tent Map: Novelty Within Universality

To examine the universality of our results, we briefly refer to a GCM of a slightly different class. A logistic or circle map has an expanding or contracting part depending on x. Indeed the slope $(|-2ax|)$ of the logistic map $f(x) = 1 - ax^2$ is smaller than 1 for $|x| < 1/(2a)$ where orbits contract. An orbit expands or contracts with time, and the orbital instability fluctuates in time. On average it is expanding, when chaos exists. Then what happens for a GCM if the orbit in each map is always expansive? One typical example is given by the tent map

$$f(x) = a(0.5 - |x - 0.5|). \tag{4.36}$$

If $a > 1$, the orbit is always expansive. Here we briefly discuss the GCM (4.1) with this tent map instead of the logistic map.

From the linear stability analysis, a coherent state with $x_n(i) = x_n(j)$ is stable if the condition $(1 - \epsilon)a < 1$ is satisfied. Indeed, our system falls on this coherent attractor if this condition is satisfied. When this condition is not satisfied, there is no clustering with a small number of clusters. Since each map is expansive everywhere, it is almost impossible to imagine a mechanism to maintain coherence between two elements. Indeed, none of the elements are synchronized for $(1 - \epsilon)a > 1$.

The phase diagram for the globally coupled tent map is given in Fig. 4.40. As stated above, there is no ordered state with a few clusters. Instead, we have band structures in the model [Kaneko 1995].

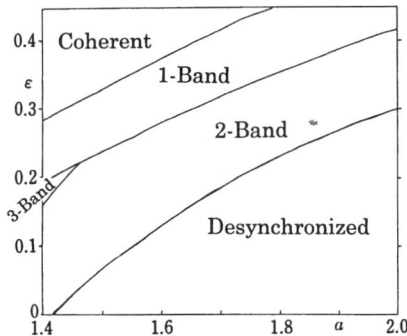

Fig. 4.40. Rough phase diagram of the globally coupled tent map, obtained from the observation of a few attractors for each parameter, incremented by 0.01. (From K. Kaneko, *Physica D* **86** (1995a) 158, with the permission of the publishers.)

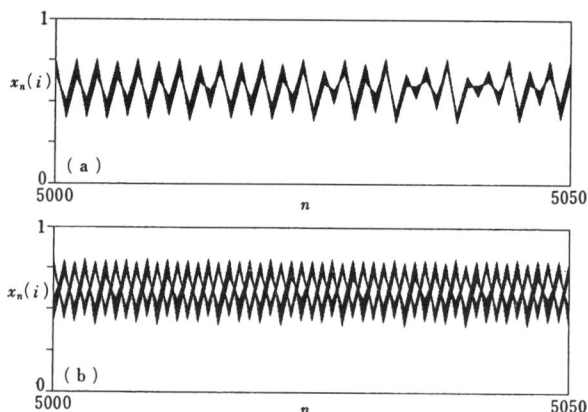

Fig. 4.41 a,b. Time series of all elements $x_n(i)$ are overlaid for the globally coupled tent map. $N = 100$. **(a)** $a = 1.55$ $\epsilon = 0.3$, for time steps from 5000 to 5050 (a single-band state), **(b)** $a = 1.75$, $\epsilon = 0.3$, for time steps from 5000 to 5050 (a two-band state).

For the single-band phase, there is strong correlation among elements, although the synchronization is not perfect. The number of clusters is N, but all elements are strongly correlated as can be seen in Fig. 4.41a. When the coupling strength is decreased, the attractor has a two-band structure as in Fig. 4.41b. In the two-band state, elements split into two groups with the same oscillation phases as the period-two band.[11] Here many attractors coexist depending on the number ratio with regard to each band, similarly to the two-cluster case. If the coupling strength is decreased or a increased, the system enters a desynchronized phase, without any band structures.

It is quite certain that there is some coherence among the elements, leading to nonfixed mean-field dynamics, in the band phases where elements are not synchronized. In the single-band phase, the mean field has a clear component of the low-dimensional map, that is the original tent map. In the two-band phase, the band which an element belongs to is fixed (i.e. an attractor), and cannot switch. In both cases, the variance of the mean-field fluctuation does not decrease at all with the increase of N, in contrast with the case discussed in Sect. 4.7.1. It is also clear that the rejected mechanisms (a), (b) in Sect. 4.7.2 hold here. In the single-band case, there is clear low-dimensional motion, in contrast with Sect. 4.7.2 (a), while there is frozen correlation among elements in the two-band case, in contrast with Sect. 4.7.2 (b).

Then is there some kind of collective dynamics in the ('completely') desynchronized phase? First, the variance of the mean-field fluctuations decreases with $1/N$ up to a huge size (say $N = 10^7$) [Kaneko 1995]. Could perhaps

---

[11] In the transition regime between single- and two-band dynamics, we have also observed a state with intermittent switching between the two.

the fixed-point solution of the self-consistent Perron–Frobenius equation be stable in our globally coupled tent-map case?

This problem has recently been studied by several researchers [Just 1995; Ershov and Potaopov 1995; Morita 1996; Nakagawa and Komatsu 1998; Chawanya and Morita 1998]. The results are rather interesting. First, Ershov and Potapov proved that fluctuations of the mean field of the order of $\exp(-/\epsilon^2)$ remain. Although the decrease of the MSD continues up to a huge size, this stops at some size and tiny, but finite, fluctuations remain even in the $N \to \infty$ limit. By studying the self-consistent Perron–Frobenius (SCPF) equation (in Sect. 4.7.3; with the tent map) several tongue-like structures and bifurcations of the mean field were discovered by Nakagawa and Komatsu [1998] (see also Morita [1996]). It is now confirmed that some coherence remains and that the mean-field dynamics is not attracted to a fixed point. Both the (low-dimensional) torus-like motion and the high-dimensional motion coexist [Nakagawa and Komatsu 1998; Chawanya and Morita 1998]. These features are common with the globally coupled logistic map case. However, the scaling form of the MSD of the mean-field amplitude is singular as $\exp(-1/\epsilon^2)$, in contrast with the ϵ^2 behavior in the logistic case. (Indeed the amplitude of the mean-field dynamics is much smaller than in the logistic case.) The origin of this singular behavior is now understood by the marginal stability of the fixed-point solution of the SCPF equation by Chawanya and Morita [1998]. Indeed their analysis of the SCPF equation clarifies the origin of the collective dynamics in the globally coupled tent map.

5. Significance of Coupled Chaotic Systems to Biological Networks

5.1 Relevance of Coupled Maps to Biological Information Processing

While in the next chapter neural information processing will be discussed from the viewpoint of chaos, here we briefly survey the significance of the CML and the GCM to biological information processing and to the complexity of living systems. We will also mention their applicability to engineering problems. Some of the conclusions of the current chapter will be important for the dynamic brain processes to be discussed in Chap. 6 [Tsuda 1990a].

In biological information processing, it may reasonably be assumed that the number of degrees of freedom of the nonlinear dynamics is rather high. Accordingly, a novel viewpoint for biological information processing by chaos needs to be considered in order to open a new conceptual framework which can incorporate these degrees of freedom. In Japan, Tomita [1985] and Yamaguti [1986] discussed the possible relationship between chaos and living systems, and inspired many researchers that followed.

On the other hand, a stable system with a large number of degrees of freedom can have many fixed-point (or stable) attractors to which memory can be assigned. This correspondence between attractors and memory is one of the fundamentals of the 'static' aspects of neural networks. However, in spite of the ability to store given data, such systems cannot generate new memory or perform dynamic information processing. Conversely, chaos with a small number of degrees of freedom has the possibility to create new information. However, it is limited in regard to the dynamic processing of a large amount of stored memories. A coupled chaotic system has the merits of both approaches, and also offers many possibilities as a new information-processing system. We believe that some of these new possibilities will be relevant to the dynamic processes of the brain, and will have useful applications in biological and engineering information processing.

The characteristic features of a chaotic network, shown in Chap. 4, provide a new viewpoint for biological-network systems in general. First, the formation of synchronized clusters demonstrates how identical elements can be differentiated and form groups of similar (same) dynamic behaviors. In

general, there can be three types of clusterings: phase, amplitude and oscillation frequency. In the examples in Chap. 4, the differentiation by the
oscillation phase is the most relevant to the clustering, and then the difference
in amplitude is brought about by the difference in the number of elements
in each cluster. Although clear differences in frequency do not appear in
the present GCM, there are some differences in the frequency components
following the differences in the dynamic behaviors of each cluster. In considering a biological network, it is necessary properly to use the three types
of clustering: phase, amplitude and frequency. For example, concerning the
synchronization in a neural system to be discussed later, clustering by phase
has attracted the most attention recently, but clustering by amplitude and
frequency will also be important in the future. In the study of the cell-
differentiation model, to be mentioned in the last part of this chapter, phase
differentiation of metabolic oscillation occurs first, and is fixed by the differentiation by frequency and amplitude later [Kaneko and Yomo 1995]. Here it
is important to note that, while differentiation by phase can be continuously
shifted, the differentiation by amplitude leads to several discrete types, distinguished clearly by intensity. Hence, there is a tendency to generate digital
information.

Second, chaotic itinerancy can provide a prototype of dynamics to form
and reconstruct a relationship among elements. Also, hidden coherence in
the desynchronized phase of the GCM leads to complex collective dynamics.
These two concepts are related to the problem of how a biological network
can generate and maintain diversity, and how such a state with diversity can
be robust against external disturbances. The significance of the network of
chaotic elements to the biological-network system will be discussed later in
this chapter.

5.2 Application of Coupled Maps to Information Processing

5.2.1 Memory to Attractor Mapping and the Switching Process

(1) Information storage with many coexisting attractors:

In the frozen chaos of the CML, at least 2^N attractors coexist. In the
GCM, if the partition distribution into clusters is given by $(N_1, N_2, \cdots, N_k)$,
then there are $(N!/N_1!N_2! \cdots N_k!)$ attractors, depending on the partition. In
general, it may be impossible to distinguish and use all of them, in particular
those with small N_k. At any rate, a huge number of attractors coexist. If
an attractor corresponds to a memory state, a huge amount of memories is
available. Although it may be hard to use these attractors as a long-term
memory for storage and retrieval, they can be used as short-term memory.
In this case, dynamic switching among memory states is necessary, as will be
discussed in item (3). So far we have considered cases in which elements have

an identical parameter and an identical coupling. By extending our coupled maps to a case with heterogeneous parameters and couplings, it may be possible to design a coupled map with the ability to store long-term memory, and retrieve the memory from the mean field.

(2) Hierarchical memory:

It is important to note that the attractors are organized hierarchically. In a state around the onset of frozen chaos in a CML, there are many attractors coexisting hierarchically, where domains with smaller sizes exist within larger domains, as is discussed in Fig. 3.5. In a GCM, the cluster is coded hierarchically as a tree structure. This hierarchy is coded by the degree of synchronization of oscillations among elements, as is determined by the number of elements in the cluster (see the tree structure in Sects. 4.5.2 and 4.5.3). By using the attractors with this hierarchical structure for memory construction, categorization with a tree structure is possible. In other words, the proximity among memories is formed hierarchically, reminding us of our own way of categorization by classifying many items hierarchically. Here, chaos is essential to form this hierarchical complexity, since the chaos leads to successive splittings of clusters.

(3) Switching among attractors by inputs to elements:

When using attractors for memories as in (1) and (2), it is important to find a method for switching from one memory to another. By applying an input to a single element, i.e. by changing $x(i)$ to $x(i) +$ input, we can make our system jump from one attractor to another. Here we adopt the GCM and discuss the simplest case in which only one element is changed externally at one time step [Kaneko 1990c].

First we consider the simplest case in which only two clusters exist. As mentioned in Sect. 4.2, attractors are coded by the number of elements belonging to one cluster. By choosing one element belonging to one cluster and putting an input to the element externally, we can increase or decrease the number of elements in the cluster. Therefore, it is possible to change the cluster distribution as $(N_1, N - N_1) \rightarrow (N_1 + 1, N - N_1 - 1) \rightarrow (N_1 + 2, N - N_1 - 2) \cdots$. However, as mentioned in Sect. 4.4, the number of elements in a cluster cannot be larger than some given threshold. If we try to increase the number beyond the threshold, an intermittent 'posi–nega' switch occurs, as shown in Fig. 5.1. The intermittent transient in the phase switch reminds us of our oscillatory mental state when we look at some of Escher's figures or Necker's cube. By looking at them, we wonder which is the "figure" or the "ground". Before we decide from a higher level, our mental state changes intermittently. We have to decide which is which (corresponding to a switch to a different attractor), to understand the figure.

The switching is also possible even if the cluster number is larger than two. If clusters are hierarchically organized, the switches are also hierarchically organized. We have switches between the lower-level clusters or the higher level clusters. A switch at a higher level affects its lower-level clusters. On the

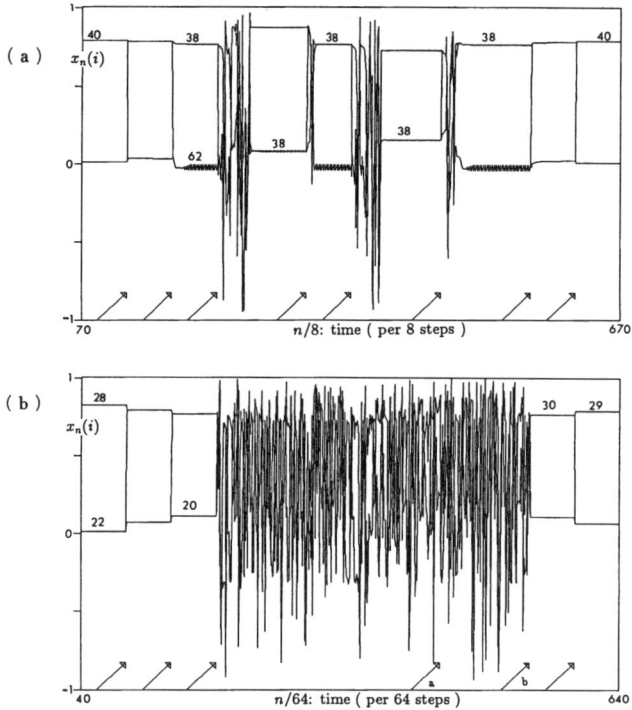

Fig. 5.1a,b. Time series with switches among attractors. $x_{2n}(i)$'s for all i are plotted as a function of time. If there are only 2 lines, the system has fallen to 2-clusters at the time step. Arrows indicate inputs as described in the text. Numbers on the time series of $x(i)$ denote the size of the cluster. (a) $a = 1.95$, $\epsilon = 0.3$, and $N = 100$. $N_{\text{thr}} = 62$. Here the 2-cluster state is stable only up to $N_1 = N_{\text{thr}} = 62$. If an input is applied to increase the number of elements beyond N_{thr}, there appears transient intermittent chaos, and then the system is attracted to a 2-cluster attractor with $N_1 < N_{\text{thr}}$. (b) $a = 1.93$, $\epsilon = 0.3$, and $N = 50$. By increasing N_1 beyond $N_{\text{thr}} = 31$, there appears intermittent chaos, which itself is an attractor, and remains even if we iterate over longer time steps. In the input at "a", we have tried to increase N_1 from 31 to 32, but our state has come back to the original attractor with $N_1 = 31$. By the input to decrease N_1 from 31 to 30, we can eliminate the intermittency (as shown by the arrow "b"). (From K. Kaneko, *Physica D* **41** (1990c) 137, with the permission of the publishers.)

other hand, a switch at a lower level does not affect the higher-level clusters up to some threshold. We can regularly change the number of sub-clusters, as in the case for the switch in 2-clusters.

If we try to increase the difference in the numbers of elements in the sub-clusters beyond a threshold, two types of switches are possible depending on the parameters and inputs. In the first type, an intermittent phase switch between two sub-clusters occurs, without any change to the other clusters in the tree structure (inter-cluster switch). The tree structure of the attractor is

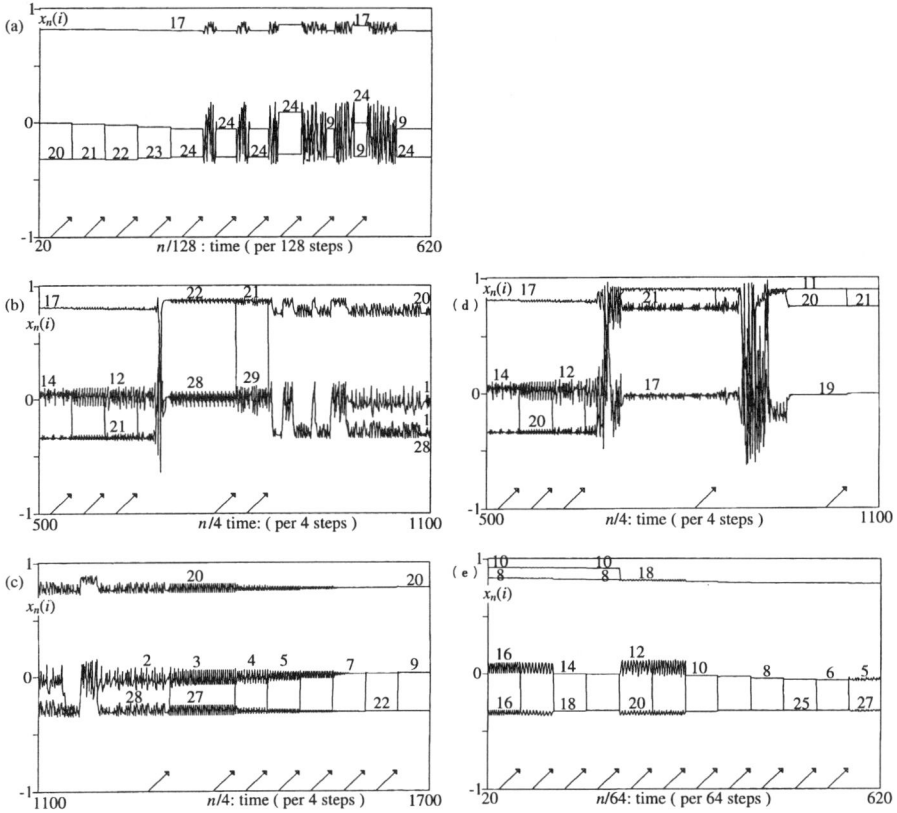

Fig. 5.2a–e. Time series with switches among attractors. Calculated and plotted in the same manner as in Fig. 5.1. Numbers on the time series of $x(i)$ denote the size of the cluster. (a) $a = 1.76$, $\epsilon = 0.2$, $N = 50$. There appears intermittent switching at two sub-clusters without destroying the upper level structure of clusters, (b) $a = 1.83$, $\epsilon = 0.2$, $N = 50$, (c) continued from (b): By an input to change the clustering from [21, 29] to [20, 30], there appeaars long chaotic transient (during (b)–(c)), until the system is attracted to a 3-cluster attractor with [20, 2, 28], (d) another switching process for the same parameter set as (b)–(c), with slightly different input values, (e) continued from (d). (From K. Kaneko, *Physica D* **41** (1990c) 137, with the permission of the publishers.)

preserved. In Fig. 5.2a the switching at a lower level does not affect the higher level. However, as the number of elements in one cluster goes beyond some threshold, a switch that changes the tree structure appears, as shown in Fig. 5.2b–d. In this case, a switch in a lower level propagates to the higher level. This is understood as a "chaotic revolt" in Chap. 4. As the difference between the sizes of the two sub-clusters is increased, the instability by chaos becomes larger, till it destroys the higher-order structure. Note that if memories of one category are stored into three-cluster states and memories of another category are stored into two-cluster states, we can successfully retrieve the memories of both categories by switching from a two-cluster state to three-cluster state as shown in Fig. 5.2.

In switching to an attractor with a different number of clusters, the state after the switch is not determined uniquely because the state must pass through a strong chaotic state. A slight difference in the input or the timing of the input (the value of $x_n(i)$ at that time) changes the state reached after the switch (see Fig. 5.2b). Conversely, when a switch occurs without changing the number of clusters, an input within some range of values yields the same switch, and there is no sensitive dependence on the exact input value. Thus, it is possible either to use a regular switch or an (apparently) probabilistic switch depending on the situation.

In general, in the transient process to change the cluster structure, strong chaotic oscillation often appears. When such oscillation appears, the transient chaotic state is used as a retrieval tool for the search among memories. This chaotic search was studied by Freeman [1987] for the collective potential of the olfactory bulb and was studied by Tsuda *et al.* in his nonequilibrium neural network [1987]. Similar chaotic search mechanisms in artificial neural networks were studied by Nara and Davis [1992, 1994].

In this item, switches involving only two or three clusters are studied. For a case with many clusters hierarchically organized, there can be several levels of switches in the hierarchy.

5.2.2 Chaotic Itinerancy and Spontaneous Recall

In the previous section, switching by external inputs was discussed. In contrast, in a partially ordered phase or in spatiotemporal intermittency, the orbit switches from one ordered state to another spontaneously. Here we discuss its relevance to information processing.
(1) Spatiotemporal intermittency and chaotic itinerancy as mechanisms for spontaneous retrieval of memories:

In the spatiotemporal intermittency of the CML and in the partially ordered phases of the GCM, an orbit itinerates chaotically among several ordered states which can be regarded as destabilized attractors. If each ordered state, for example an almost-clustered state, corresponds to a memory, then it is possible to obtain spontaneous recalls of memories by chaos, as was first proposed by Tsuda *et al.* [1987]. Since the order of jumping from one memory to another is governed by chaotic dynamics, it has both some degree of regularity and randomness. Thus, a somewhat flexible rule is formed for the jumping process, which plays an important role for a spontaneous recall and binding of memories by it. The chaotic itinerancy found in a model of the brain and its significance to information processing will be discussed in the next chapter.

(2) Hierarchical dynamic memory and dynamical categorization:

In the partially ordered phase of the GCM, the hierarchical structure in Sect. 5.2.1 (2) spontaneously changes. It itinerates over memory states of the same category (up to a lower order), and then switches spontaneously to a memory at a different level (category). Such dynamics will be useful as a dynamic memory.

(3) Partial coherence as a mechanism of grouping or feature detection:

In the GCM, several clusters with synchronization are formed. Let us assume that these clusters work as units of feature detection. For example, elements in a cluster represent a continuous object. This idea of grouping was first proposed by Gray, Singer and others [Gray *et al.* 1989; Eckhorn *et al.* 1988] for a function of correlated activity discovered in the visual cortex. At first, they proposed feature detection by periodic oscillations of spike trains of simple cells and the synchronization between them. Indeed a similar idea had already been proposed by Shimizu *et al.* [1985]. According to the data by Singer *et al.*, however, the auto-correlation decays within two or three periods. If the data are of mathematical significance, they imply that the oscillation is not periodic and that the synchronization is partial. Indeed, we believe that their oscillation is chaotic, and we also take the possibility seriously into account that partial synchronization as in the GCM and chaotic itinerancy occurs in the visual cortex. This conclusion is not only drawn from the analysis of experimental data but also from the theoretical consideration that feature detection must have dynamic and diverse aspects. Indeed, there are severe limitations when trying to cope with diverse and dynamical external inputs in terms of simply periodic dynamics. Furthermore, it is difficult for elements with periodic dynamics to synchronize and desynchronize successively according to external inputs within a short time scale. (Since the limit cycle oscillation is stable, it is not easy to desynchronize once the oscillations are synchronized.) On the other hand, partial synchronization and the accompanying chaotic itinerancy have the potential to respond to diverse inputs. It is also easy to change the number of synchronized elements successively depending on the inputs (see Chap. 6).

The relevance of a state between order and disorder to biological systems has recently been discussed from several viewpoints as the edge of chaos or homeochaos, as will be discussed again in Sect. 5.5.

Chaotic itinerancy is the conceptualization of dynamics whose effective number of degrees of freedom varies. Since the upper levels are formed and destroyed spontaneously in chaotic itinerancy, it provides an example to overcome the antithesis between the top-down and bottom-up approaches discussed in Chap. 1. Although emergence is often used incorrectly (from our viewpoint) as a type of dissipative structure [Nicolis and Prigogine 1977] or as an example of the slaving principle by Haken [1979], chaotic itinerancy provides a framework for emergence through chaotic revolt against the slaving principle.

5.2.3 Optimization and Search by Spatiotemporal Chaos as Spatiotemporally Structured Noise

Recently, algorithms for optimizing searches with the help of the concepts of statistical mechanics have made some progress. For example, in simulated annealing [Kirkpatrick *et al.* 1983] a search in phase space has been carried out by using some equivalent of thermal noise. The search with noise, however, often destroys the phase-space structure. In contrast, a search with spatiotemporal chaos may be carried out without destroying the structure of the phase space. Since the strength of spatiotemporal chaos differs depending on elements, it is possible to perform a search by separating the determined and undetermined parts successively. Since an orbit of spatiotemporal chaos does not necessarily cover the entire phase space, the search may omit unnecessary parts in phase space when the search problem is suitably coded in the phase-space structure of the dynamics. Thus, a search with high-dimensional chaos by coupled maps is expected to have a large potential for optimizing searches. A successful example in this direction was proposed by Nozawa [1992] (as mentioned in Chap. 1), by combining the GCM and neural networks.

5.2.4 Local–Global Transformation by Traveling Waves – Information Creation and Transmission by Chaotic Traveling Waves

As mentioned in Sect. 4.3, chaotic traveling waves transform local dynamics to global features. When attractors with different traveling-wave velocities coexist, one can switch from one attractor to another with a different speed by adding a local input at one lattice site. With this switch, the local input is transformed by the global information. This is easily understood by recalling the local phase slip that affects the global traveling wave. By a local input, the number of phase slips is changed, which changes the speed of the traveling wave over all lattice points. Thus the local input is transferred to all lattice points. On the other hand, when chaotic modulation is applied to a traveling wave, chaos is transmitted in the opposite direction of the wave. Hence bidirectional information transmission may be possible simultaneously.

5.2.5 Selective Amplification of Input Signals by the Unidirectionally Coupled Map Lattice

As discussed in Chap. 3, in the unidirectional coupled map lattice (chain), a disturbance is amplified in the downstream direction. There we have observed that an upstream input is amplified down-flow selectively depending on the frequency of the input. The amplification rate can reach 10^{10} and the frequency dependence is very sharp. Combining several coupled map chains it may be possible to construct a system with selective amplification of various inputs. Such a mechanism of amplification may be adopted in our neural system. In the next section, we study the information mechanics of coupled map chain systems from a different viewpoint.

As an input is amplified down-flow, the spatiotemporal pattern at the downstream is changed successively. As a result, a tiny input at the upstream can smear out the chaos of the downstream (Fig. 4.29). Thus, a CML with unidirectional coupling may also be applicable for the control of chaos.

5.3 Information Dynamics of a CML with One-Way Coupling

Here we discuss a CML with unidirectional coupling in order to investigate its information-processing structure. The coupling of the CML here is a little bit different from that of the CML discussed in Chapt. 3. In dynamic information processing in the brain, a combination of excitatory and specific inhibitory couplings is generally observed. The coupling form discussed here was chosen such that some key features of the neural system are taken into account. In this sense, this section provides a basis for discussing the relevance of the CML to information processing in the brain. For the dynamics of each element, we choose a one-dimensional map $x_{n+1} = f(x_n; s)$, with $f(x)$ a nonlinear map, n a discrete time step, and s a bifurcation parameter of the dynamical system. Here $f(x)$ is not necessarily the logistic map. Although there are various ways to couple elements unidirectionally in a chain, the actual coupling form does not make an essential difference with regard to the resulting information structure.

In particular we consider the following CML, consisting of N elements,

$$x_{n+1}(i) = f(x_n(i); s(i)) + Dx_n(i-1) - D'x_n(i), \tag{5.1}$$

where $D = D'$ for $1 < i < N$, while $D = 0$ at $i = 1$ and $D' = 0$ at $i = N$. A specific form of $f(x)$ will be given later [Tsuda and Shimizu 1985].

Here we study how information is deformed and transmitted along the chain after inputs are externally applied to the CML (Fig. 5.3). In Sect. 2.7, we have mentioned that the information dynamics in a chain depends on the choice of the map $f(x)$ based on a study by Matsumoto and Tsuda [1987, 1988]. Here, we adopt the BZ map (2.1), discussed in Chap. 2, as $f(x)$ for the elementary dynamics.

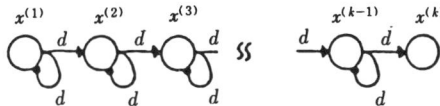

Fig. 5.3. The circles represent the maps f. A unidirectional coupling is adopted. A self-inhibitory coupling is added except to the map at the right edge. (I. Tsuda and H. Shimizu, Complex Systems (ed. H. Haken, Springer-Verlag, 1985) 240.)

172   5. Significance of Coupled Chaotic Systems to Biological Networks

At the left end $i = 1$, we introduce information by coupling it to a new dynamical system. Including this new information source, the equation of the whole CML becomes $x_{n+1}(i) = f(x_n(i), s(i)) + Dx_n(i-1) - D'x_n(t) + EY_n$ with $Y_{n+1} = G(Y_n)$. Here G is a nonlinear map.

First we discuss the case when the map $G(x)$ shows chaos. Here the result shown below is essentially independent of the choice of $G(x)$. It is basically true, even if Y is a random sequence. For example, Fig. 5.4 shows a result for the case that we chose the BZ map for $G(x)$. The numerical result here is analysed by the information theory in Sect. 2.7.

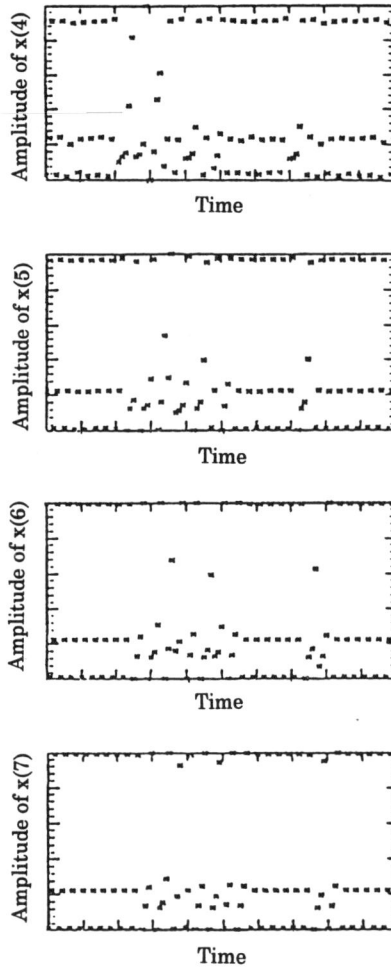

Fig. 5.4. The transmission of a chaotic burst from $x(4)$ to $x(7)$. $D = 0.12$ and $E = 0.2$. (I. Tsuda and H. Shimizu, Complex Systems (ed. H. Haken, Springer-Verlag, 1985) 240.)

Recall that there is almost no information loss in Fig. 2.9a. A similar result is obtained here. Information generated by the source dynamics $G(x)$ is transmitted along the chain at a constant transmission speed. The transmission speed is dependent on the bifurcation parameter $s(i)$ and the coupling constant D. In a BZ system, since the fluctuation of information flow is large and can shuffle information from neighboring elements, information creation at each element can overcome information loss. Microscopic information is efficiently transmitted to the neighboring site through the shuffling.

Next, we employ a periodic dynamics for $Y_{n+1} = G(Y_n)$. In other words, information in the form of a periodic oscillation is put into the chain (see also Sect. 2.3.3). As shown in Fig. 5.5, a complete image of the periodic oscillation is copied to the right end of the chain, in spite of the chaotic dynamics at intermediate lattice points in the CML. The period, amplitude and waveform of the periodic oscillation is copied to the endpoint of the chain. (It might be interesting to interpret this phenomenon as pattern recognition by chaotic dynamics.) To achieve this type of information processing, the following conditions are required.

Condition 1: To copy an oscillation state of period n, some element in the CML must have a period-n solution, and the input has to be applied to that element.

Condition 2: When the input is not applied, all elements fall onto a periodic state with some period by controlling the coupling parameter D, while the oscillation phases from element to element are shifted by a constant ratio (a metachronal phase shift).

Under the conditions 1 and 2, it is also possible to bifurcate the CML connection as in Fig. 5.6, so that the system forms a branched network. In this case, as many copies of periodic inputs as desired can be constructed. It is also possible to transmit input information to distant locations in space.

By designing a loop at the edge of a chain, memory as to the input information can be stored in this loop. Various patterns can also be memorized simultaneously, as is illustrated in Fig. 5.6. Once memory is established within a chain, it is robust against external noise. Therefore, noise-tolerant dynamic memory is constructed as a result of the information dynamics of this CML. The dynamic processing of information in the CML is based on a computation process in chaos.

Here we have discussed the information-processing ability of a CML that uses the BZ map as the local element. Chaos of the same type as in the BZ map (chaos with large fluctuations in information flow) is found in the nerve cells of mollusks [Hayashi et al. 1982] and in the olfactory bulb processing the olfactory information in rabbits [Freeman 1987; Skarda and Freeman 1987]. It is also interesting to note that BZ-type chaos has been discovered in a neural-network model for the successive retrieval process in mammalian brains [Tsuda et al. 1987; Tsuda 1990a, 1991a, b, 1992a, c, 1994a]. Coupled

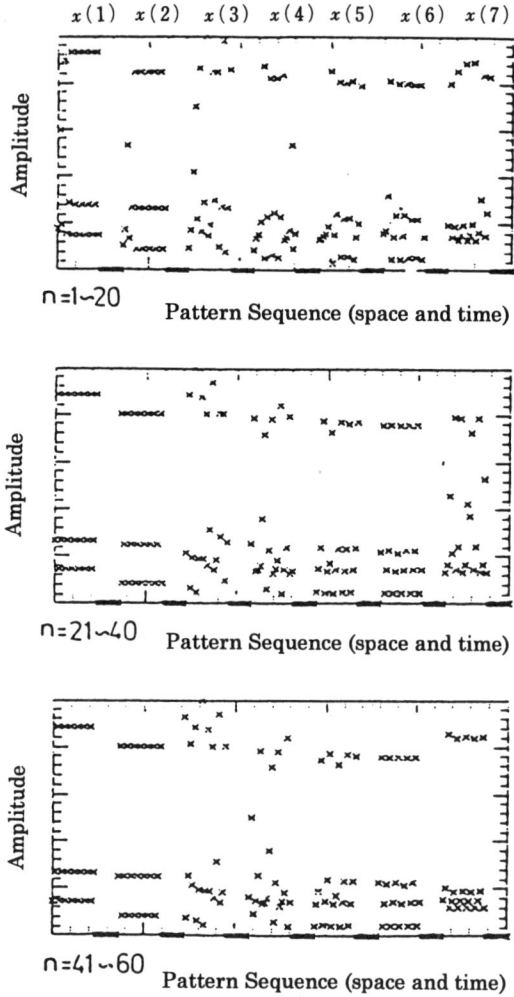

Fig. 5.5. A copy of the period-4 oscillation at $x(7)$ when the period-4 oscillation is input at $x(3)$ which showed a period-4 oscillation before coupling. In this figure, n indicates the discrete time. The amplitude of the period-4 oscillation appearing at $x(7)$ is almost the same as that of the input oscillation. The case for $D = 0.12$ and $E = 0.2$ is shown. (I. Tsuda and H. Shimizu, Complex Systems (ed. H. Haken, Springer-Verlag, 1985) 240.)

chaos of BZ-type will be relevant to information processing of the brain and computation machines.

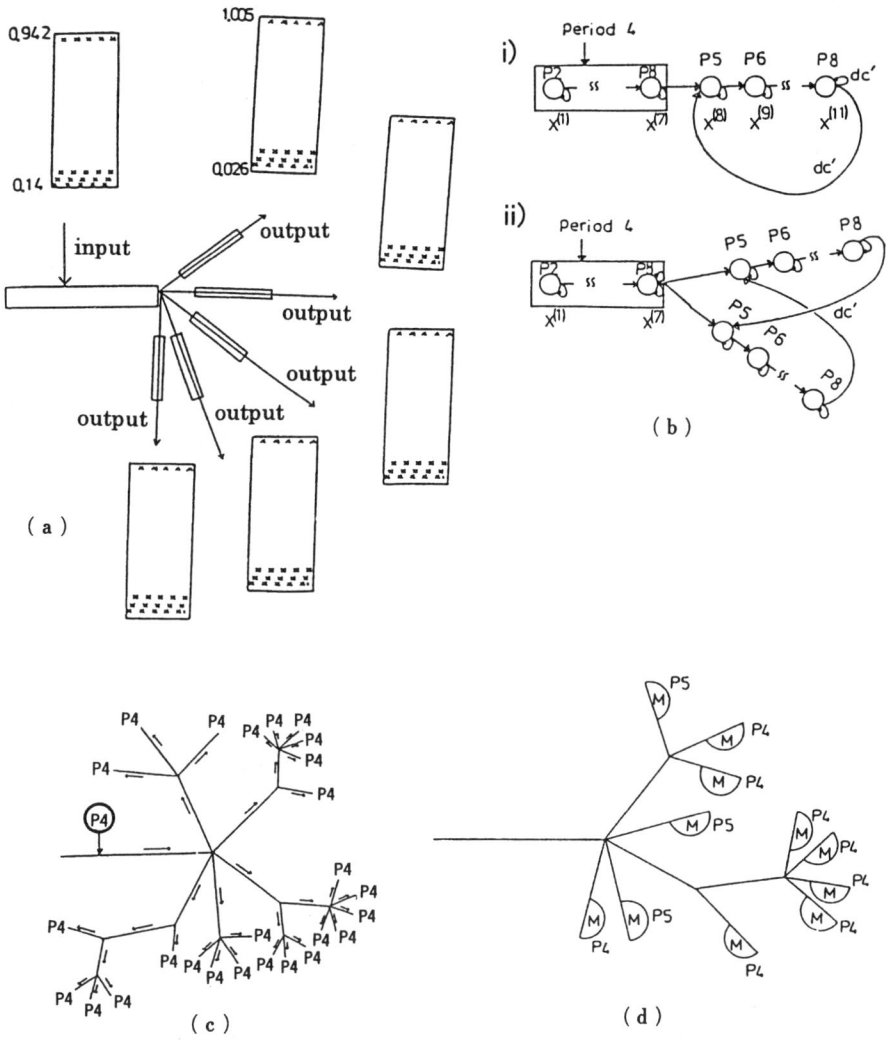

Fig. 5.6. (a) An arbitrary number of copies is possible. To obtain this, as in this figure, the edge of the system is bifurcated by coupling the subsystem at the right side of the element having the input with the edge of the system. The case of the coupling constant at the bifurcation point $d_c = 0.05$ is shown ($D = 0.12, E = 0.2$). (b) A way of constructing dynamic memory: pn shows the period-n oscillation. (c) An arbitrary number of memories is also possible by making a branched channel. (d) An arbitrary number of memories of different periods (here periods 4 and 5) can also be made. (I. Tsuda and H. Shimizu, Complex Systems (ed. H. Haken, Springer-Verlag, 1985) 240.)

5.4 Design of Coupled Maps and Plastic Dynamics

So far we have studied coupled maps with nearest-neighbor, uni-directional, and global couplings. In general, by choosing suitable coupling forms, it may be possible to design coupled maps with various information-processing abilities. Still, we have not clearly addressed the question yet of how one can design such coupled maps, or how one can train coupled maps to learn certain inputs.

In traditional neural networks, the learning process is carried out by adjusting the coupling strengths. There, the 'Hebbian learning' method is adopted, where a coding is assigned to the firing patterns $x(i)$, while the plastic change and memory are assumed to be assigned to the interaction strength between two elements i and j. It is not yet certain whether the Hebbian learning method is adopted in real neural systems, but at least the learning method works in some situations. Still, such a learning method may not be relevant to coupled maps. Rather, a new type of learning mechanism may be required here. In this respect, the learning process with global and competitive dynamics of synapses by Edelman [1987], or globally coupled evolutionary dynamics by Eigen and Schuster [1979] may be useful.

As a preliminary step, we first study a GCM with heterogeneous coupling. Here we extend the previous GCM by allowing for inhomogeneous parameters by elements, instead of identical parameters for all. There are two possibilities here: either we choose a distributed $a(i)$ instead of the constant nonlinearity a, or we choose a distributed $\epsilon(i)$ instead of the constant coupling ϵ. Indeed, both give similar results. Here we show the latter case [Kaneko 1994a] given by

$$x_{n+1}(i) = (1 - \epsilon(i))f(x_n(i)) + \frac{\epsilon(i)}{N}\sum_{j=1}^{N} f(x_n(j)). \qquad (5.2)$$

In Fig. 5.7, we have plotted a snapshot of $x_n(i)$ for a system with a homogeneously distributed coupling over $[\epsilon_{\min}, \epsilon_{\max}]$ (i.e., $\epsilon(i) = \epsilon_{\min} + (\epsilon_{\max} - \epsilon_{\min})(i/N)$). As can be seen, the clustering structure depends on the coupling strength at each element. Elements with a large coupling ($\epsilon > 0.22$) form a single synchronized cluster, while the number of clusters increases successively with the decrease of the coupling strength at the element. Such clustering bifurcation looks similar to the phase change in Chap. 4. However, this is not a trivial extension. The clustering change in Chap. 4 is a bifurcation that occurs when changing the system parameters, while the change here is included in a single network system, where all the elements therein are connected by a unique mean field. Still, "internal" bifurcation among elements occurs here. At the edge parameter region between clustered and desynchronized states ($\epsilon \approx 0.12$ in Fig. 5.7), the motion includes chaotic itinerancy. Desynchronized bursts emitted from the elements with smaller $\epsilon(i)$ flow to elements with larger $\epsilon(i)$, where clustering can change in time.

The above result suggests the possibility that external information can be embedded into several clustered states, by changing the coupling ϵ in accordance with the input. So far, in most neural-network studies, the information is coded in the firing pattern $x(i)$, while the learning process is carried out

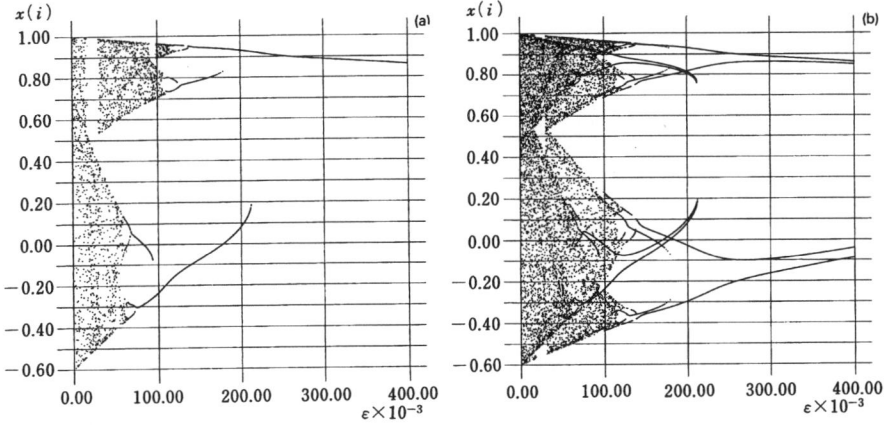

Fig. 5.7 a,b. Snapshot of the globally coupled map (5.2) with distributed coupling values. ($a = 1.6$, $N = 10^5$). The coupling $\epsilon(i)$ is ordered as $\epsilon(i) = 0.4(i/N)$. (a) snapshot of $x_n(i)$ at the time step $n = 5000$ is plotted as a function of i. (b) Successive snapshots $x_n(i)$ at the time steps 5001, 5002, 5003, and 5004 are overlaid. (From K. Kaneko, *Physica D* **75** (1994a) 55, with the permission of the publishers.)

through adjustment of the coupling strength between the elements i and j. On the other hand, it was proposed recently that the information is coded in the correlation between the elements $x(i)$ and $x(j)$ [Malsburg 1986; Aertsen and Vaadia 1992; Tsukada 1992; Aertsen 1993; Aertsen *et al.* 1994; Fujii *et al.* 1996]. Then it may be interesting to pursue this "converse" limit for the code and learning. In other words, let us assume that the learning rule is assigned to the coupling $\epsilon(i)$ at i, and not to the interaction between i and j [Kaneko 1994a]. Thus we take the following simple example

$$x_{n+1}(i) = (1 - \epsilon_n(i))f(x_n(i)) + \frac{\epsilon_n(i)}{N} \sum_j f(x_n(j)), \qquad (5.3)$$

where $\epsilon_n(i)$ is in-(de-)creased according to inputs at the i-th element. To be specific, we take the following dynamics: first increase the coupling to the mean field, according to the input $s_n(i)(> 0)$ by

$$\epsilon'(i) = \epsilon_n(i) + \gamma s_n(i), \text{ with } \gamma > 0 \qquad (5.4)$$

and then rescale the coupling so that the average of $\epsilon(i)$ is conserved:

$$\epsilon_{n+1}(i) = \epsilon_0 \times \frac{\epsilon'(i)}{\sum_j \epsilon'(j)}. \qquad (5.5)$$

Indeed the latter equation is introduced only to suppress the over-growth of $\epsilon_n(i)$, and may be replaced by other forms. This gives global competition for the coupling term. What we need here is a mechanism for (a) the suppression of an indefinite increase of the coupling $\epsilon(i)$ and (b) competition among elements for an increase in their coupling.

When common inputs are applied to elements $i_0 \leq i \leq i_1$, the degree of synchronization of the oscillations $x_n(i)$ for $i_0 \leq i \leq i_1$ increases. Depending on the input strength, two elements i, j with common inputs often synchronize completely. After inputs are eliminated, a pair of (almost) synchronized elements remains coherent or correlated. We have thus achieved clustering according to inputs [Araki 1994]. This clustering is often preserved even if the coupling $\epsilon_n(i)$ is later restored to be homogeneous.

The formation of clustering according to inputs works well if the parameters a and ϵ_0 are chosen such that the number of clusters is not far from the expected cluster number by inputs. If the number of clusters at the corresponding a and ϵ_0 values is smaller than that required by the inputs, some of the groups are fused into the same cluster. The relationships between some input groups are self-organized by the internal dynamics. Generally, the above clustering is formed by a balance between the internal dynamics and the external inputs.

There is a recent report of an interesting experiment by Hayashi and Ishizuka [1995], where the synchronization among neurons is increased when a rat is continuing a task with a feedback to the brain. In our model this corresponds to the growth of $\epsilon(i)$ for all elements. When further continuing the task, the synchronization reaches its maximum, and the rat's brain goes into an epileptic state, where the ability to process information is lost. In our model complete synchronization would be attained if the suppression term ((5.5)) were not included. To avoid such a catastrophe, we have to introduce some mechanisms to suppress the increase of ϵ in our model. We may expect that such a mechanism exists in the real brain, and that its breakdown leads to epileptic states.

5.5 Construction of Dynamic Many-to-Many Logic and Information Processing

It is important to construct a theoretical framework for studying situations where many 'causes' are mutually related and lead to several 'effects', i.e. situations where a one-to-one causal relationship cannot be extracted.

The flexible dynamics of the hierarchical structures discussed in Sect. 4.5 will also be relevant to the study of the brain. For example, the existence of functional modules is often assumed there. Although it is certainly convenient to assume the existence of modules which have certain functional roles at some levels, it is questionable whether these are in the form of fixed and rigid structures. It might be better to assume that the module has an approximate structure which is valid only within some limited time span, and to study the formation and transformation of the module in terms of dynamical relationships.

This problem of many-to-many relationships is also relevant to the design of a computer. For example, as a computer gets smaller, it becomes more difficult to assume that each element of the device will work independently. Global electromagnetic interaction among the elements is no longer negligible. In the trend of the present technology, engineers try to minimize electromagnetic interaction among elements. However, it will be necessary in the future to design a computer that positively makes use of such global interaction.

Studying dynamic many-to-many relationships will be a basis in designing such computers. In other words, we need a logic where many causal relationships run in parallel interfering with each other. This logic may have some things in common with the present parallel computer. However from our standpoint, in contrast with the use of independent elementary processes in a parallel computer, we will take interactions into account positively. Interference among elementary processes may sometimes lead to logically incorrect operations. This interference, however, can lead to a logical jump in the process. Such a jump might arise from some relation between the internal dynamics and the external inputs, such as the synchronization between the two. With this jump, one can construct information processing including its creation, that is not passively determined by external inputs.

A related motivation can be seen in the study of probabilistic logic by von Neumann [1956]. He discussed how accurate computation could be possible by an ensemble of unreliable elements. Conversely, by a logic that works under nonlinear many-body dynamics, we aim to construct an information-processing system that assigns a higher priority to continual operation even if some errors occur than to perfect accuracy.

In Chap. 1, we discussed some recent studies on real-number computation theory. A computer based on a CML may possibly perform parallel computation on real numbers. The present computer of von Neumann type has the same computability level as a cellular automaton. It is possible that a CML-based computer has a higher computability. It is an important problem for future study to construct a CML-based computer and to reveal its computational ability.

5.6 Implications to Biological Networks

The issues outlined in Chap. 4 will be important not only to brain and computer information processing but also to biological networks in general. This topic was already mentioned in Sect. 1.6, and some applications to ecological networks and cell differentiation were discussed in detail elsewhere [Kaneko and Ikegami 1998]. Since this topic is essential to complex systems, we will briefly discuss it here.

From our point of view, there are three interesting issues concerning the evolution of biological networks [Kaneko 1994c]: (1) how does a higher level in an organization emerge? Examples include the origin of life, the origin of eucaryotic cells, and the origin of multicellular organisms; (2) the origin and maintenance of diversity; (3) the origin and evolution of complexity. We briefly discuss why clustering and chaotic itinerancy are possibly relevant to these issues.

5.6.1 Prototype of Hierarchical Structures

In the partially ordered phase of the GCM or a frozen random phase of the CML, attractors form hierarchical structures spontaneously. In a GCM, there are sub-clusters within a cluster at a larger scale, and smaller-scale sub-clusters within these, and so forth. This was already mentioned in connection with categorization, but it may also be relevant to the origin of hierarchical structures in general, observed for example in ecological and social systems.

It is known that such hierarchical structures appear in spin-glass systems, which have been shown to display "ultrametricity" [Mezard *et al.* 1987]. As mentioned already, the GCM and spin glasses have some features in common. However, as noted in Chap. 4, there are two differences: (1) in a spin glass, since the interactions are given in advance at random, systems are not uniform, whereas in our example, identical elements have identical interactions. In spite of this uniformity, the elements spontaneously form a hierarchy; (2) the hierarchical structure is not fixed. The oscillation of each element continues, and the degree of synchronization among the elements may change in time, as e.g. in chaotic itinerancy.

In ecological, economical and social systems, there is global competition among dynamical elements. Thus the problem therein can be regarded as belonging to the same class as the issues discussed here. In this sense, the hypothesis "chaos gives a source of the hierarchy of natural and social systems" is worthwhile examining seriously in each individual field [Kaneko 1990c].

5.6.2 Prototype of Diversity and Differentiation

A typical origin of diversity is seen in cell differentiation. By cell division, each cell reproduces itself with differentiation and forms a network and a cell society. Nonlinear dynamics is involved in biochemical reactions within each cell, and cells interact with each other. On the other hand, in the GCM, identical elements form several clusters that show different dynamical behaviors. Thus, the clustering may provide a principal mechanism for the differentiation of identical elements. Indeed, along this line, a theory of cell differentiation was proposed by Kaneko and Yomo [1994, 1995, 1997, 1999]. The model consists of intracellular chemical-reaction dynamics, cell-to-cell interaction through a medium, and cell division. As for the internal reaction dynamics, auto-catalytic reactions among the chemicals is chosen. Such auto-catalytic reactions are necessary for producing chemicals in cells for reproduction [Eigen and Schuster 1979]. Autocatalytic reactions often lead to nonlinear oscillation in chemicals. Here we assume such oscillations in the intra-cellular dynamics.

As the interaction mechanism, diffusion of chemicals between cells and their surroundings is chosen. The surrounding medium is assumed to be spatially homogeneous to avoid the complication of spatial-pattern formation. In other words, each cell is coupled to all the other cells. For some models, we assume a nutrition chemical which is actively transported into the cells. The rate of active transport depends on the concentration of some other chemicals in the cell. For other models, only diffusion is taken into account.

The cell divides according to its internal state. In a number of models, we assume that some products are accumulated through chemical reactions, and that the cell divides into two when the accumulated product rises above some threshold. In other models, the cell volume depends on the amount of chemicals in the cell. A cell divides into two when the volume becomes twice the original volume.

Of course, there is a large variety of choices for the chemical-reaction network. The observed results, however, do not depend on the details of the choice, as long as the network allows for cell division, i.e. for growth in the number of cells. Note that the network is not constructed to resemble an existing biochemical network. Rather, we try to demonstrate that some important features in biological systems are the natural consequences of them being systems with internal dynamics, interaction, and reproduction. From the study we extract a universal logic underlying this class of models.

From a dynamical systems point of view, the model has a novel feature not included in the globally coupled map. Here the number of degrees of freedom varies via cell division. When a new cell is born, we need additional degrees of freedom to indicate the chemical concentrations of the new cell.

Based on the results of extensive simulations, Kaneko and Yomo [1994, 1997] have proposed the concept of "isologous diversification". Starting from a single unit, we have found the following scenario for the development of a society of diversified units.

(1) Synchronous oscillations of identical units:

Up to a certain number of cells (which depends on the model parameters), cells dividing from a single cell have the same characteristics. Although each cell division is not exactly symmetrical due to the accompanying fluctuations in the biochemical composition, the phases of the oscillations in the concentrations, and as a result, the timing of the cell division, remain synchronous for all cells. Synchronous cell division is also observed for the cells in the developmental process of mammals up to eight cells.

(2) Differentiation of the phases of oscillations of the internal states:

When the number of cells rises above a certain (threshold) value, the state with identical cells is no longer stable. Small differences introduced by the fluctuations start to be amplified, until the synchrony of the oscillations is broken. Then the cells split into a few groups, each having a different oscillation phase. Each cluster has a different phase, and the cells belonging to each cluster are identical in phase. This diversification in the phases, however, cannot be called cell differentiation, because the time average of the biochemical concentrations reveals that the cells are almost identical. At this stage, only in the phases of oscillation are the cells different. The mechanism of this clustering has already been discussed in Chap. 4.

(3) Differentiation of the amplitudes of internal states:

With the further increase of the cell number, the average concentrations of the biochemicals over the cell cycle become different. The composition of the biochemicals as well as the rates of the catalytic reactions and the transport of the biochemicals become different for each group. The orbits of the chemical dynamics plotted in the phase space of biochemical concentrations lie in a distinct region within the phase space, while the phases of the oscillations remain different for cells within each group.

Hence distinct groups of cells are formed with different chemical characters. Each group is regarded as a different cell type, and the process to form such types is called differentiation. In biological terms, this third stage is no other than the division of labor of several biochemical reactions in the cell, since the use of chemical resources is different for each group.

Disparity in chemical activities is sometimes observed here. The speed of division of active cells is also differentiated by cells. The active cells may correspond to germ cells, while others correspond to somatic cells. Here, somatic cells are also differentiated according to the concentrations of the contained chemicals.

(4) Transfer of the differentiated state to their offspring by reproduction:

After the formation of cell types, the chemical compositions of each group are inherited by their daughter cells. In other words, chemical compositions of cells are recursive over divisions. To see this recursivity, we have plotted the "return map", that is the relation between the chemical averages of the mother and daughter cells. In the return map, the recursivity is represented by points lying around the diagonal ($y = x$) line. Initially, chemical compositions change by divisions, but later they settle down on almost fixed values. The biochemical properties of a cell are inherited by its progeny, or in other words, the properties of the differentiated cells are stable, fixed or determined over the generations.

It is important to note that the chemical characters are "inherited" just through the initial conditions of chemical concentrations after the division. We have not explicitly implemented any external mechanisms to get such cellular memory. The determination of cells occurs at this stage, because daughters of one cell type preserve their type. After several divisions, initial conditions of units are chosen such that the next generation is of the same type as its mother cell. Thus a kind of memory is formed. This memory lies not only in the internal states but also in the interactions among the units.

(5) Hierarchical organization:

As the cell number increases, further differentiation proceeds. Each group of cells differentiates further into two (or more) subgroups. Thus, the total system consists of units of diverse behaviors, forming a heterogeneous society.

The most interesting example here is the formation of stem cells [Furusawa and Kaneko 1998]. This cell type, denoted as 'S' here, either reproduces the same type or forms different cell types, denoted for example as type A and type B. Then after division events $S \to S, A, B$ occur. Depending on the adopted chemical networks, the types A and B replicate, or switch to different types. For example $A \to A, A1, A2, A3$ is observed in some network. This hierarchical organization is often observed when the internal dynamics has some complexity such as chaos.

The differentiation here is "stochastic". The choice for a stem cell either to replicate or to differentiate looks stochastic as far as the cell type is concerned. This stochastic branching is accompanied by a regulative mechanism. When some cells are removed externally during the developmental process, the rate of differentiation changes so that the final cell distribution is recovered. To such regulated differentiation, the two faces of chaos, determinism and stochasticity are essential.

As the units reproduce, the competitive interaction among them becomes stronger, resulting in successive diversification of their behavior. Generally speaking, identical elements tend to become diversified through the interplay of internal dynamics and interaction, and through reproduction. The first three stages listed above are the consequences of dealing with globally coupled dynamical systems, as has been discussed. The emergence of the fourth stage, however, is due to the reproduction of units. The initial conditions after reproduction are selected such that the units keep the same character. We believe that this emergence of recursivity or memory is an important feature of coupled dynamical systems with reproduction, and thus essential to information flow and memory in biological systems.

A theory which clarifies the logic for cell differentiation was proposed [Kaneko 1998a, b]. The developmental process comes through diversification, formation of discrete types that are recursive, and rule generation. It is also interesting to note that the appearance of germ-line segregation, stem cells, determination, and tumor cells are a natural consequence of a system with internal dynamics, interaction and reproduction. It is stressed that the development process in this scenario is rather robust against molecular fluctuations or macroscopic perturbations such as damage or removal of cells. Several other features in cell society are also naturally explained by the theory [Kaneko and Yomo 1997, 1999; Furusawa and Kaneko 1998; Kaneko 1998a, b].

5.6.3 Formation and Collapse of Relationships

As was discussed for the globally coupled map, the collective motion of an
ensemble of elements often displays behavior different from the dynamics of
each individual element. In the GCM, even if an individual element shows
chaotic motion, the mean-field dynamics can show a different type of chaos
or, in some cases, quasiperiodic motion. Such collective motion implies some
cooperative motion among the elements, and is seen in biological systems. For
example, one type of ants (*Lepthorax acevorum*) repeats active and resting
periods, in turn. Time series of the activity of a single ant and an ensemble
of ants have been studied by Cole [1991]. The time series of a single ant is
found to show chaos with a dimension of about 2.5. On the other hand, the
average activity of an ant colony shows a regular rhythm. Indeed the average
dynamics is quasiperiodic on a three-dimensional torus. Although the form
of coupling among the ants is unknown, this report seems to provide us with
an example for collective motion of chaotic elements as in the GCM (see
also Goodwin [1994]). In the model of formation of ant paths with passive
motion by pheromones (as discussed in Sect. 1.8), such collective dynamics is
not possible. Hence, this experiment demonstrates collective motion due to
active elements such as chaos.

The motion of the above ants is, of course, observed in a rather ideal
situation. Ants in the real world exhibit more complicated motion, by sometimes assembling together and then separating. In the motion of fishes and
birds, clusters are formed spontaneously and then collapse. The phenomenon
of chaotic itinerancy is significant in the study of such cluster dynamics. In
fact, in a 'coupled map gas' model [Shibata and Kaneko 1995] in which each
element has internal dynamics and moves around in space while interacting
with other elements, clusters repeatedly form and collapse. Often clusters
exchange their elements, and move around the space. In this coupled map
gas, dynamics to change relationships among elements are formed through
the change of synchronization in motion.

5.6.4 Clustering in Hypercubic Coupled Maps; Self-organizing Genetic Algorithms

Let us discuss the dynamics of relationships, by choosing the population
dynamics model mentioned in Sect. 3.5. There a coupled map on a hypercubic
lattice is studied, given by $(1 - \epsilon)f(x_n(i)) + (\epsilon/K)\sum_{j=1}^{K} f(x_n(\sigma_j(i)))$, where
$\sigma_j(i)$ is a "type" whose j-th bit is different from species i and ϵ a mutation
rate. The process of mutation is characterized by a diffusion process in the
space of genes. If the "gene" space is represented by a bit space (such as
$i = 0010111$) as is often the case for genetic algorithms [Holland 1986], the
single-point mutation process is given by a flip-flop $0 \leftrightarrow 1$. The population
(density) of each type i is given by $x(i)$ and the mutation process provides
the diffusion in hypercubic bit space.

In this model, we have again found the formation of synchronized clusters as in Chap. 4 (i.e., $x_n(i) = x_n(j)$ for two elements i and j in the cluster). In the present case, the split into clusters is organized according to the hypercubic structure. Examples of such clusters are given below.

(a) 1-bit clustering

Two clusters with synchronized oscillation are formed. Each of the clusters has $N/2 = 2^{k-1}$ elements, determined by the bit structure. For example, elements may be grouped into two clusters with bit sequences **0*** and **1*** (* meaning that the symbol there is either a one or a zero), each of which has 2^{k-1} species. This clustering is formed by cutting the k-dimensional hypercube along a hyperplane.

In the genetic algorithm, irrelevant bits are initially considered as "don't care" bits represented by "*". Here, such bits are spontaneously created during the temporal evolution.

(b) 2-bit clustering

Depending on initial conditions and parameters, both the number of clusters and the number of bits relevant to clustering can be larger than in case (a). An example is a 2-cluster state with 2 relevant bits under the XOR (exclusive-or) rule. Here the elements split into the groups (i) **10*** or **01*** and (ii) **00*** or **11***, for example.

(c) Parity-check clustering

Elements split into two groups according to the parity of the number of 1's in each bit representation. For example, elements split into two clusters as follows (i) 000, 011, 101, 110 and (ii) 001, 010, 100, 111, for $k = 3$. The clustering thus implies a parity check. It is a hypercubic version of the (one-dimensional) zigzag or (two-dimensional) checkerboard pattern [Kaneko 1989b].

Besides these examples, attractors with many clusters are also found. Most of these states are constructed by combining the above clustering schemes. For example, 4 clusters with two relevant bits are found as a direct product state of case (a). Here the hypercubic space is cut by two hyperplanes. Elements split into four clusters, for example, coded by 01*****, 10*****, 11*****, and 00*****.

Here we have to note that not all partitions are possible in the present case, in contrast with the GCM case. Even if we start from an initial condition with an arbitrary clustering, the synchronization condition ($x(i) = x(j)$ for i, j belonging to the same cluster) is not satisfied at the next step for most such initial conditions. Not all possible partitions can be a (stable or unstable) solution of the evolution equation.

As is discussed, the present result opens up the possibility of automatic genetic algorithms. Relevant bits are spontaneously formed. Furthermore, when chaos in each element is sufficiently strong, we have found a chaotic itinerancy state where the relevant bits change according to the temporal evolution. For example, elements spilt into two clusters according to the bit

structure 0**** and 1****, for a while, and then switch to a clustering with
0 and **1**, and so forth.

With the introduction of external inputs to each element, it is also possible
to have a clustered state depending on the external information [Kaneko
1994a]. Relevant information is extracted through this process spontaneously,
and stored as a relevant bit in the clustering.

5.6.5 Homeochaos

Coupled chaos provides a source of diversity and complexity. In connection
with high-dimensional chaos, Kaneko and Ikegami [1992] proposed a hypothesis of homeochaos for dynamic maintenance of diversity.

As mentioned in Chap. 1, the origin and maintenance of complexity and
diversity are important when considering the evolution of a biological network
such as an ecosystem. So far we have shown that clustering in the GCM can
be a source of diversity. As for the origin of complex behavior, evolution to
the edge of chaos has often been discussed. The hypothesis is that the state
between chaos and order has a high information-processing ability and that
evolution to it occurs. Although this hypothesis may be effective for explain-
ing the evolution of complex behavior, it cannot account for the maintenance
of diversity. For problems involving many degrees of freedom with complex
relationships among elements, how can diversity and complexity stably be
maintained?

For the coupled map studies in the previous sections, the parameters of
our models are given in advance, and we have investigated the effect on the dynamic behavior when changing their values. In some instances in a biological
evolution, the system parameters can reasonably be assumed to change with
time (with a slower time scale). Here we briefly discuss an extension of the
hypercubic coupled map to population dynamics by also allowing mutation
of the mutation rate. As mentioned, the mutational process is described by
a diffusion process in the gene space, represented by a bit space for simplicity. Kaneko and Ikegami considered an example of population dynamics
with hosts and parasites by studying the following coupled map where the
population of the hosts is denoted by $x(i)$ and that of the parasites by $y(i)$:

$$x_{n+1}(i) = (1 - \epsilon)f(x_n(i), y_n(i)) + \frac{\epsilon}{k}\sum_{j=1}^{k} f(x_n(\sigma_j(i)), y_n(\sigma_j(i))), \qquad (5.6)$$

$$y_{n+1}(i) = (1 - \epsilon)g(x_n(i), y_n(i)) + \frac{\epsilon}{k}\sum_{j=1}^{k} g(x_n(\sigma_j(i)), y_n(\sigma_j(i))), \qquad (5.7)$$

where f, g give the local population dynamics, while $\sigma_j(i)$ is a species whose
j-th bit is different from the species i (with only one bit difference), and

k is the total bit length of the species (totally there are 2^k types for each group) [Kaneko 1994a]. The functions f and g are chosen to be $f(x, y) = c_h x \exp(-\beta y)$ and $g(x, y) = c_p x(1 - \exp(-\beta y))$ for the May–Hassell model [Hassell et al. 1991], while some other variants are also adopted [Ikegami and Kaneko 1992].

In the above model, we have again found the formation of synchronized clusters. By choosing the parameters of nonlinearity in the map f, g and the mutation rate suitably, chaotic itinerancy of clustered states is again observed. The dynamics shows successive change of clustered states.

Next we added the mutation of mutation rate to the model by assuming that the same type i can have different mutation rates (which also change by mutation). The mutation rate of the mutation is given by the mutation rate itself. Since the null mutation rate is not changed further by the mutation, it may be expected that the system evolves to a state with a null mutation rate. Indeed, if there is no coupling among the two groups (i.e., $f(x, y)$ does not depend on y, nor $g(x, y)$ on x) or the parasite does not exist at all ($y(i)$ is set to zero), the mutation rate finally falls to zero with time. This is also true if the growth of type i depends on the fitness landscape (i.e, the growth function f depends on i).

However, the decline in mutation rate does not occur when there are interactions between hosts and parasites. In this case, the mutation rate is sustained at a finite level, which depends on the interaction strength. Hence a variety of types coexists for hosts and parasites.

Then what type of dynamics is selected through the change of mutation rates? Here we have found that the system shows chaotic itinerancy. In other words, the system alternates between desynchronized and synchronized states. Indeed, over a fairly long time span, the system forms the following feedback process [Kaneko and Ikegami 1992; Ikegami and Kaneko 1992; Kaneko 1994a].

(i) Increase of mutation level leads to synchronization of population oscillation;

(ii) synchronization leads to the decrease of mutation level;

(iii) decrease of mutation level leads to the split of synchronized clusters;

(iv) split of clusters is associated with irregular temporal dynamics and leads to increase of mutation level, and then to (i).

Through this feedback process the system stays around the ordered and desynchronized states, and chaotically itinerates over several clustered states.

The same behavior is observed, when the growth rates in the functions f and g depend on the type i, and the fitness has a rugged landscape. Summing up the results, we have found that the population of each type shows chaotic oscillation, but the amplitude is much smaller than in the case for a single pair of host–parasite dynamics. For example, if the mutation rate is set to zero, the dynamics of each type of host/parasite is disconnected, and it can show strong chaos for some pairs. When the mutation rate is sustained at a high level, however, chaotic instability is shared by almost all degrees of freedom. Indeed the dynamics leads to weak high-dimensional chaos.

The mutation rate corresponds to the coupling parameter ϵ in the previous sections. Thus the change of mutation rates leads to the change of the clustering of the population dynamics. The diversity, allowing for the coexistence of many types, is sustained dynamically through the interaction. By generalizing these numerical results we proposed the "homeochaos" hypothesis for sustaining dynamic stability maintaining diversity.

The following four points are the essence of homeochaos:

(i) Weak chaos: Homeochaos is supported by weak chaos. Quantitatively, the maximal Lyapunov exponent is slightly positive, and close to zero. The amplitude of oscillations is not too large. This weak chaos, for example, is essential to avoid a too violent change or extinction in the population dynamics.

(ii) High-dimensional chaos: Homeochaos is represented by high-dimensional chaos. "High-dimensional" here means that the number of relevant degrees of freedom is large. There are many positive Lyapunov exponents, although they are close to zero.

(iii) Chaotic itinerancy: The system itinerates over several ordered states successively in time. In the above example, each ordered state is composed of only a few clusters, and the chaotic itinerancy is sustained by a feedback mechanism which changes the number of degrees of freedom. Indeed item (iii) is closely connected with (i) and (ii). If chaos were too strong, the oscillations of many elements would be desynchronized, and many positive Lyapunov exponents of large magnitudes would exist. If chaos were completely suppressed, clustering with small numbers of clusters would follow. Chaotic itinerancy with partial clusterings can provide weak and high-dimensional chaos, as postulated by (i) and (ii).

(iv) Dynamic stability against external perturbations: Homeochaos provides dynamic stability for a complex network. The robustness of our homeochaos can easily be seen by introducing an external perturbation to the population dynamics. When the dynamics follows low-dimensional strong chaos, the population often drops to a level very close to zero. By applying an external perturbation, the population may go to zero, and extinction follows. On the other hand, the weak chaos in homeochaos makes the oscillation amplitude much smaller, and the population remains far from zero. Hence extinction is avoided.

Indeed this robustness is attained through (i)–(iii). The suppression of strong instability is supported by (i), and by distributing the strong chaotic instability to many modes with the process (ii). When the system becomes unstable with a rather chaotic motion, a switch to a different ordered state occurs through the mechanism (iii), eliminating the instability again. In an ecological system, the coexistence of many species is not easy, as has been discussed by May [1973] with regard to the instability of static states in a random network. Thus it is important to search for a dynamical mechanism which allows for diversity in a system with interacting population dynamics.

Our proposal here is that the coexistence of many species is dynamically sustained as homeochaos. A typical example of a complex ecological network can be seen in the species of the tropical rain forest. The ecology there consists of a huge number of species. Several studies on the tropical rain forest have shown that its complex ecological dynamics is in a dynamic rather than a static state [Connel 1978; Kikkawa 1990], although detailed dynamical studies are not yet available. It is hoped that future studies will measure the population dynamics of the species in the tropical rain forest, which, we believe, is in a homeochaotic state.

In real ecological data, it is of course not easy to estimate the Lyapunov exponents. Recently Ellner and Turchin [1995] have examined several data sets of annual populations of some animals in fields (within Europe and Northern America), as well as the population of diseases such as measles in some cities, and estimated the Lyapunov exponents. They have not found many instances that provide evidence for positive exponents, but their data do suggest frequent appearance of exponents near zero. Although it is difficult to draw any definite conclusions from their analysis, it is interesting to note that the dominance of weak chaos in their estimates is consistent with our homeochaos hypothesis.

An important factor in the evolution dynamics is the change in the number of degrees of freedom themselves. When a new species appears, its population must be included as a new variable. Thus the number of variables can change in time. In conjunction with those cases of chaos where the number of degrees of freedom can change, the concept of "open chaos" was proposed [Kaneko 1994a, c]. Although for a system with a varying phase-space dimension one cannot use the definition of chaos in its strict sense, the orbital instability similar to chaos leads to changing numbers of degrees of freedom in a class of "open" dynamical systems. This is why we call such a class 'open chaos'.

As the population dynamics becomes chaotic, the probabilities for extinction and emergence of new species increase. After a regime of rapid change, the population is reorganized into a stationary state. The system switches alternately between a stationary state of regular dynamics with a fixed number of degrees of freedom and a transient regime with chaotic dynamics and a change in the number of degrees of freedom. This "meta-dynamics", analogous to chaotic itinerancy, supports stability with diversity, since the instability is weakened by the change in the number of degrees of freedom successively. The meta-dynamics again leads to homeochaos.

5.6.6 Summing Up

Our view on living organisms in the last subsection can be summarized as follows [Kaneko 1998b]. Organisms, if successful in reproduction, start to have stronger interactions with each other. Then, even when starting with the replication of identical organisms, their character starts to be differentiated by the nonlinear dynamics within each organism and the interactions. Thus a society with diversity is formed. Such diversity is sustained as homeochaos. This state makes diverse and flexible responses against external disturbances or inputs by making use of chaotic itinerancy. Along this line, Kaneko and Yomo are trying to extract a universal principle underlying cell biology [Kaneko and Yomo 1997, 1999; Kaneko 1998a, b].

6. Chaotic Information Processing in the Brain

6.1 Hermeneutics of the Brain

Let us call a machine involving simple stimulus–response rules a Cartesian automaton, and call a Cartesian automaton possessing a control system including feedback systems a Craik automaton [Johnson-Laird 1983]. By the studies of Karl von Frish, it turned out that a honey bee calculates the positions of flowers in relation to the position of the sun, and correctly sends such information to other bees. This seems to be a kind of sophisticated information processing, but its processing mechanism is essentially the same as that of a Craik automaton. The reason is as follows. Once the control system is set up, it performs correct calculations and outputs the results correctly, based on the present state and the purpose. In other words, the same result is output, provided the conditions are the same.

As seen not only in bees but also other insects, this type of information processing is also a general characteristic of the present digital computers. Actually, a Turing machine is a Craik automaton [Johnson-Laird 1983]. On the contrary, information processing in the human brain apparently differs from these automata. One of the differences is that the brain does not directly respond to external stimuli, but responds to images and symbols created inside the brain. The response to symbols depends on the history of experience, the present internal state and the present external stimuli. Let us call this history-dependent response an *interpretation*. The main concern in this section lies in the process of interpretation. What dynamics represents this interpretation process?

The meanings of the signals generated inside the brain can at present not be determined because neither the coding scheme nor the framework for the information representation are known. We have to investigate what meaning is for the brain. This has been pointed out by Shimizu [1992]. We have to see the system from within, but we have not yet obtained any appropriate framework to do so. Thus, we are obliged to interpret the signals. In this respect, this is a typical problem of endophysics as proposed by Rössler [1987]. In general, the problem of 'interpretation' is secondary in exophysics (conventional physics), but in endophysics and also in information sciences, it is essential, because the question of what information is meaningful for the system concerned and how it makes sense of the world is not solved in advance.

In this context, Tsuda [1984] proposed a 'hermeneutics of the brain',[1] stimulated by Marr's theory for vision [Marr 1982]. Marr's proposition can be summarized as follows. In the internal representation of visual information, starting from the gray-level's description and ending at the level of the construction of a three-dimensional model via a $2\frac{1}{2}$-sketch, visual information can be represented by symbols. Even in early vision where the gray-level's description may work, visual information can be represented by *primitives* as a symbol. Hence the brain does not directly map the external world. From this proposition follows the notion of the 'interpreting brain', i.e. the notion that the brain must interpret symbols generated by itself even at the lowest level of information processing. It seems that many problems related to information processing and meaning in the brain are rooted in the problems of the mechanisms of symbol generation and their interpretation.

In this respect, Tsuda has investigated the role of chaos for symbol generation, and proposed the notion of *chaotic hermeneutics* [Tsuda 1984; Nicolis and Tsuda 1985; Tsuda 1985, 1987a, b; Tsuda 1990a, 1991b]. Let us now ask ourselves how one can know the *activity* of the brain. One interesting option is to measure the spatiotemporal patterns of neural networks in the brain. According to the findings by Freeman and his colleagues, such spatiotemporal patterns are chaotic. It will be plausible to think that the chaos observed underlies the correlated dynamics of the brain and the body [Freeman 1995]. Here, chaotic dynamics provides a dynamic representation of the states of the neural networks. On the other hand, one cannot directly measure the meaning of such a representation. One simply interprets the neural activities, conceiving the process of 'mind' constructing logics and simulating the neural activities, whose process itself is interpretative.

In general, the level where chaotic dynamics works differs from the level where logics work [Freeman 1995], but we think what underlies logics may also be chaos. Chaos can be viewed as a symbol generator if one provides a code in phase space. Determined by the value of the bifurcation parameters and depending on the initial conditions, a chaotic dynamical system generates a symbol sequence. It is Markov in some cases and nonMarkov in others. Chaotic dynamical systems provide a grammar for the generation of symbol sequences. Could a symbol exist inside the brain? Indeed, we can use symbols as the means to represent internal activities of the brain, but this as such does not provide evidence for the presence of symbols inside the brain. Rather, a symbol is what exteriorizes the internal activities by means of activities in somatosensory and motor cortices. By the presence of this externalization the feedback from symbols to the brain activities will be

---

[1] Other theories on hermeneutics of the brain have been proposed by Arbib and Érdi. See for this [Arbib *et al.* 1998] and references cited therein. The readers are also recommended to refer Érdi and Tsuda [2000].

possible via somatosensory and motor cortices once more. Thus, an image of symbols will be embedded into the internal activities of the brain. Symbol generation with chaos may be incorporated in such a process. In this context, the relation between logics and chaos would be interesting. Nicolis and Tsuda [1985] put forward this problem, and recently Tsuda and Tadaki [1997] have proposed a theory for the interface between logics and chaotic dynamics and for the self-organization of symbol sequences.

Understanding the dynamics of the brain is equivalent to understanding the way the brain recognizes the world. Here it is valuable to consider a conventional question 'can the brain understand the brain itself?'. In our 'hermeneutics of the brain', it is stated that the way the brain recognizes the world is interpretative. The interpretation process is, in particular, clearly seen in vision. Suppose one looks at a table with a half-eaten apple, books randomly piled up, and a cup half-hidden by books. A half-eaten apple possesses only incomplete information as an apple, but one can recognize it as an apple. A cup half-hidden by books also provides us with only incomplete information, but one rapidly judges it to be a cup. Thus the brain transforms the incomplete information to the complete one concerning the object, supplementing the supposed lost information, and then interprets with the descriptions what the object means. Since this supplement cannot be determined uniquely, the process of interpretation must be repeated. If it converges, one feels to have understood the object well. The lack of uniqueness of supplement may lead to errors in interpretations. An equivocal figure or "trick" art is what utilizes this nonuniqueness.

The idea that the visual world in the brain is what is interpreted by the brain itself has been embodied by a neural-network model of the coupled-oscillator type [Shimizu and Yamaguchi 1987]. Shimizu further developed the notion of supplement mentioned above in the theory that the essence of interpretation is indeterminacy of information, which gives rise to a principle of the generation of meaning.

The supplement of information accompanied with interpretation strongly depends on one aspect one must take into account before starting interpretation, namely *preunderstanding* of the object. Preunderstanding depends on how one has recognized the world so far and how one wants to act on it. The brain can never be independent of the world, so that it cannot observe the world from without but *participates* in it from within. In this situation, epistemologically, the brain is obliged to interpret the world. The process of understanding such a brain is also interpretative.

The notion of the interpretation we are here dealing with overlaps in most parts with the so-called hermeneutics developed as the art for interpretation of texts such as the Bible, history, classical philosophy, and so on, but some parts differ. To clarify the difference, we will briefly review hermeneutics in the next section and will touch upon the difference in Sect. 6.3.

6.2 A Brief Comment on Hermeneutics (the Inside and the Outside)

Since it is not our concern to give a total description of hermeneutics, we restrict ourselves to the problems of the inside and the outside. For an exhaustive description, including scientific hermeneutics, see Noé [1993]. We think that the study of hermeneutics in complex systems studies will be essential when reexamining the conventional scientific methods and considering the methodology of social sciences in complex systems studies. The study by Noé precisely shows the situation where hermeneutics will become necessary also for natural sciences.

Hermeneutics is retroactive to the interpretation of the Bible, where it is inevitable to ask the meaning according to God of words in the text in which the words are written by humans. In ancient times, a poet was an interpreter for revelation. In ancient Japan, Hitomaro Kakinomoto was such a man [Matsuoka 1986]. In this century, philosophical hermeneutics was founded by Diltey [1924], Heidegger [1927], Gadamer [1976] and others as the art for understanding [Pöggeler 1977].

In the interpretation of literature, the meaning of individual sentences cannot be determined unless the meaning of the literature is *pre*understood. Such an indication of meaning is called a preunderstanding (Vorverständnis). Preunderstanding is based on the author's philosophy, individual history, understanding of life, the contents of references, and so on, namely a related whole history. On the other hand, a book consists of individual sentences which consist of individual words. Here, correlations between the structure and meaning of sentences emerge, and interactions between the interpreter with what is interpreted bring about the so-called hermeneutic circle. One should note that a hermeneutic circle differs from a vicious circle. The latter occurs in the irrational condition that the consequence is included in the premise, whereas the former occurs in the rational condition of the preunderstanding being altered in the interpretative process. This alteration accompanied with interpretation is called a hermeneutic spiral [Pöggeler 1977]. Hermeneutics becomes via this circle similar to dialectics if one views it as a developmental process of life.

The characteristic of hermeneutics as an art lies in the separation of interpreter and things interpreted. Primarily, the necessity of interpretation lies in the impossibility of grasping a whole. In hermeneutics, the situation that the interpreter is inseparable from the world is avoided, because the interpreter tries to understand the meaning of the world, based on an understanding of the relations between the parts of the world. One exceptional case seems to be Heidegger's notion of *commitment*, which indicates the participation of the interpreter in the world. Even in such a case, however, it seems to imply a participation from without, and the assumption of transcendental beings.

6.3 A Method for Understanding the Brain and Mind – Internal Description

The hermeneutic method for understanding the brain and mind Tsuda has proposed is based on the notion that the brain and mind cannot necessarily be described from without, and must be described from within. The brain maps the image of the world into itself, but the mapped image is altered by the activity of the brain. Furthermore, the brain observes and describes the world from within, but that world is embedded into the brain. We must describe a modality of the brain where the description itself becomes internal. Let us explain this in more detail by an analogy with a formal system.

As mentioned in Sect. 2.2, Gödel explicitly showed one feature of a formal system inevitably appeared when the formal system describes itself [Hofstadter 1979; Nagel and Newman 1964]. It is undecidability. Introducing a way of describing a whole system by itself, Gödel could prove the incompleteness theorem. Gödel's method was applied to the study of computability by machines. By asking undecidable problems, computability was well defined. If the machine computation is restricted to some specific problems, an essential impossibility cannot be seen. On the other hand, an essential impossibility clearly appears in the machine computing all of what is written by a finite algorithm. Thus, in a finite description, the description on the whole of a system from within leads to undecidability.

External descriptions lead us to truth in the case that the descriptions never have an essential influence. The results obtained by external description will asymptotically approach the truth if the influence by the description can be controlled to be arbitrarily small. If distinct behaviors are obtained by a slight change of description, such a description cannot ensure that it approaches the truth.

Theories in physical sciences are justified by experiments, whose process has contributed to find the truth. If the experiment, however, distorts the image of world qualitatively, the experimental data alone could not reveal the truth. In this case, the world concerned should include the experimental apparatus and even the way the experiment is conducted. This is a similar situation to the one in quantum mechanics, though the apparatus is discriminated as a c-number from a so-called q-number in the conventional framework (it might be useful to remember von Neumann's discussion on "abstract ego"). Thus the observer who describes the world will be involved in the world, therefore the choice of the coordinate of such an internal observer becomes essential. An internal observer might give the least influence, adopting an appropriate coordinate. If such an interfacial coordinate is found, it will give a new external description which does not essentially influence the world [Rössler 1994].

Let us think of a standing object. You are asked to gaze at it. If you move your head on both sides, your eyes move in the opposite directions of your head. This reflex is called a *vestibulo-ocular reflex*. The internal mechanism of this reflex was investigated in detail by Ito *et al.* [1980]. By this reflex, the brain can judge, in spite of the head movement, that the object is standing, because the retinal image is standing. On the other hand, when you are asked to pursue by eyes a moving object, with a fixed head position, your eyes move smoothly to pursue the object. This is called a *smooth pursue*. Here, the retinal image is standing, but the brain judges that the object moves. This indicates that the retina cannot be an interface in the above sense. In the vestibulo-ocular reflex, the brain concludes that the object is standing, based on the data of the standing retinal image, the head movement and the reflexive movement of the eyes. In the case of smooth pursuit, it concludes that the object moves, based on the data of the standing retinal image, the standing head and the smooth movement of the eyes. Discrimination of these two movements is performed by some calculation on the condition of movement of head and eyes.

How could the premise of the vestibulo-ocular reflex be obtained? When you shake your head to both sides violently, how do you see an object? We have experienced the movement of an object in the opposite direction of the head movement. Why could the reflex not be fixed in such a way? Can we conclude that such a reflex is irrational since one can judge by touch that the object is standing? The present technology of virtual reality proved indeterminacy of sensation including touch, as well as psychopathological and neurophysiological evidence. The question "what is reality?" is not trivial but essential, so that we have to keep asking.

The brain must have started computations, based on some assumptions. In the vestibulo-ocular reflex, a somatic sensation might have had priority over the visual sensation, thereby the reflex might have been fixed to acquire the perception of standing images. Here another question arises. Could the brain have known the "reality", from within, that the object is standing? The brain seems to have acquired a method to observe the world from without in spite of being inside the world.

In this aspect, the first procedure seemingly necessary for understanding the brain will be to find a canonical coordinate which may regulate both internal and external descriptions. A computational theory with a canonical coordinate leads to a method of interpretation for the brain functions. Probably the only method for finding internal descriptions is to construct the world concerned. A medium for construction may be computers, because one can control and describe the world to be constructed inside computers from without, and construct it internally via programming.

6.4 Evidence of Chaos in Nervous Systems

The brain is constructed by several spatial hierarchies: the level of macromolecules such as proteins (10^{-8} m), the level of synapses (10^{-6} m), the level of neurons (10^{-5} m), the level of neural networks (10^{-4} m), the level of maps

$(10^{-3}$ m), the level of systems $(10^{-2}$ m), and the level of central nervous systems $(10^{-1}$ m). It should be noted that the hierarchy addressed here roughly represents anatomical spatial hierarchies and it does not mean the corresponding functional hierarchy. Each hierarchy does not work alone, but works with mutual relations. Every hierarchy must commit to the emergence of higher functions.

In order for the functional hierarchy to coincide with the spatial hierarchy, a temporal hierarchy will need to accompany it. It is, however, quite difficult to find the temporal hierarchy in a functional manifestation. Actually, for each hierarchy a corresponding characteristic time scale is not apparently seen. Time scales are seemingly mixed up. For instance, only a few hundred spikes of a neuron whose duration is of the order of 1 s can evoke an intracellular reaction of macromolecules (the 2nd, or the 3rd, or the n-th messengers, and even DNA) sustaining from around a day to a week (e.g. LTP in the hippocampus). This is a typical characteristic seen in complex systems. Furthermore, the primary process of mental phenomena is closely related to a dynamic process of neural networks which sustains about a hundred milliseconds, and a moving object can be perceived as the same one even if the velocity is different, because of a scale transformation in time. Thus it would be plausible to think that the spatial levels addressed above are not independent, but intertwine over several time scales.

Probably, relating to this mixing of time scales, chaos can be observed in almost all spatial hierarchies. Actually, several investigations have proven chaos in some of the higher hierarchies, though no research has been done yet at the level of macromolecules and synapses: the onchidium giant axon [Hayashi and Ishizuka 1987], the squid giant axon [Aihara *et al.* 1985], the pyramidal neuron in hippocampal CA3 in rats [Hayashi and Ishizuka 1990, 1995], the hippocampal neural network in intracranial self-stimulations in rats [Hayashi and Ishizuka 1995], the neural network in rats' and rabbits' olfactory bulb [Freeman and Skarda 1985; Freeman 1987; Skarda and Freeman 1987; Yao and Freeman 1990], the olfactory system [Kay *et al.* 1995], the circulatory system influenced by the central nervous system [Tsuda *et al.* 1992], and the human brain [Babloyantz 1986; Destexhe *et al.* 1988; Layne *et al.* 1986; Basar 1990]. This fact implies the possibility of chaos appearing in a single neuron and also in networks of neurons whose number could be up to ten billion. Let us here call the chaos in nervous systems *neurochaos*.

Theoretical studies have predicted and confirmed neurochaos. The Hodgkin–Huxley equation, which has a physiological basis, of the squid giant axon reveals chaos if it is periodically forced [Aihara *et al.* 1989]. A modified Hodgkin–Huxley equation which takes into account the effect of ion channels bringing about slow changes of the membrane potential shows various bifurcations and chaos [Hayashi and Ishizuka 1992]. Freeman *et al.* made a neural-network model for the olfactory activities and obtained chaos and bifurcations similar to those found in experiments [Yao and Freeman 1990]. Furthermore, it is known that a network with asymmetric couplings can produce chaos [Parisi 1986; Hopfield 1986; Amit 1986].

Our concerns here are the origin of neurochaos, the information structure of neurochaos, and its biological significance.

6.5 The Origin of Neurochaos

Three cases on the origin of neurochaos can be considered [Tsuda 1991b].
(1) Chaos at a higher level directly stems from chaos at a lower level. Then, a spatial cascade of chaos can appear.
(2) Chaos at some higher level is independent of the lower levels' chaos and it stems from an enhancement of the damped oscillation via feedback from further higher levels' activities.
(3) Chaos at a higher level stems from the self-organization of lower levels' activities with the help of fluctuations.

In order to clarify the problem, let us consider chaos at the level of neural networks. This level is the lowest level at which all the above three possibilities should be taken into account.

If the first possibility occurs, the mathematical framework dealing with this problem would be as follows.
(A) The case of continuous space and time

The formulation by partial differential equations:

$$\frac{\partial S}{\partial t} = f(S) + D\frac{\partial^2 S}{\partial x^2}, \tag{6.1}$$

with the state variable $S = (s_1, s_2, \cdots, s_n)^t$ and nonlinear functions $f = (f_1, f_2, \cdots, f_n)^t$ in n-dimensional space, the space coordinate $x \in \mathbf{R}^d$ ($d = 1, 2$ or 3), and the diffusion coefficients $D = (d_{ij})$ ($i, j = 1, 2, \cdots, n$). It is also assumed that the dynamical system $dS/dt = f(S)$ reveals chaos.
(B) The case of continuous time and discrete space

The formulation by n-coupled nonlinear differential equations:

$$\frac{dS}{dt} = f(S) + Cg(S), \tag{6.2}$$

where C is a matrix denoting coupling strengths c_{ij}, and g is a coupling term. It is also assumed that $dS/dt = f(S)$ reveals chaos.

It is expensive to analyse these systems with a computer in a reasonable amount of time. But, there is a much more economical alternative. According to Kaneko [1984b, 1989a, 1990a, c], a partial differential system of some type could be approximated by a coupled difference system with discrete time. Then we obtain the third case.
(C) The case of discrete space and time

The formulation by the CML, or the GCM:

$$S(t+1) = f(S(t)) + Cg(S(t)), \tag{6.3}$$

where t is a discrete time, and it is assumed that $s_i(t+1) = f_i(s_i(t))$ reveals chaos.

The last formulation was discussed in detail in Chaps. 3 and 4. Here, it is significant to note that the globally coupled map (GCM) is equivalent to a fully connected neural network under some assumption. If, in the GCM

$$s_i(t+1) = w_{ii} f_i(s_i(t)) + \sum_{j=1,\neq i}^{n} w_{ij} f_j(s_j(t)), \tag{6.4}$$

we substitute $y_i = f_i(s_i)$, the representation of a conventional neural network is obtained:

$$y_i(t+1) = f_i \left(w_{ii} y_i(t) + \sum_{j=1,\neq i}^{n} w_{ij} y_j(t) \right). \tag{6.5}$$

Here, w_{ij} is interpreted as the coupling strength from neuron j to neuron i, $y_i(t)$ as the i-th neuron's state at time t, and f_i as a nonlinear transformation expressing the input–output relations of the i-th neuron. Then, the original variable s in the GCM is interpreted as a membrane potential of the neuron. Usually, in the case of neurons, f is chosen to be a sigmoidal function, but here we regard f as a chaotic dynamical system. Could it be that there is a neuron following this type of f?

If we combine an excitatory neuron and an inhibitory one, we can construct a function f possessing a maximum value at some intermediate value of input. Furthermore, let us take into account interacting excitatory and inhibitory receptors inside the membrane of a single neuron. Next, suppose the excitatory receptors are strongly activated and the inhibitory ones weakly activated for excitatory inputs. Excited inhibitory receptors may slightly inhibit the excitatory receptors. Approximate the effect of this feedback inhibition by up to the second order of input. Then, the membrane potential may change in the form $ax - bx^2$ for the input x, where a and b are positive. Thus, the input–output relation f of a single neuron may allow chaotic behaviors of the membrane potential, and hence in spike sequences too.

There are other cases where f can be regarded as a chaotic dynamical system: e.g. in the presence of a slow channel, or when there is a refractory period, that is, a delay. According to Hayashi and Ishizuka [1987], when providing a monosynaptic stimulus to the onchidium neuron, such a neuron's activity becomes chaotic. This has been observed in interspike intervals and thus in the membrane potential. The "effective input–output relation" of such a neuron must follow a chaotic dynamical system. This might imply that the dynamics at the trigger zone follows the chaotic dynamics. A similar mechanism is seen in neurons with a refractory period. Let us think of the following neuron model (Caianiello–Delluca equation, or Nagumo–Sato equation):

$$y_{n+1} = f(x_n), \quad x_n = -\alpha \sum_{j=0}^{n} b^j y_{n-j} + u_n - \theta, \tag{6.6}$$

where $0 < b < 1$, θ is a threshold, n a discrete time, and u_n an external input. Here, f is taken as a sigmoidal function, that is

$$f(x) = \frac{1}{1 + \exp(-\beta x)}, \tag{6.7}$$

where β (> 0) is a parameter for the determination of the slope of the map at the origin. In Caianiello's equation and also in Nagumo's analysis, f was taken to be the threshold function, i.e. the Heaviside function:

$$f(x) = \begin{cases} 1 \ (x \geq 0) \\ 0 \ (x < 0). \end{cases} \tag{6.8}$$

In order to make the discussion more realistic, we here adopt (6.7) instead of (6.8).

From (6.6),

$$x_{n+1} = -\alpha f(x_n) + bx_n - b(u_n - \theta) + u_{n+1} - \theta$$

follows. We assume that the inputs u_n are independent of time, namely $u_n = u_{n+1} = u$. Then, we obtain $x_{n+1} = -\alpha f(x_n) + bx_n + (1 - b)(u - \theta)$. Putting $a = (1 - b)(u - \theta)$, we obtain

$$x_{n+1} = -\frac{\alpha}{1 + \exp(-\beta x_n)} + bx_n + a. \tag{6.9}$$

This dynamical system possesses a chaotic solution when β is sufficiently large and b is not too small.

Aihara showed the presence of chaos in this type of neuron dynamics, based on the experimental data of a squid giant axon [Aihara 1990a]. Nagumo and Sato showed, adopting (6.8) as the function f, that neural responses obey a devil's staircase (Cantor function as a singular function) in the parameter space of a. Following (6.8), however, chaos cannot be created. If one adopts the following piece-wise linear function in place of (6.8), the overall system becomes a chaotic dynamical system as shown in (6.11) [Aihara 1990a, b]:

$$f(x) = \begin{cases} 1 & (x \geq \epsilon) \\ (1/\epsilon)x & (0 \leq x < \epsilon) \\ 0 & (x < 0), \end{cases} \tag{6.10}$$

$$x_{n+1} = \begin{cases} bx_n + a - \alpha & (x_n \geq \epsilon) \\ -(\alpha/\epsilon - b)x_n + a & (0 \leq x_n < \epsilon) \\ bx_n + a & (x_n < 0). \end{cases} \tag{6.11}$$

The map, (6.11), is a piece-wise linear approximation of the map, (6.9). One should note that the above Nagumo–Sato equation does not directly display observable chaos, but has a chaotic solution. According to Hata [1982, 1998], the domain of chaotic solutions constitutes a Cantor set. This type of "topological" chaos[2] can be observed if the Nagumo–Sato equations are coupled in such a way as in the CML discussed in Chap. 3. Actually, Crutchfield and Kaneko [1987, 1988] showed for this system that the quasistationary supertransient described in Sect. 3.3.1 (7) appears, and that chaos can be observed as long as the system size is sufficiently large. Moreover, the Nagumo–Sato equation coupled in the form of a neural network also reveals observable chaos.

Since biological neurons possess a refractory period, the alteration of the membrane potential, that is, the pulse amplitude at the trigger zone which gives rise to the alteration of the spike intervals, can be described by chaotic dynamical systems, even if the input–output relation of the neuron is represented by a monotone function like the sigmoidal function. This does not, however, directly justify the interpretation of (6.4) and (6.5). On the other hand, there is still the possibility that a neuron is found whose observation allows such an interpretation. Furthermore, adopting the interpretation that the state variable in (6.5) represents the state of a neural network, (6.5) can be reinterpreted as describing the chaotic transformation of the network states. Then, (6.5) reveals a network of neural networks. It will not be straightforward whether a network of neural networks is functionally equivalent with a neural network of the same size. If the neural networks are described by an order parameter, they would differ. Then, if the order parameter is a chaotic dynamical system, (6.5) will describe a megaro-network, namely a network of a larger scale than macroscopic scales.

The second possibility has already been proposed by Freeman [1987], in relation to the dynamic behaviors that he found in the rabbit and rat olfactory bulb. According to Freeman, the neural network in the bulb constitutes a damped nonlinear oscillator. By the delayed feedback excitation from the prepiriform cortex, the oscillator can be enhanced, and chaos can appear in some regions of feedback strength. Actually, Yao and Freeman [1990] succeeded to simulate, in this framework, chaos and more coherent motions as they found in the series of experiments. In general, a delayed feedback system can reveal chaos since it is embedded into infinite-dimensional dynamical systems. A typical system is Ikeda's optical system [Ikeda *et al.* 1980].

The third possibility may be related to the notion of noise-induced chaos. This is the possibility that a neuron behaves simply, but chaos appears at the network level by the interactions. Among others, the simplest one is the case of asymmetric couplings. In the classical model for associative memory formulated by Kohonen, Amari, and Hopfield, symmetric couplings ($w_{ij} = w_{ji}$) were adopted. After these works, Hopfield [1986], Parisi [1986], Amit [1986], and Sompolinsky *et al.* [1988] showed the presence of chaos in asymmetric networks with $w_{ij} \neq w_{ji}$. As we will mention later, since chaos in such a

---

[2] Refer to the footnote on p. 42.

network has nothing to do with any learned patterns, it is less functional, simply implying "I don't know" states. We will here discuss a more interesting case which will be valuable both biologically and theoretically. This is the case that chaos is created in a noisy environment.

In general, there are two kinds of stochastic neural dynamics as follows:

$$\text{(I)} \quad y_i = f(U_i + \sigma), \tag{6.12}$$

where y_i is the activity of the i-th neuron, U_i a dendritic potential of such a neuron, and σ a noise term.

$$\text{(II)} \quad y_i = f^{p_i}(U_i), \tag{6.13}$$

where f^{p_i} is probabilistically chosen from a set of possible dynamics, which may be defined as follows:

$$\begin{cases} y_i(t+1) = f(U_i(t)) & \text{with probability } p_i \\ y_i(t+1) = y_i(t) & \text{with probability } 1 - p_i. \end{cases} \tag{6.14}$$

In the dynamics of (6.12), the external noise is additive, which may stem from a leaked current of adjacent neurons, giving rise to a random fluctuation of the dendritic potential. Thus the noise term reveals a "dendritic noise". On the other hand, in the dynamics of (6.13), the probabilities are provided for a choice of two dynamical systems, f and the identity map. The following two cases can be considered for the physiological origin of this type of stochastic dynamics.

(a) The case of stochastic emission of synaptic vesicles [Tsuda 1990b]. This stochastic release differs from spontaneous emission of vesicles. The latter is usually ineffective for firing because it gives only a tiny potential change of the order of μV, but the former can be effective due to a much larger influence of the order of mV.

(b) The case of a series of stochastically incoming input pulses. Because of the presence of a distributed synaptic delay for each pulse, whether the postsynaptic membrane potential at each time is above or below the threshold can be stochastic.

Thus, in the dynamics of (6.13), a randomness is embedded in the synaptic connections, that is a "synaptic noise" is taken into account.

The stochastic neurodynamics of the type of (6.12) has been investigated with a model such as the Boltzmann machine [Hinton et $al.$ 1984]. If we take f as the threshold function with zero threshold such as $f(z) = 1$ for $z > 0$ and $f(z) = -1$ for $z < 0$, the probability of the neuron i being active ($f = 1$), $P(U_i + \sigma > 0) = P(U_i > -\sigma)$ is given by $P_i = \int_{-U_i}^{\infty} \exp(-\sigma^2/2s^2) d\sigma / \int_{-\infty}^{\infty} \exp(-\sigma^2/2s^2) d\sigma$ if the noise term σ is given by a Gaussian distribution. Therefore,

$$P_i = 1/\sqrt{2\pi s^2}\chi(U_i), \tag{6.15}$$

where s is the standard deviation, and χ an error function. This is equivalent to the Boltzmann machine's algorithm. However, chaos has not been observed for this algorithm. In general, when a neuron follows a simple threshold dynamics, a neural network consisting of the same kind of neurons only or with symmetric couplings only cannot show chaotic behaviors.

On the other hand, the dynamics of (6.13) can reveal chaos, and thus be one of the origins of neurochaos. It should, however, be noted that this dynamics does not always produce chaos, but could do so if a specific neural network structure such as seen in the brain is taken into account. In the same structure, the dynamics of (6.12) does show random and disordered behaviors.

The stochastic dynamics of (6.13) is not known so widely, but shows curious behaviors. Moreover, it can provide a new direction of research for chaotic dynamics as well as neurodynamics, extending a dynamical system to a wider functional space. Let us briefly comment on some other aspects of this dynamics.

6.6 The Implications of Stochastic Renewal of Maps

6.6.1 Chaotic Game

Extending the stochastic renewal of neurodynamics, one can consider a kind of 'game' for chaotic dynamical systems. A general form of a 'chaotic game' [Barnsley 1988; Tsuda 1991b] is defined by n-tuples $(f_1, f_2, \cdots, f_n)$ of dynamics and corresponding n-tuples $(p_1, p_2, \cdots, p_n)$ of assigned probabilities. As a stochastic dynamics the following is defined:

$$(f_1, f_2, \cdots, f_n), (p_1, p_2, \cdots, p_n), \tag{6.16}$$

where each dynamics f_i can be controlled by a bifurcation parameter (or parameters), whereby fixed points, limit cycles, tori, chaos, or in the case of higher dimensions chaotic itinerancy [Tsuda 1990a, 1991a, b, 1992a; Kaneko 1990c, 1994a; Ikeda et al. 1989; Davis 1990] may appear. At each k-times interval, according to the assigned probability, the corresponding dynamics is chosen and executed. The last value of the previous dynamics (the value k time steps earlier) is taken as the initial condition of the newly chosen dynamics. By this procedure, the state of the game at each time step is defined. For each state, n choices are allowed. If one thinks of the initial state of the game as a whole phase space spanned by n-tuples of dynamical systems, each state is its subset. The orbits in phase space represent a chain of states of the game, and hence reveal the on-going states.

The overall dynamics would exhibit various new phenomena according to the choice of the n-tuples $(f_1, \cdots, f_n)$ of the dynamics, the n-tuples $(p_1, \cdots, p_n)$ of the assigned probabilities and a tuning parameter k.

This game is a generalization of the random fractal by Falconer [1986] and the random iterations by Tanaka and Ito [1982], which is one of skew-product transformations. In the random fractal, f is restricted to a contracting map and was effectively used as an algorithm for reconstructing an original image by fractal images.

6.6.2 Skew-Product Transformations

The stochastic renewal of dynamical systems is viewed as a dynamical system in extended phase space. The idea is to redefine probabilities by means of chaotic dynamical systems. Here, we formulate it for the case of neurodynamics [Tsuda 1991b, 1994a]. Define a neural network by n-dimensional dynamical systems on the interior of an n-dimensional hypercube, including boundary, $X = [-1, 1]^n$:

$$U_i(t+1) = h_i(U_1(t), U_2(t), \cdots, U_n(t)), \tag{6.17a}$$

or

$$U_i(t+1) = g_i(U_1(t), U_2(t), \cdots, U_n(t)), \tag{6.17b}$$

where $i = 1, 2, \cdots, n$.

Suppose the probability p for stochastic renewal is generated by a chaotic dynamical system. We consider the product space of such a phase space and that of the neurodynamics. The variables working on the latter space depend on those working on the former space, and not vice versa. Hence the product is a skew product [Cornfeld et al. 1982].

What chaotic dynamical system should be selected depends on the probability structure demanded. Here, we show a simple example for it, using a Bernoulli shift, $B(p, 1 - p)(y \mapsto y/p \ (0 < y \le p), \ y \mapsto (y - p)/(1 - p) \ (p < y < 1))$. The domain of the definition of the Bernoulli shift is $Y = [0, 1)$. The skew product is defined on the space $T = X \times Y$.

a. Bernoulli model

$$T(U(t), y(t)) = \begin{cases} (h(U(t)), y(t)/p) & (0 < y \le p) \\ (g(U(t)), (y(t) - p)/(1 - p)) & (p < y < 1). \end{cases} \tag{6.18}$$

Here, the neurodynamics h or g is coupled with the Bernoulli shift $B(p, 1-p)$. If the decision point which decomposes phase space Y into two nonoverlapping subregions labeled, for instance, 0 or 1, is $y = p$, the symbolic dynamics generates independent symbol sequences of any size.

b. Markov model

$$T(U(t), y(t)) = \begin{cases} (h(U(t)), 2y(t)(\mathrm{mod}\, 1)) & (0 < y \le p) \\ (g(U(t)), 2y(t)(\mathrm{mod}\, 1)) & (p < y < 1). \end{cases} \tag{6.19}$$

Here, the neurodynamics h or g is coupled with the Bernoulli shift $B(1/2, 1/2)$. If the decision point is selected as $y = p$, the symbolic dynamics generates n Markov symbol sequences, thus the generated symbols are no longer independent, except for $p = 1/2$.

6.7 A Model for Dynamic Memory

A correlation method is a typical method to store patterns, events, and their series in neural networks. Usually, a state for the storage is represented by an attractor, in particular, by a fixed point. The representation by a limit cycle has also been proposed [Nara and Davis 1992, 1994]. By just considering these attractors, however, the problem for *memory* is not necessarily solved. One of the most fundamental functions of the brain is a memory process which allows learning, recalling of what has been and is being learned, and seeing reality. If we would not have memory, we could not see reality.

In Greek mythology, Mnemosyne who is a wife of Zeus is the goddess of memory, and the nine Muses are the goddesses of art. This conveys that the memory *process* is inevitable for creativity and imagination. Moreover, Proust describes in *A la Recherche du Temps Perdu* that a taste stimulation can trigger a series of memories which had been thought to be lost. Here, we review a series of our studies relating to similar memory processes.

In most models for an associative memory, a direct rule for association is used. A typical model is used in the studies by Amari, Nakano, Kohonen, and Hopfield [Amari 1978; Fukushima 1979]. There are many metastable states in these models. It is possible that a deep valley among the metastable states represents a learned pattern, and that a shallow valley is a parasitic mode which is unrelated to the learned patterns. The purpose of this system is to output one metastable state corresponding to the learned pattern. Thus only one association is anticipated for one pattern. There are no dynamic transitions among the learned patterns. The system is fixed to continually output one pattern after a relaxation process. In these systems, another rule for an order of associations is necessary in advance if we want to avoid this adherent state. A system just following a given rule does, however, lose flexibility.

There is another weak point in this system. If the number of patterns to be learned becomes large, a recall of the stored patterns becomes impossible, namely, that a large correlation among the patterns voids the ability of recall.

In these models, and also in the model we review here, a Hebbian learning rule is adopted. According to this learning algorithm, we usually face difficulty in the retrieval process. If some pattern is more strongly learned than some other one, the recall of this pattern is dominant. In order to acquire a correct recall and increase the memory capacity, various attempts have been made. A reduction of parasitic modes by a forced forgetting algorithm is effective. Here, forced forgetting indicates that a synaptic strength is forced to be weakened when the system becomes a steady state. This procedure is done many times for randomly selected initial conditions. It is known as inverselearning.

This algorithm works for synaptic weights related to embedded patterns in order to equalize them. This makes the probability of recall for each pattern equal for a fuzzy input. In spite of the advantage of this algorithm, the recalled pattern is sometimes an unlearned pattern and in the worst case a learned pattern is deleted.

Let us call a memory represented by a fixed point a first-order memory. The above facts imply that a perception process cannot be related to a cognitive process only by the storage of a first-order memory. For this, at least a mechanism is necessary for dynamically linking the first-order memories, and thereby a rule for linking is created. According to the studies by Tsuda *et al.* since 1987, it turns out that such a created rule can be described by chaos, and chaos has biological significance via this organization process. Then, a dynamically linked behavior of a first-order memory can be thought of as a second-order memory. The emergence of the second-order memory may be used for an increase in the capacity of the first-order memory. This was realized by the game-theoretical neural networks introduced in the previous section.

6.8 A Model for Dynamically Linking Memories

The model here is based on the structure of mammalian cerebral cortices and also on the neurophysiological data for stochastic emission of synaptic vesicles under stimuli [Tsuda *et al.* 1987]. Some simplifications were made out of theoretical considerations. A neuron as an element of a network is particularly supposed to obey a convergent dynamics. A skeleton of the model is shown in Fig. 6.1.

The neural dynamics we took is as follows ((6.20)–(6.25) (or replacing (6.24) by (6.24′), or (6.25) by (6.25′)) [Tsuda 1992a].

$$S_i(t+1) = f^p \left(\sum_{j=1}^{2m} C_{ij}(t) S_j(t) + d_i R_i(t) - \delta_i \phi(t) \right), \qquad (6.20)$$

where $\phi(t) = x(t_1), t_1 = \max_{t>s}\{s \mid x(s) = x(s-1)\}$, $\delta_i = 0$ for $i = 1, \cdots, m$ and 1 for $i = m+1, \cdots, 2m$, and $x(t)$ is given by $x(t) = \sum_{j=m+1}^{2m} C_{ij}(t) S_j(t)$,

$$R_{i+l}(t+1) = f^p \left(\sum_{j=1+l}^{m+l} e_j S_j(t) \right), \qquad l = 0 \text{ or } m, \qquad (6.21)$$

$$C_{ij}(t+1) = C_{ij}(t) + \varepsilon S_i(t) S_j(t) \text{ for } 1 \le i, \ j \le m \qquad (6.22)$$
$$\text{or for } m+1 \le i, \ j \le 2m,$$
$$C_{ij}(t+1) = C_{ij}(t) + \varepsilon' S_i(t) S_j(t) \text{ for } 1 \le i \le m \text{ and } m+1 \le j \le 2m$$
$$\text{or for } m+1 \le i \le 2m$$
$$\text{and for } 1 \le j \le m. \qquad (6.23)$$

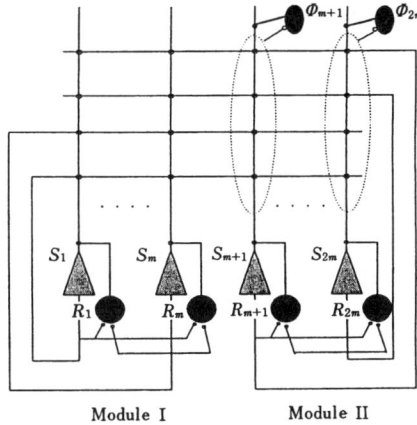

Fig. 6.1. A model network based on a typical cortical structure. It is still question-able if memory dynamics in the cerebral cortex or in the hippocampus is performed on this type of network. The network construction for the model study is made, based on the hypothesis that a memory process will be performed on the network structure common to all areas, since it underlies all sensory information processing. (Reprinted from I. Tsuda, *Neural Networks* **5** (1992a) 313, with the permission of Elsevier Science.)

After the learning is completed in the initial stage, we take

$$C_{ij}(0) = \sum_{\mu} S_i^{(\mu)} S_j^{(\mu)}, \tag{6.24}$$

where the vector $S^{(\mu)}$ is the μ-th stored pattern. When the learning is per-formed from the initial stage, we take

$C_{ij}(0) =$ random numbers over some positive to negative values.   (6.24′)

The probability law adopted in S_i and R_i is given by

$$y(t+1) = \begin{cases} \mathbf{1}^p(x(t)) = \begin{cases} +1 & \text{if } x(t) > 0, \\ -1 & \text{otherwise} \end{cases} & \text{(with probability } p) \\ \mathbf{1}^p(x(t)) = y(t) & \text{(with probability } 1-p) \end{cases} \tag{6.25}$$

or

$$y(t+1) = \mathbf{1}^p(x(t)) = \begin{cases} \tanh(\gamma x(t)) & \text{(with probability } p) \\ y(t) & \text{(with probability } 1-p) \end{cases} \tag{6.25′}$$

where y denotes S or R and γ denotes the steepness of the sigmoidal function.

The physiological neurons for the S and the R units are the pyramidal cells and the spiny stellate cells, respectively. Synaptic connections $\{e_j\}$ of the S

units with the R units are supposed to stem from axon collaterals of the pyramidal cells. Since the distribution of axon collaterals is random [Szentagothai 1975] and there are intervenient inhibitory neurons, we assumed that the value of e_j took a quasirandom number distributed uniformly over $[-\alpha, \alpha]$. Synaptic connections $\{d_j\}$ of the R units with the S units are more specific, but similarly assumed to take the quasirandom values distributed uniformly over $[-\beta, \beta]$, since the spiny stellate cell takes a synaptic connection with a basal dendrite of the pyramidal cell via spines possessing some distribution [Crick and Asanuma 1986] and also via the intervenient inhibitory small basket cells [Szentagothai 1975, 1983]. The recurrent connections of the S neurons stem from the recurrent connections of the pyramidal cells via their axon collaterals and inter-regional feedback connections among pyramidal cells.

The negative feedback by the ϕ unit was introduced in a specific fashion. If the collective activity of the S units in the module II ($i = m + 1 \sim 2m$) approaches a steady state, then the ϕ unit starts to work. The ϕ unit keeps the value of the dendritic potential of the apical dendrite of the pyramidal cells till another steady state appears. When another steady state appears, the ϕ unit takes another value corresponding to the potential of the pyramidal cells at that steady state. This algorithm seems to be justified, if there is a differential operation detecting a steady state of activity of the pyramidal cells, which seems to be possible since the neuron in the cortical layer II is compact. The axonal tuft cell in layer II or Martinotti cell in layer VI may have such a function. Compared with the number of pyramidal cells in the cortical layers II and III, there is a relatively small number of axonal tuft cells. Hence we introduced the effect of the ϕ unit in the module II only.

The model network we deal with here consists of two types of subnetworks: a symmetrically coupled network, and an asymmetrically coupled network. A symmetrically coupled network is related to memory storage, and an asymmetrically coupled one has nothing to do with memory itself but makes the overall system be in a nonequilibrium state. This combination of symmetric and asymmetric couplings allows neural networks to be temporally unstable, keeping stability due to convergent dynamics.

As shown by (6.25) or (6.25'), this neural network is a game-theoretic one. As mentioned in Sect. 6.6, this type of system can be described by a dynamical system, introducing a skew-product transformation. The introduction of the probabilities in this form avoids the decomposition of the system into a dynamical part and a noise part in an additive form. The additive noise in neurodynamics, which is usually used in model studies, may stem from a dendritic noise due to a leaked current. On the other hand, a stimulus-induced stochastic release of synaptic vesicles is resposible for a multiplicative noise, which may be reduced to the stochastic renewal of neurodynamics treated here. Which type of stochastic process is essential depends on the complexity of the system.

Fig. 6.2. The representation of memory structure on a supposed potential in a typical static model for associative memory. Each memory is represented by a stable fixed point or a stable limit cycle.

Now, let us consider to embed a vector consisting of the states S_i of the S-unit as a first-order memory in this network. By Hebbian learning ((6.22) and (6.23)), this memory state is represented by a stable fixed point in the state space spanned by the states of the S-unit. Hence, for the learning of m patterns, at least m basins corresponding to m memory representations are needed. This is a usual situation for an associative memory. A typical case is shown in Fig. 6.2.

Choosing some point in this space as an initial condition represents an input of a pattern in the neural network. The orbit starting from such an initial condition converges to a fixed point which is a representation of some memory state. This convergent process represents a retrieval process of memory.

In our network, however, a convergent dynamics only does not necessarily work. As shown in Fig. 6.3 schematically, the presence of interactions between the state space made of R- and ϕ- unit's states and S-state space dynamically alters the landscape, and the stochastic renewal of dynamics temporarily destabilizes the fixed points. The orbit stays at a stable fixed point or its neighborhood or can escape from the basin of such a fixed point along a thin unstable manifold like a whisker to another fixed point. As long as external stimulation is absent, this chaotic transition repeats eternally, thus the network can access *automatically* correct memory states. A parasitic mode is naturally used in this transition process.

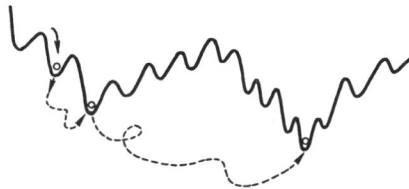

Fig. 6.3. The representation of a dynamic memory process on a supposed potential which can be dynamically varied (the case of chaotic itinerancy of memory). In this model, the state of the network chaotically changes among memories which are represented by a fixed point. The orbit in phase space stays at a fixed point for a while. After that, it approaches another fixed point along a generated unstable manifold. In this process, an itinerant motion appears. Unless the network is forced by a strong external input, this itinerant process eternally sustains.

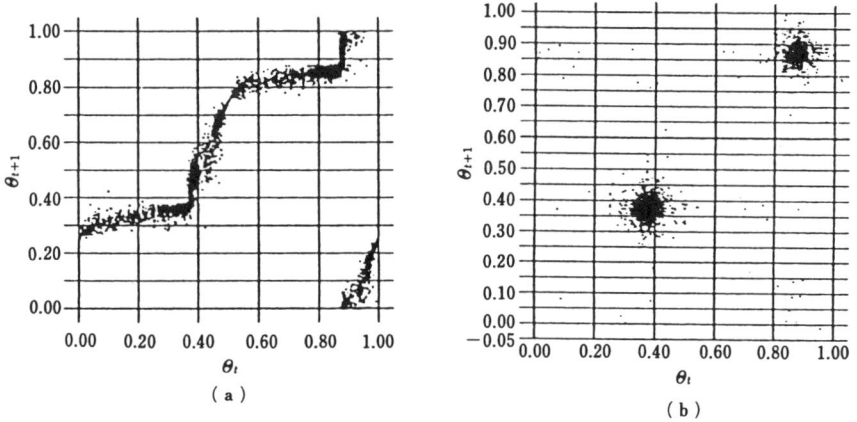

Fig. 6.4 a,b. The transition rule of an itinerant process in the space of the network state. A deterministic law is apparently seen in the form of a one-dimensional map (**a**). For a comparison, the return map is depicted in (**b**) in the case that the transition is forced by random noise with uniform measure in the same network. In (**b**), the stochastic renewal of the map used in (**a**) is replaced by adding a noise.

The transition obeys a dynamical rule which is self-organized in the system. Defining the activity of a subnetwork by $(1/m)\sum_{i=1}^{m} S_i(t)$, and investigating its return map, it turns out that the orbit is distributed in the neighborhood of the closed curve and moves almost clockwise. In other words, the return map itself is almost two-valued but one of two branches is deterministically selected. Then, choosing an origin and a phase appropriately, another return map with respect to the phase θ_t is drawn. We found it chaotic. This is shown in Fig. 6.4a. The dominant part of this map can be approximated by the following circle map:

$$\theta_{n+1} = f(\theta_n), \quad f(\theta) = \theta + A\sin(4\pi\theta) + C(\mathrm{mod}\,1). \tag{6.26}$$

This map self-organized in the network provides a rough rule to link memory representations.

The network behaviors approximated by (6.26) do not depend on the detailed parameter values, but depend on the network structure and the probabilities p of the game. Since we adopt the network structure mimicking a physiological one, our interest here is the dependence on this probability, which is shown in Fig. 6.5.

If $p = 1/N$, where N is the number of neurons, that is, the system size, then the way of introducing probabilities in the present model is equivalent to that of Hopfield's associative-memory model. In other words, in the Hopfield model, only one neuron is selected randomly and independently at each time step and an object for update while the other neurons are set up to take the same values as in the previous time step. Hence it is equivalent to the

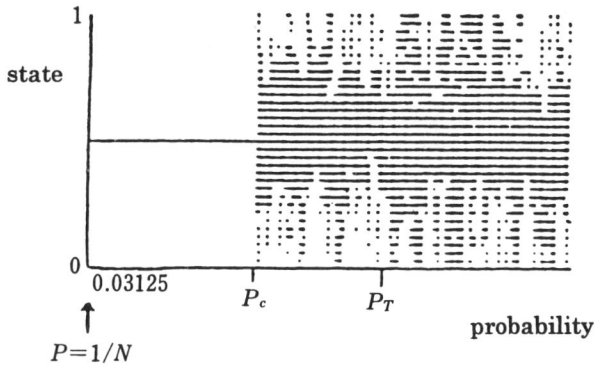

Fig. 6.5. The dependence of a network state on the probability p of the renewal of dynamics.

dynamics with probability $p = 1/N$. Let us call this point on the probability axis a Hopfield point.

As seen in Fig. 6.5, the present model behaves like the Hopfield model at the Hopfield point, that is, it outputs only a steady state representing a memory state which is close to the input. The Hopfield-model-like behavior continues up to the point $p = P_c$. Beyond this critical point, the dynamic linking state mentioned above appears. This chaotic state continues until just prior to $p = 1$. At $p = 1$, a multi-stable state consisting of limit cycles, each of which links fixed points representing a memory state, appears.

It will be meaningful to judge which state has a physiological significance. As mentioned above, the introduction of the probability p stems from a stochastic process of quantal emissions of synaptic vesicles under stimulations. From this, one can calculate what value of p is physiologically allowed. In the actual case the value of p may differ for each synapse, but here we discuss the simplest case that the value p is equal for every synapse. Thus we obtain

$$p(N) = 1 - (1 - 1/N)^N. \tag{6.27}$$

Supposing the system size is large enough,

$$p(N) \geq \lim_{N \to \infty} p(N) = 1 - e^{-1} = 0.63212 \cdots \tag{6.28}$$

is obtained. The number $0.63212 \cdots$ is a universal constant. Thus it turns out that our network is physiologically meaningful in the domain $p \geq 0.63212 \cdots$. This second critical point of the probability is indicated as P_T in Fig. 6.5. Empirically, we find $P_T \geq P_c$. Hence, a physiologically significant state in this network is in the chaotic state only if we use a stochastic renewal of neurodynamics. Thus the allowed probability values are determined by physiological conditions. Taking this into account, the chaotic linking process of memories is determined by the construction of the network. The computations of physical quantities like the diffusion coefficient, the probability distributions of residence time at memory states, and so forth show the criticality. This implies that the dynamic process of linking memories may be viewed as a kind of self-organized criticality [Bak *et al.* 1987].

The chaotic linking of memories has been widely observed after Tsuda's finding. A neural network consisting of chaotic neurons with refractoriness shows a similar chaotic linking [Aihara *et al.* 1990c], though in this case the itinerancy is transient and an asymptotic state is one of steady states which could be a memory state. In the case that a memory is represented by a limit cycle, the chaotic transition appears among memory states for a sufficient size of the average number of connections [Nara and Davis 1992; Nara *et al.* 1993; Nara and Davis 1994; Nara *et al.* 1995].

The dynamical state appearing in chaotic-linking processes differs from conventional ones. The phenomenon treated here is a chaotically itinerant phenomenon in the sense that the past history of orbits remains for linking. A dynamical system's state to describe it will be as follows.

Let us imagine we have many attractors in a high-dimensional phase space. Each attractor is separated and has its own basin of attraction. Then, the asymptotic behavior is represented by one of these attractors, depending on the initial conditions. What happens when the system is destabilized? If the instability is strong enough, many chaotic modes appear, and consequently the system goes toward a turbulent state, that is, a quite noisy macroscopic state. Then, any features of the original attractors disappear. If, however, the instability is not so strong, an intermediate state between order and disorder can appear. The dynamics may be regarded as an itinerant process which assures a transition among states. These states were at the beginning an attractor, but now are no longer an attractor. In this case, a crucial characteristic is that some original features of the attractor remain in spite of the generation of unstable manifolds. This itinerant process often becomes chaotic. We called a destabilized attractor an *attractor ruin*, and the overall behavior *chaotic itinerancy*. In this situation, a unique attractor is a collection of attractor ruins and itinerant orbits connecting each attractor ruin. This was called an *itinerant attractor*.

By this reason, the dynamic state found in the nonequilibrium neural networks was named a chaotic itinerancy as we mentioned in Chap. 4.

6.9 Significance of Neurochaos

Let us here discuss the significance of chaos in neural systems. We first describe it based on the facts which have been obtained in both experimental and computational studies, and later touch upon some possible functional roles, too. The significance of neurochaos has been clarified through studies related to the chaotic linking of memories described in the previous section. The computational results are summarized below [Tsuda 1992a, c].

1. Let's assume the situation that a pattern is input in the network while memories are chaotically linked. If the pattern is close to some memory state, in other words, the input pattern is inside a basin of attraction of some memory state, then the network state becomes a weak chaotic one whose orbits move around such a memory state. In other words, the memory state concerned, or the pattern close to it, is output. On the other hand, if the input pattern is far from any single memory state, e.g. when the input pattern overlaps several memory states or when the input is in the basin of a state other than a memory-related one, then the output becomes a chaotic itinerancy whose orbits itinerate chaotically among all the memory states. This implies that the chaotic itinerancy plays the role of *novelty filter*. Actually, a similar dynamical behavior has been observed in neurophysiological experiments in mammals [Freeman 1992, 1994, 1995].

2. The network in the previous section includes a stochastic recurrent net as a memory-related subnet, which is reduced to a Hopfield net for the Hopfield point of the probability. Hence, the comparison of the memory capacity with that of the Hopfield net makes sense. Suppose the neural network has learned patterns up to the critical capacity of memory in the sense of the Hopfield net. Suppose the neural network learns a new pattern when the chaotic linking of memories appears in the network. One can see the learning effect in two-dimensional parameter space, where one parameter is the time for presentation of such a pattern to the network, and the other is an increment of synaptic weight per unit time. The following three phases were observed in this space.

S-phase: The network does not learn the input pattern, but preserves all memories.

C-phase: The network does learn the input pattern, and preserves all memories. After learning, the recall is the dynamic process with chaotic itinerancy including the additionally learned pattern. Thus, in this phase, the linking among memories is reorganized.

D-phase: The network learns the input pattern, but this learning destroys all memories.

Additionally, between C- and D-phases, there appears a hierarchy of phases, where the learning of the input pattern destroys one memory, two memories, three, four, etc. But, these phases are narrow in parameter space.

By this computer simulation, it turns out that learning is possible in the dynamic retrieval process of memories represented in the form of chaotic itinerancy. Furthermore, when the same experiment is conducted for the case without chaos but with additive noise, only so as to ensure the transitions, the C-phase completely disappears in the learning process of the new pattern [Tsuda 1994a]. This indicates that chaotic itinerancy enhances the learning capability of the network.

The mechanism of the increase of critical capacity of memory is still unknown, but one may conceive the following situation which might provide a hint for the mechanism. Because of the chaotic itinerancy, a strong correlation among linked memories is generated, which gives rise to the learning of *orbits* in phase space, in spite of the Hebbian learning which contributes to the learning locally. The learning of orbits may avoid the destruction of memories, since memory is represented by a point in phase space.

In this section, we showed that neurochaos can subserve for the dynamic linking of memories, the classification of novelty of stimulations, the acceleration of learning, and the reorganization of memory space in terms of "super"-learning. Thus far, we have mentioned only some cases for the biological significance of neurochaos, but we conceive many other kinds of significance. Neurochaos may underly the generation of internal images with the help of chaotic itinerancy in memory space, the dynamic preservation of the perceived world, creative actions and imagination, as well as the dynamic maintenance of the metabolism of neurons. It seems that in complex systems possessing dynamic interactions among nonlinear elements, a higher-order complexity which chaos possesses is inevitable for the generation and maintenance of higher functions.

6.10 Temporal Coding

The coding of information by means of neural networks has been studied, based on the notion of spatially combining neural activities. The "grand-mother-cell" hypothesis that a single neuron in a higher level represents a grandmother as the result of a specific combination of a feature representation by each single neuron, and the "distribution" hypothesis that information is distributedly represented on a whole network are in particular well-known, as mentioned in Chap. 1. Another notion that has been proposed is that the brain uses an intermediate representation or each one of them case by case. Among others, a sparse coding has recently been highlighted [Miyashita 1993]. The idea is that if the number of active neurons is small in the network representation of the information, the memory capacity increases, giving rise to an effective coding. This was verified in the network of some part of the temporal cortex [Miyashita 1993]. These coding schemes are called *spatial coding*. The notion of rate coding, in which information is encoded by spike density, underlies this spatial coding.

On the other hand, the importance of *temporal coding* has been recognized. Actually, many experiments suggest that dynamic information processing at various levels of functions is conducted. The activities of neurons and/or neural networks alter in the process of perception and memory, depending on external and/or internal stimuli. Since every perception process is accompanied by memory processes, the network state in the perception process is observed as a nonstationary process. In conventional experiments, the temporal coding of information has been overlooked, since only integrated data over a time axis or only averaged quantities were taken into account. However, as the viewpoint that neural networks can be described by dynamical systems was established, the studies of a dynamic character of the network have made progress. Furthermore, since the hypothesis was proposed that a coincidence of spikes indicates a functional connection, the notion of *temporal coding* has been highlighted. On the other hand, since the necessity of the alteration from static to dynamic aspects has been discussed in various fields of biology, as mentioned in Sect. 1.5, the focus on the dynamic aspects of the brain [Tsuda 1984, 1990a, 1991b, 1996], including the notion of temporal coding, can be viewed as a paradigm shift along this trend.

The hypothesis of dynamic binding of features in perception is one of the key notions in temporal coding [Shimizu *et al.* 1985; Malsburg 1986; Koerner *et al.* 1987; von Seelen *et al.* 1988; Gray *et al.* 1989; Reitboeck *et al.* 1990; Koerner and Boehme 1991; Aertsen 1993; Tsukada 1992; Tsuda 1987b, 1991a, b, 1992a, b, 1993, 1994a; Fujii *et al.* 1996]. If the spikes of different neurons which encode the same feature are synchronously input to another neuron concerned, or if the underlying oscillations of neurons are entrained, this neuron instantly may link to such a feature. Furthermore, if different neurons encoding distinct features by spikes synchronize to other neurons concerned, such neurons may link to different features. Thus, a binding problem which is difficult to solve by rate coding can be solved by a *coincidence-detecting neuron* [Fujii *et al.* 1996].

Freeman clarified in collective activities of the rabbit olfactory bulb that a spatial pattern of neuroactivities generated by a spatial entrainment of nonlinear oscillators subserves for olfaction, and alters, depending on the sequence of inputs [Freeman 1987, 1992, 1994]. Freeman also found that the collective activity in this case is *chaotic* (see, for instance, Skarda and Freeman [1987]; Freeman [1995]).

Stimulated by Freeman's finding, Gray *et al.* conducted the experiment in the cat visual cortex, and found oscillatory phenomena in orientation- and direction-specific neurons [Gray *et al.* 1989]. They hypothesized that a dynamic link of features can be performed by the synchronization. This idea is similar to the one that Shimizu *et al.* had proposed with a model for pattern recognition (which was called *holovision*) [Shimizu *et al.* 1985].

Recently, Singer *et al.* conducted further experiments and found a coincidence of neurons in the visual cortex. They observed a synchronization in the LFP (local field potential) of neurons also in the case that they possess different orientation specificity, but a synchronization in impulses only among those neurons possessing a similar orientation specificity. They think that this indicates that not neural oscillation, but synchronization among pulses, is responsible for functional connectivity, and oscillations are simply the product of such a coincidence.

Whether the oscillatory activities in the visual cortex are functional or not is still in dispute, but the analysis in terms of information quantities by

Richmond *et al.* [1987a, b, 1990a, b] seems to be important. Below, we would briefly like to comment on this.

It seems that the brain encodes information by the *behavior* of neural activities in order to act adaptively and creatively. Here, we see the necessity that chaos participates in the information processing of the brain. Since the memory process must be accompanied with any perception process, it would be meaningful to study the possibility of "chaos representing information in its dynamic behaviors" relating to the memory process.

So far, many temporal codings have been proposed. Here, we classify them into three classes. The first one is the notion of *coincidence detector* (see, for instance, Aertsen [1993]; Aertsen and Vaadia [1992]; Aertsen *et al.* [1994]; Fujii *et al.* [1996]). As mentioned above, this is the notion that the binding among different features and/or different modalities is due to the synchronization of spikes in some neuron from other connecting neurons. Let us call it the 1st-TC.

Second, we observe the case that the information is encoded by a quantity, like a temporal change of the spike density. More precisely, according to Richmond and Tsukada, it is the case that the information is encoded by the second or higher statistics of spike sequences. We call it the 2nd-TC.

Furthermore, even though the coding scheme of external information at the neuronal level is the 1st-TC or the 2nd-TC, one can conceive computation processes by network dynamics, regarding the dynamic representation of information by dynamical systems. In other words, the information representation is extended to a time axis in the former two cases, but in the latter the representation of the computational process is extended to a time axis. It would provide a better understanding for the latter case to conceive the case that a transition rule of dynamical systems is again represented by dynamical systems. Let us call the third case the 3rd-TC.

Typical examples of the 2nd-TC are seen in the works of Richmond and Tsukada. Here, we briefly introduce Richmond's experimental results. The temporal alteration of the spike density of neurons in the primary visual cortex and in the inferior temporal cortex related to vision can be viewed as a waveform. They used two-dimensional patterns formed by a direct product of Walsh functions as visual stimuli. The Walsh functions constitute a system of completely orthonormal functions. The dominant component of the waveform of a single neuron as a response to the Walsh stimuli was extracted, using a principal-component analysis. It turned out that the mutual information between the principal components and the stimuli is much larger than the mutual information between the stimuli and the long-term average of the spike density representing the neural activity. Furthermore, the mutual information between the principal components and the stimuli was compared with the cases when independently changing each parameter of the stimulus, and when changing all the parameters simultaneously. It was found that the mutual information in the latter case is larger than the mutual information summed over a kind of parameter in the former case. From these calculations, Richmond *et al.* concluded that it is more rational to think that visual stimuli are encoded in the waveform of the neuron than in the state of the neuron or the neural network, and that the waveform can encode simultaneously multiple characteristics of the stimulus.

The meaning of the use of Walsh functions can be understood as follows. Since any two-dimensional pattern can be represented by Walsh functions, one can state that the obtained results are not for specific patterns but for general ones. By this function, one can give at once various types of patterns over all receptive fields. The latter is consistent with the finding of dynamic receptive fields consisting of temporally defined subfields by Tsukada *et al.* [1983] in the retinal ganglion cells, and by Dinse *et al.* [1990] in the primary visual cortex.

The notion of dynamic receptive fields is of importance for a dynamic interpretation of brain functions, since the conventional notion of receptive fields is static and encounters difficulties when trying better to understand the dynamic brain. It has been observed that the dynamics of subfields sustains a few hundred milliseconds, and that an active region inside an entire receptive field alters according to the dynamics. The static receptive field can be redefined as a spatiotemporal average over an entire receptive field and over a few hundred milliseconds.

Related to the 2nd-TC by Richmond, the reports by Dinse *et al.* [1990, and references cited therein] on the presence of dynamic tuning of orientation and direction selectivity of neurons in the primary visual cortex would be important. In addition to the static neurons found by Hubel and Wiesel, Dinse *et al.* observed neurons which alter the specificity within a few hundred milliseconds after the onset of stimulations. Tsuda predicted the presence of this type of dynamic neurons in 1987, and made a neural-network model. Then, computer simulations showed the reorganization to alter the neuron's specificity and its other functions [Tsuda, unpublished].

One question arises for the Walsh-pattern analysis by Richmond. This analysis is a linear one just like Fourier analysis, and hence it is difficult to evaluate the computational results when it is applied to a system where nonlinearity is essential. In his analysis, it was assumed that the relation between the response of a neuron and the stimulus is linear, but it has not been assured.

Next, we proceed with the discussion about neuro-oscillators in the cortex, which was simulated by Shimizu *et al.* According to the mechanism of genesis of oscillations, this is an example of the 1st-TC or 3rd-TC. If entrainment among neuro-oscillators is essential, it will belong to the class of the 3rd-TC. If synchronization among spike sequences of neurons is essential, it will be in the 1st-TC. If the presence of inherent neuro-oscillators is assumed, the coding of, for instance, orientation in terms of phases or frequencies allows entrainment among nonlinear oscillators representing neurons with a similar orientation selectivity. Conversely, if the information is encoded in phases in a spike sequence, synchronization among spike sequences can occur, which may give rise to synaptic learning in a short time, and consequently synchronized neurons can constitute a large ensemble. The occurence of this case implies oscillations as a by-product.

It would be interesting to think that chaos is related to this information processing [Tsuda 1993]. Whichever, entrainment or synchronization, is essential, these would be effective for rapid perception and judgment. But, in such a case, *off*-entrainment or *de*-synchronization is also necessary. If the degree of entrainment or synchronization is proportional to the connection strengths of neurons, the time necessary for it is of the order of the inverse of the connection strength. On the other hand, off-entrainment or de-synchronization can be determined by the degree of chaos. It is measured by the largest Lyapunov exponent. Hence, the time necessary for off-entrainment or de-synchronization is estimated by the inverse of the largest Lyapunov exponent. As another possibility of off-entrainment or de-synchronization, one can think of the role of noise. But, it is not effective for the following reason. Since the orbital separation by noise is realized in at most a power of time, being different from chaos, an infinite time is necessary for off-entrainment or de-synchronization in principle. Actually, Sompolinsky and Crisanti showed with a computer simulation that a rapid off-entrainment or de-synchronization cannot occur with noise [Sompolinsky *et al.* 1988]. Thus the significance of temporal coding in Singer's experiments lies in the rapid judgment by means of dynamical processes [Tsuda 1993]. Kaneko and Tsuda [1991, unpublished] investigated the alteration of entrainment in coupled circle maps (see also Sect. 4.8), where the maps are coupled globally in one direction and in another direction coupled only with the nearest neighborhoods. Under the condition that the elementary map shows chaotic behavior, the elements with the inputs are almost synchronized, and the synchronized region moves smoothly according to the movement of the inputs. Thus, we observed a rapid de-synchronization in the chaotic dynamical system.

A typical example of the 3rd-TC is seen in the dynamics of the rabbit olfactory EEG found by Freeman and his colleagues. Olfactory information is represented by a spatially synchronized pattern of the olfactory bulb. But this spatial pattern depends on the history of motivated learning. Furthermore, the activity of the bulb's network can be described by a multiple-wing chaotic attractor generated via chaotic itinerancy in the process of learning. It is, in particular, important to note that the olfactory activities do not encode odorants, but rather that odor information is generated inside the brain, depending on the past states of perception and the present motivation. In the olfactory system, it has actually been shown that such a chaotic process plays the role of a *novelty filter*. Thus the perception process of odor is represented by chaotic dynamical systems. This is the reason why we classify it as being different from the 1st- and the 2nd-TCs. The 3rd-TC may be called a dynamic *representation* of meaning, rather than using the term *coding*.

As already described, Tsuda found a dynamic process of memory concerning the external information encoded in a spatial pattern of activity of nonequilibrium neural networks. In this process, the dynamic linking of memories is represented by a relatively low-dimensional chaotic dynamics. In other words, the rule for the linking itself is chaos, which is generated in the network. Such a chaos subserves for a novelty filter and for the enhancement of learning. It is also possible to obtain the transition between memories by using an external noise, but in such a case one cannot expect causality for the linking of memories since noise breaks the dynamical causality. Thus chaotic activities in the brain can play functional roles, preserving causality on macroscopoic scales [Tsuda 1994b].

6.11 Capillary Chaos as a Complex Dynamics

6.11.1 Significance of Capillary Pulsation in the Brain Functions

In this section, we discuss chaos found by Tsuda *et al.* [1992] in human capillary vessels and its relation with brain dynamics. It is also one of the themes of complex systems.

The recent development of dynamical systems' theory enables us to interpret systematically complex behaviors in physical and even in biological systems. In particular, the technique of embedding [Packard *et al.* 1980; Takens 1981] of observed data into finite or infinite dynamical systems may force a change in the analysis of random motions. Before the development of such a technique, one tried to calculate the average, the standard deviation and, if necessary, higher moments of an observed random variable as well as a probability distribution function. Whereas, in the present, one may try to analyse random motions as an entity, not as a decomposed one as above, by extracting an implicated order in the form of a nonlinear smooth manifold, i.e., geometry [Campbell *et al.* 1984].

By the embedding technique, the presence of deterministic chaos has been clarified in many complex systems. In particular, evidence of deterministic chaos in human brain and heart has been obtained in several experiments by adopting this technique to the data of an electro-encephalogram (E.E.G.) [Babloyantz 1986; Layne *et al.* 1986; Rapp *et al.* 1989] and an electrocardiogram (E.C.G.) [Babloyantz and Destexhe 1988; Goldberger *et al.* 1988].

Tsuda *et al.* found another human chaos: the capillary chaos. Recent studies on applications of dynamical systems' theory to biological systems raise expectations of the discovery of novel indications for the process of recovery of health. Such appropriate indications could be utilized in care and cure [Rapp 1980, 1989]. It is worth studying the correlation of the peripheral data with controlled conditions by reconstructing the dymanics of the peripheral, since the peripheral activities vary easily with changes of mental or physical conditions.

One of the crucial problems for human brain research is how to get an indication for the recovery of mental health, i.e., a care and rehabilitation problem. The motivation of the study of human capillary vessels is to find a good indicator for the rehabilitation of a psychiatric patient as well as for a daily self-care. This also leads to a study of the peripheral. If the patient could be aware of his current condition by means of an appropriate indicator that is simple and, in particular, appears as a visual image, such an indicator would be available for use in the process of healing and rehabilitation. This idea could be generally applied to daily health-care.

Motivated by this fundamental idea of self-care, Tsuda *et al.* recorded a time series of the pulsation of capillary vessels, and found chaotic pulsation in a finger's capillary vessels in both normal subjects and psychiatric patients. They observed the geometry of an attractor constructed from the time series of a single variable, i.e., the peripheral blood pressure, and they also calculated the Lyapunov exponents from the experimental data, which can express the degree of orbital instability. The results proved practically the presence of deterministic chaos in human capillary vessels. Forms of the chaos depended on the mental or physical conditions of the subjects; here the term "form" indicates a triple of geometry, orbital instability, and information.

Furthermore, cardiac activities, i.e., the beating of the heart simultaneously measured by electro-cardiography with a pulsation of the capillary vessels, also exhibited deterministic chaos, whose forms were, on the contrary, almost independent of the subject's condition. This leads to a hypothesis on autonomic nerve innervation. The hypothesis is based on the experimental evidence indicating that central nervous systems, peripheral cardiovascular systems, and metabolic systems are inseparable.

They also studied the ability of the observed chaos in detecting information fed from outside. This ability was measured by mutual information. From this information-theoretical point of view, we obtain a scenario on the role of chaotic processors for a dynamic maintenance of living systems.

6.11.2 Embedding Theorems

Although several problems which we face in data analysis were pointed out, the fundamental significance of the embedding technique has not been weakened. Accordingly, we first summarize embedding theorems, and then introduce the work of Tsuda *et al.* on chaotic hermeneutics in mental health-care, where only the fundamental technique of embedding was adopted.

Let $\varphi : M \to M$ be a diffeomorphism or a flow (vector fields) on a compact manifold M.

Here, an observable is assumed to be a smooth function $g : M \to \mathbf{R}$, where φ, g are at least a C^2 function. Let us ask as follows:

"In the observation of the function $t \mapsto g(\varphi^t(x)) \equiv G(t)$ for time evolutions φ^t of some dynamical system, how does one know the information of the original dynamical system?"

Takens provided one possible answer for this question [Takens 1981]. It is the following embedding theorems.

Theorem 1 Given an m-dimensional manifold M, for a dense pair (φ, g) in functional space,

$$
\begin{aligned}
\text{a mapping } \Phi_{(\varphi,g)} &: M \to \mathbf{R}^{2m+1} \\
\Phi_{(\varphi,g)}(x) &= (g(x), g(\varphi(x)), \cdots, g(\varphi^{2m}(x)))
\end{aligned} \tag{6.29}
$$

is an embedding.

Theorem 2 For a dense pair of vector fields X and g on an m-dimensional manifold M,

$$
\begin{aligned}
\text{a mapping } \Phi_{(X,g)} &: M \to \mathbf{R}^{2m+1} \\
\Phi_{(X,g)}(x) &= \left(g(x), \left.\frac{dg(\varphi^t(x))}{dt}\right|_{t=0}, \cdots, \left.\frac{d^{2m}g(\varphi^t(x))}{dt^{2m}}\right|_{t=0} \right)
\end{aligned} \tag{6.30}
$$

is an embedding.

For experimental data $\{G(i)\}_{i=1}^{N}$, a large error happens when applying (6.30) hence (6.29) is usually applied and the embedding theorem is adopted for data $\{G(i), G(i+k), \cdots, G(i+2mk)\}$ with an appropriate value of k [Packard et al. 1980; Shaw 1984].

One should note that the data could be embedded in space with lower dimension than $2m + 1$ since the theorem assures a sufficient condition.

6.11.3 Experimental Systems

The setup of the experiment is shown in Fig. 6.6. The data were recorded from the surface of the bulb of the left forefinger by detecting light reflected by the vascular tissues of the infrared ray with wavelength 940 nm emitted from the light emitting diode (LED). The peripheral blood pressure was measured by the photo-coupler attached to the inner surface of the cuff which fixed the measurement place. The light emitted from the infrared LED was reflected from the vascular tissues and detected by the photo-transistor. Detected light intensity which was transformed to an electric signal by the photo-transistor was stored in computer through the A/D converter after being amplified by 10 000 times. In the measurement, a sampling frequency is 200 Hz with 12-bit resolution.

A forefinger of the left hand was consistently chosen for all the subjects. Tsuda et al. also detected almost the same output for the other seven fingers except thumbs; the thumbs exhibit a slightly different data set. They tried to measure from toes, but did not obtain an apparent oscillatory activity.

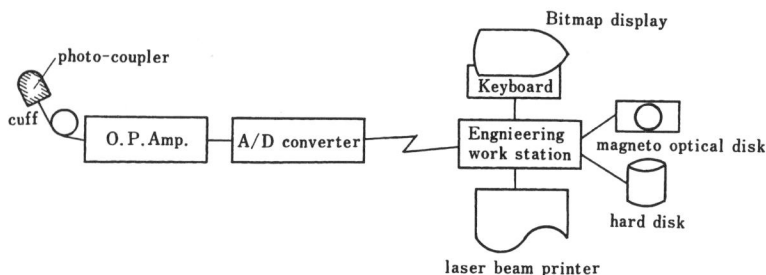

Fig. 6.6. A schematic drawing of the experiment. (From I. Tsuda *et al. Int. J. Bifurcation and Chaos* **2** (1992) 313, with the permission of the publishers.)

If the finger cannot move, a variable part in the data is reasonably derived from the motion of the blood flow. It is, however, questionable if the effect of the motion of the bulb could be removed. To check this, another measurement was made, where the data were taken on the surface of the nail. The result was the same, except for a slight decrease of the output intensity.

In each capillary vessel, blood flows in pulsatile fashion, namely it is not a smooth movement. For the spatial extension of the present measurement place, it will be a plausible estimation that it is within a few millimeters square. Thus the apparatus can measure the time series of the collective pulsation of the capillary vessels, namely, the peripheral blood pressure. The data were taken from 20 normal subjects and 15 psychiatric patients.

6.11.4 Reconstruction of the Dynamics

A typical time series is shown in Fig. 6.7a. An oscillation with a period of about 1 s is dominant, which is a reflection of the cardiac activity. In a long time observation (about 100 s in the present case), however, both amplitude and period fluctuate over a rather wide range. The dynamical structure from these experimental data was reconstructed, by adopting the method of embedding.

For a variable $x(t)$ denoting a time series of the peripheral blood pressure, new variables $y(t) = x(t + \tau), z(t) = x(t + 2\tau), w(t) = x(t + 3\tau)$, etc. are defined, where τ is the order of the correlation time. In the following embedding, $\tau = 25$ ms is adopted. In the embedding into three-dimensional phase space, they observed some complicated structure of the reconstructed attractor, but could not obtain a consistent topology with possible vector fields of three-dimensional dissipative dynamical systems. This indicates that at least the fourth dimension is needed to satisfy the topological consistency; hence the embedding into four-dimensional phase space $(x(t), y(t), z(t), w(t))$ was made. Fig. 6.7b shows one example, where the orbits in a new $(x'(t), y'(t), z'(t))$-coordinate are shown with a more bright curve as w becomes larger, taking the parallel projection of a supposed four-dimensional object.

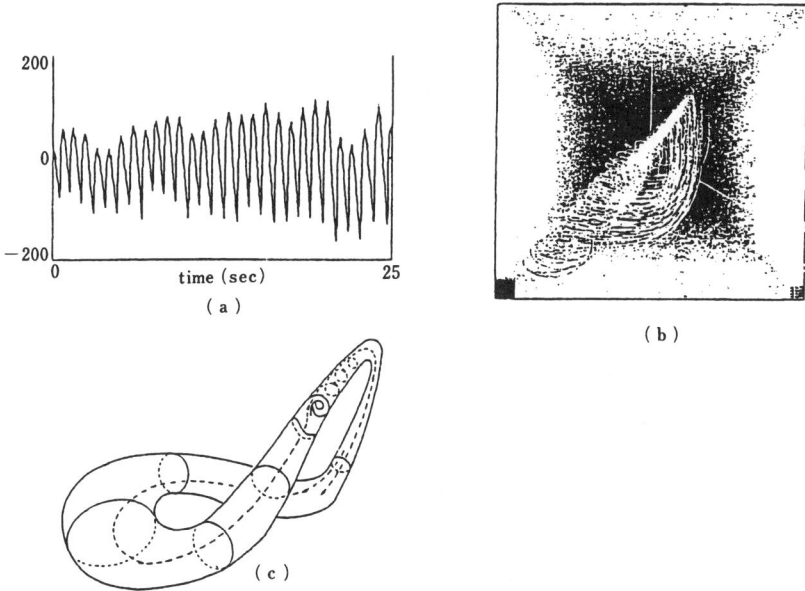

Fig. 6.7. (a) A typical time series of capillary pulsation measured from the surface of a human finger (the subject H. T. in reading condition). It is considered as a motion of pulsation of capillary vessels averaged over a few cubic millimeters. (b) Three-dimensional parallel projection of orbits in four-dimensional embedding. The direction of sight of viewing the orbits, $n = (3/4, 1/4, 1/4, \sqrt{5/4})$. The larger the value of $w(t)$ is, the more bright is the orbits depicted. The background shows the contrast. (c) A model for a topology of capillary chaos. (From I. Tsuda et al. Int. J. Bifurcation and Chaos 2 (1992) 313, with the permission of the publishers.)

The projection is given by the following equation:

$$x'(t) = n_2 x(t)/A - n_1 y(t)/A,$$
$$y'(t) = n_1 n_3 x(t)/AB + n_2 n_3 y(t)/AB - A z(t)/B, \qquad (6.31)$$
$$z'(t) = n_1 n_4 x(t)/B + n_2 n_4 y(t)/B + n_3 n_4 z(t)/B - B w(t),$$

where $A = (n_1^2 + n_2^2)^{1/2}, B = (1 - n_4^2)^{1/2}$, and $\mathbf{n} = (n_1, n_2, n_3, n_4)$ is a unit vector representing the direction of sight of viewing the orbits. The fourth axis w was rotated to coincide with the direction of that unit vector.

Figure 6.7c shows a possible model for the skeleton of the attractor. It is reasonable to think of the three-dimensional torus as the ground state of the peripheral blood pressure. This is because the main dynamical components of the peripheral blood pressure will be the three distinct oscillations: the heart rhythm, the respiration cycle, and the hormone cycle of the fast type. If these three oscillatory components are incommensurate with each other in frequency, a three-dimensional torus is obtained in phase space. If there are nonlinear interactions among these components, this torus could be destabilized and chaos can appear [Ruelle and Takens 1971]. Thus capillary chaos could be a strange attractor generated from a torus, following the Ruelle–Takens scenario. A similar model in the four-dimensional embedding has also been proposed in the local E.E.G. of mammalian brain [Freeman 1987; Yao and Freeman 1990].

6.11.5 Calculations of Lyapunov Exponents

Since the observed attractors are very nonuniform, conventional methods for the measurement of the attractor's dimension such as the Grassberger–Procaccia method [Grassberger and Procaccia 1983a, b] are inappropriate for such attractors. In general, even when the overall attractor is nonuniform, the Grassberger–Procaccia method is applicable to the Poincaré sections if uniformity of invariant density on the sections is presumed [Schaffer *et al.* 1988]. In the present case, however, it was practically impossible, because of the difficulty in recording indefinitely many data of the peripheral blood pressure, assuring its stationarity. The maximum number of data recorded was about 20 000 sampling points by the measurement with 5 ms sampling time. This number is too small to assure the invariance of the probability density of orbits on three-dimensional Poincaré sections. Furthermore, referred to Mandelbrot's elaborate work on fractal geometry appearing in nature and fractal dimension, the correlation dimension may also become fractional also in the case of nonchaos, for instance, colored noise, stochastic processes like Levy flight, etc. [Mandelbrot 1977]. The presence of a fractional dimension does not always indicate the presence of chaos.

Since the presence of a positive Lyapunov exponent indicates the orbital instability, the Lyapunov exponent can be a measure of deterministic chaos. However, even adopting this exponent, it is unable to decisively determine whether the data are chaotic or not. Furthermore, the algorithms proposed so far for the estimation of the Lyapunov exponents also have decisive weak points such as the impossibility of discriminating chaos and noisy limit cycles in critical systems, and the possibility of the appearance of spurious positive exponents when the embedding dimension is much higher than the system's dimension [Eckmann *et al.* 1986]. The former becomes serious in the case that the nonuniform stability on some subspace transversed by the limit cycle orbit occurs, and also in the case that an excitable system such that a saddle point and a stable node which are close in phase space is influenced by noise whose amplitude is larger than the distance of these fixed points. The latter becomes destructive if the system's dimension is not known in advance. This is decisive in the case of highly nonuniform attractors, in which the precise estimation of the correlation dimension is hopeless, as in the present case.

Tsuda *et al.* practically judged whether the capillary data are chaos or not mainly with calculations of the Lyapunov exponents from experimental data, and additionally calculated the correlation dimension. The estimation of the largest exponent is relatively reliable, since the arbitrarily chosen vector

Table 6.1. The first and the second Lyapunov exponents. A bit number in the unit of 50 ms.

Subject	State	$\lambda 1$	$\lambda 2$
H.T.	resting	0.56 ± 0.013	0.14 ± 0.006
H.T.	reading (interest for the subject)	0.40 ± 0.009	0.12 ± 0.006
K.M.	resting	0.44 ± 0.008	0.18 ± 0.009
K.M.	reading (no interest for the subject)	0.41 ± 0.018	0.12 ± 0.070
K.M.	reading (interest for the subject)	0.48 ± 0.026	0.24 ± 0.015
K.M.	looking at a colorful graphics	0.40 ± 0.015	0.07 ± 0.02
S.S.	before treatment	0.37 ± 0.021	0.27 ± 0.073
S.S.	under treatment	0.48 ± 0.004	0.14 ± 0.017
S.S.	after treatment	0.41 ± 0.006	0.13 ± 0.029
K.M.	V4-induction of electrocardiogram	0.13 ± 0.019	$-$

quickly tends to the direction of the unstable manifold. This is a reason why we have taken here the Lyapunov exponent as a practical measure of deterministic chaos.

The calculation of the Lyapunov exponents from the experimental data showed the presence of a positive exponent. The results are summarized in Table 6.1. In the calculation of the Lyapunov exponents, the Wolf method was used with some modification [Wolf *et al.* 1985]. The number of the present data is greater than, but close to the theoretical lower bound of the number of the data needed to estimate a correct value of the Lyapunov exponent in the case of four-dimensional embedding. Whether the algorithm works correctly was checked, applying it to the Lorenz attractor.

If orbits are sufficiently embedded into four-dimensional phase space, it is concluded that the third exponent should vanish and the fourth exponent should take a large negative value, namely the absolute value of the fourth exponent should be greater than the sum of the first and the second exponents. Moreover, in general, if the embedding dimension is lower than the dimension of the attractor, the degree of orbital instability would seemingly decrease. This situation could give a lower value than the actual Lyapunov exponent. Thus the calculated exponents would give the lower bounds. These considerations show that the pulsation of the capillary vessel can be described by deterministic chaos. The positive second exponent indicates the correctness of our assumption that the attractor should be described in at least a four-dimensional dynamical system.[3]

---

[3] The empirical Lyapunov spectrum with other methods such as Sano–Sawada method [1985] and Eckmann *et al.* method [1986] methods was also calculated, by increasing the embedding dimension. The largest Lyapunov exponent gives the same value as in Table 6.1 within the precision. However, it was quite difficult to determine whether the second exponent is zero or positive. Both Sano–Sawada method and Eckmann *et al.* method are convenient for obtaining the whole spectrum simultaneously. It is, however, still questionable whether the tangent space is correctly spanned. In particular, the assumption that the vectors in the sphere are uniformly distributed is highly questionable in our system. An elaborated algorithm has been published [Barna and Tsuda 1993].

An additional calculation of dimension gave the value 2.9 ± 0.5. These results suggest that the capillary chaos is embedded in three- or four-dimensional Euclidean space. On the other hand, as mentioned in Sect. 6.11.4, capillary pulsations should possess main three oscillatory components, which can form a three-dimensional torus. By this inference, it will be plausible to think that capillary chaos is embedded in four-dimensional Euclidean space.

6.11.6 The Condition Dependence

For various mental or physical conditions of the subjects, the dependence of chaos on such conditions was investigated. For normal subjects, moderate conditions such as resting, calculating simple arithmetic, drinking, exhaustion, unrest, sleeping, reading, looking at pictures, etc. were adopted. For psychiatric patients, only a resting condition was used after informed consent with subjects or their families. For immature babies, only conditions within and without an incubator in a neonatal intensive care unit (NICU) were for investigation.

Attractors from recorded data were reconstructed, and the Lyapunov exponents were calculated. The forms of chaos, the geometry and size of trajectories as well as the Lyapunov exponents were rather sensitive to those conditions, possessing basal forms specific to the individual. Under the same condition for a subject, the successive measurements assured invariance of the forms of chaos.

Figures as shown below are depicted with the same sight, and with the same scale for the same series, hence variations of the form of attractor likely indicate qualitative alteration of peripheral blood pressure. Figure 6.8a shows the capillary chaos in normal subject K. M. for the respective condition, resting (upper left), reading a mathematical text in which the subject has no interest (upper right), reading a story comic in which the subject has interest (lower left), and looking at a picture which suddenly appeared on a screen (lower right). Figure 6.8b shows the capillary chaos in normal subject T. T. for the conditions, resting (left), exhaustion (middle), and drinking a little alcohol (right).

We do not here show the data, but the capillary chaos of cardiac patients and the one of the patients of brain infarction possesses a specific randomness. The case of brain infarction is of particular interest. Tsuda et al. personally reported (unpublished):

"We measured a peripheral blood pressure of a subject having slight migraine, but no other subjective symptoms. A simultaneous recording of it from right and left forefingers showed an apparently different structure of capillary chaos between both fingers. The data from the left finger showed a distinct geometry, compared with a normal subject. We conceived that either his cardiac system or some part of the right hemisphere has been damaged. The doctor in our group suggested that he sees the other doctor for a medical check. A slight brain infarction was found in the right hemisphere."

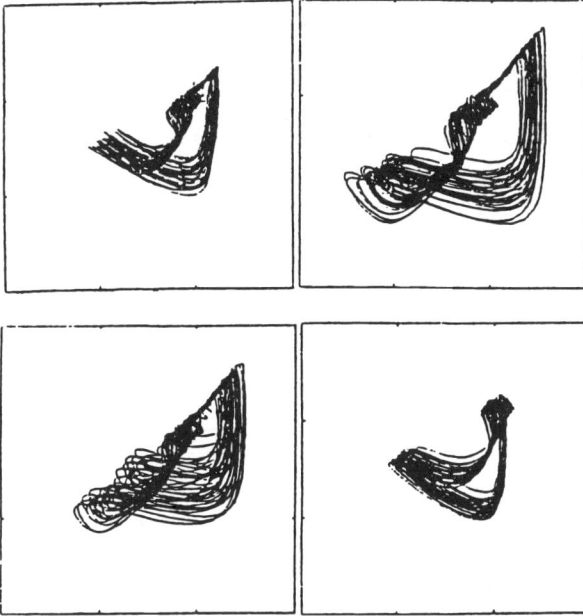

(a)

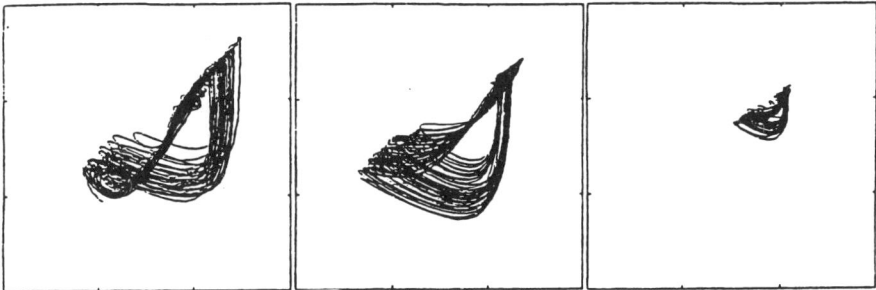

(b)

Fig. 6.8 a,b. The state-dependence of capillary chaos. The data were successively taken within an hour for the same subject. The sight is the same as in Fig. 6.7. The scale is the same for the same subject. (**a**) Normal subject K. M.: The upper left is for resting, the upper right for reading a book in which the subject has no interest, the lower left for reading a book in which the subject has interest, and the lower right for looking at a colorful unfamiliar animal which suddenly appeared on the screen. (**b**) Normal subject T. T.: From the left, resting, exhaustion, drinking a little alcohol. (From I. Tsuda *et al. Int. J. Bifurcation and Chaos* **2** (1992) 313, with the permission of the publishers.)

Figure 6.9a shows a capillary chaos for the patient M. S. of senile dementia, and Fig. 6.9b for the patient H. M. of schizophrenia. Features of chaos in Fig. 6.9a are typical for patients of senile dementia. On the other hand, it should be emphasized that none of the features in Fig. 6.9b show a specificity of schizophrenic patients. The Lyapunov exponents up to the second one are positive, the same as in normal subjects, in other words, a similar attractor can appear also in normal subjects when they are out of physical condition.

A specificity of senile dementia manifests also in the Lyapunov exponent. The first exponent is positive but has a small value, and the second one is zero within the precision. This indicates that the peripheral blood pressure in senile dementia is weakly chaotic, rather close to periodic oscillation. It should be, however, noted that whether this characteristic stems from dementia or aging of blood vessels is not determined.

Figures 6.9c,d are the data from premature newborn babies in a neonatal intensive care unit (NICU): (c) a newborn baby in an incubator, and (d) a newborn baby out of the incubator, respectively. The data in (c) show zero as the largest Lyapunov exponent; hence it could not be distinguished from

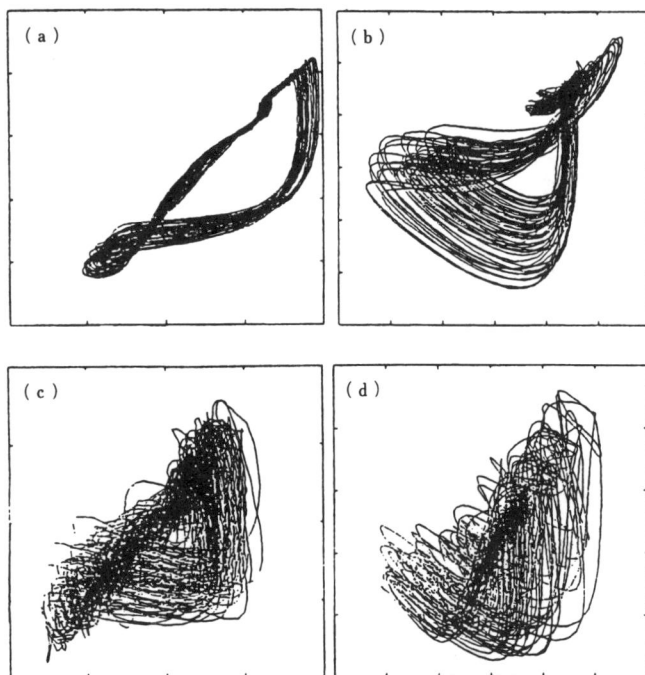

Fig. 6.9. (a) Senile dementia of Altzheimer type, M. S. (b) Schizophrenic patient H. M. (c) Immature new-born baby under treatment in an incubator. (d) The same baby as in (c) just out of the incubator. (From I. Tsuda *et al. Int. J. Bifurcation and Chaos* **2** (1992) 313, with the permission of the publishers.)

noise, and in (d) the largest Lyapunov exponent is near zero, but positive within the precision. Both data have not much structure, compared with adults' ones. It should be, however, noted that a newborn baby out of the incubator has more structure than a newborn baby in the incubator, leading to the suggestion that a normal newborn baby is already "chaotic".

Therefore, the features of the chaotic attractor reflect the degree of physical or mental activity (health) or the degree of maturity. This indicates that the feature alteration can also be an appropriate indicator in the process of the care of mental or physical diseases. We hypothesize on the process of growth and aging as follows: a capillary pulsation in a baby is rather noisy, but chaotic with a high dimension. In the growth process, it structuralizes to show a low-dimensional chaos. The most structuralization is represented by chaos generated after a collapse of a three-dimensional torus. In cases of physical and/or mental illness and aging, the degree of chaos decreases, showing a simpler structure close to a periodic oscillation. Dominance of a periodic component indicates danger in cardiovascular systems and even other systems. Only a periodic structure implies death.

What is important is that such a representation by attractors could be a good indicator of the rehabilitation process of physical and mental diseases. Actually, this method was applied to the rehabilitation process of a slightly neurotic patient in order to check whether or not the chaotic representation obtained here can be utilized as an indicator of the degree of recovery of mental health. Figure 6.10 shows reconstructed chaos at respective stages before, under, and after treatment. A conspicuous disorder in hospitalization (Fig. 6.10b, see also Table 6.1) can be considered to stem from both a self-discord of the patient and drugs for medical treatment. After recovering mental health (Fig. 6.10c) the dimension of the attractor is seemingly reduced, and there appears a complicated screw-type structure which was not so conspicuous before medical treatment (Fig. 6.10a). It should also be emphasized that

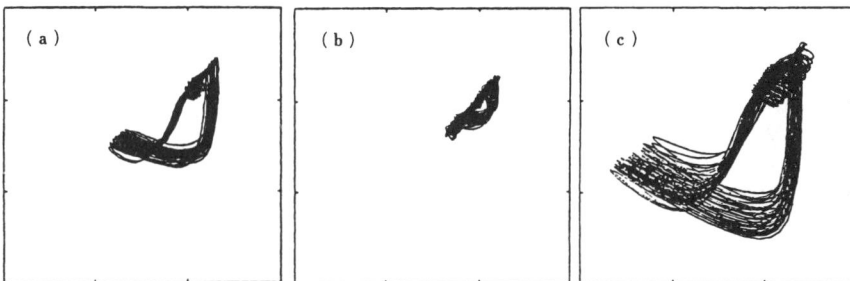

Fig. 6.10a–c. The neurotic patient S. S. (a) Before treatment. (b) Under treatment. (c) After treatment. (From I. Tsuda *et al. Int. J. Bifurcation and Chaos* **2** (1992) 313, with the permission of the publishers.)

the size of the attractor after the recovery becomes greater than that before treatment.

Readers might be suspicious about confusion of physical alteration and psychological alteration. The observations so far, however, seem to show the inseparability of physical processes and mental processes, in particular, in the peripheral.

6.11.7 Cardiac Chaos

The capillary chaos possesses a periodicity close to that of cardiac pulsations. Hence, to what extent it is influenced by cardiac pulsations is of interest. It is already known that the intervals of cardiac pulsations are chaotic. Tahara *et al.* made simultaneous measurements of the beating of the heart and the capillary motion. A long-time recording of that beating in terms of the V4-induction of the electrocardiograph also exhibited deterministic chaos in a subject without any heart disease (see also Babloyantz and Destexhe [1988]; Goldberger *et al.* [1988]), but as shown in Fig. 6.11, its topology differed very much from that of the chaos of the capillary vessels, which is quite similar to that of the upper-left pattern in Fig. 6.8a. Moreover, the topology of the cardiac chaos is insensitive to subjects and their conditions if heart diseases such as myocardial infarction, atrial fibrillation, and irregular pulse are not recognized. In Fig. 6.12, a capillary chaos of another normal subject T. Y. (a) and his cardiac attractor simultaneously recorded (b) are shown. The difference is seen in capillary chaos of K. M. and T. Y., but the cardiac attractor is almost the same in both subjects.

The Lyapunov exponent was also calculated. Since rapid velocity changes of orbits are seen in several places on the attractor, it is difficult to calculate the exponent precisely. Hence, calculated exponents should be viewed as an average local divergence rate. The first average rate was positive in both

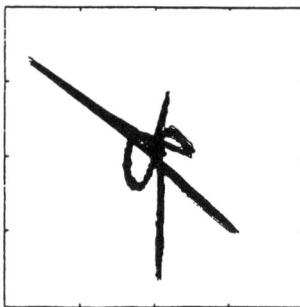

Fig. 6.11. Attractor obtained from electrocardiogram of normal subject K. M. (From I. Tsuda *et al. Int. J. Bifurcation and Chaos* **2** (1992) 313, with the permission of the publishers.)

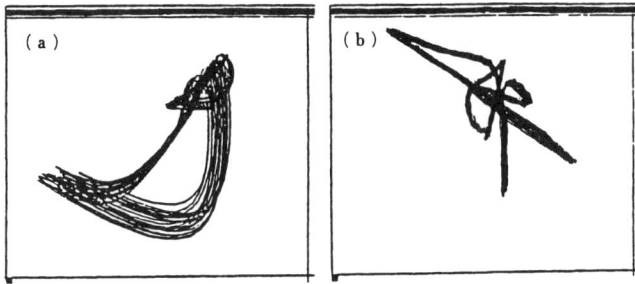

Fig. 6.12. (a) Capillary chaos of normal subject T. Y. (b) Cardiac attractor simultaneously recorded with the data in (a). (With the courtesy of Takashi Tahara.)

cases of Fig. 6.11 and Fig. 6.12b, but the second rate could not be calculated. Taking into account other reports on cardiac chaos, it seems to be plausible to think that observed attractors are chaotic.

In Fig. 6.13 is shown, a cardiac attractor of the patient of atrial fibrillation, and that of the patient of myocardial infarction is shown in Fig. 6.14a. The difference of topology is obvious. In Fig. 6.13b is shown the capillary chaos simultaneously recorded for the patient of myocardial infarction. Abnormal topology seemingly stems from the heart disease. The capillary chaos could be an indication of the degree of health, but the cardiac chaos could represent a kind of heart disease.

6.11.8 Information Structure

The cardiovascular system is an information channel [Mandell 1987] as well as the cortical nervous system. The peripheral system is, in particular, considered as a control system correlating with the nervous system. Hence, it is worth studying the information capacity of the observed chaos, namely the

Fig. 6.13. Cardiac attractor of the subject of atrial fibrillation. (With the coutesy of Takashi Tahara.)

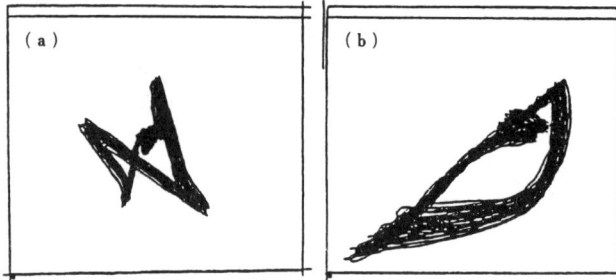

Fig. 6.14. (a) Cardiac attractor of the subject of myocardial infarction. (b) Capillary chaos simultaneously recorded with the data in (a). (With the courtesy of Takashi Tahara.)

rate for the transmission of information fed from outside. When equations of motion of the chaotic system concerned and of the input system are known, the formula of calculation was established [Matsumoto and Tsuda 1985, 1987, 1988]. In this book, it is described in relation to the dynamics of the unidirectional CML in Chaps. 2 and 5. In order to study this on the experimental data, where the equations of motion are unknown, a new algorithm is needed. A simple algorithm for the computation of mutual information between the experimental data and the other dynamical system has been proposed [Tsuda *et al.* 1992].

Let $\{\mathbf{x}(n)\}$ denote the n-th orbital point in the M-dimensional vector space. The embedding of experimental data into M dimensions allows this assignment. Let $\{\mathbf{y}(n)\}$ denote the n-th orbital point of the other data set in the M'-dimensional vector space. A set $\{\mathbf{y}(n)\}$ is supposed to have been computed in a numerical simulation of the dynamical system, or obtained in another experiment. Both sets $\{\mathbf{x}(n)\}$ and $\{\mathbf{y}(n)\}$ are numbered in the order of evolution. We consider the following type of forced system:

$$\mathbf{x}(t+1) = \mathbf{f}(\mathbf{x}(t)) + C\mathbf{y}(t),$$
$$\mathbf{y}(t+1) = \mathbf{h}(\mathbf{y}(t)), \tag{6.32}$$

where C is a matrix of coupling constants, whose elements are expressed by c_{ij}, $i = 1, \cdots, M$, $j = 1, \cdots, M'$.

Suppose that the solutions achieved in the case of $c_{ij} = 0$ for all i and j give the data sets $\{\mathbf{x}(n)\}$ and $\{\mathbf{y}(n)\}$ ($1 \leq n \leq N$). Our aim is to construct a time series $\mathbf{x}'(n)$ which closely resembles a solution of (6.32). Choose $\mathbf{x}(1)$ for $\mathbf{x}'(1)$. In each step we compute the exact evolution from $\mathbf{x}'(n)$ using the term $\mathbf{f}(\mathbf{x}'(n)) + C\mathbf{y}(n)$. However, since the effect of $\mathbf{f}$ is known only for the elements of the data set $\{\mathbf{x}(n)\}$, we substitute the nearest element of this set for the exact evolution.

The procedures can be written as follows:

$$\mathbf{x}'(1) = \mathbf{x}(1),$$
$$\mathbf{x}'(n+1) = \mathbf{x}(k), \tag{6.33}$$

where $\mathbf{x}(k)$ satisfies

$$||\mathbf{f}(\mathbf{x}'(n)) + C\mathbf{y}(n) - \mathbf{x}(k)|| = \min_{1 \le l \le N} ||\mathbf{f}(\mathbf{x}'(n)) + C\mathbf{y}(n) - \mathbf{x}(l)||.$$

Divide both $\{\mathbf{x}'(n)\}$ and $\{\mathbf{y}(n)\}$ into m cells. Find the probability $p_i (i = 1, 2, \cdots, m)$ of $\{\mathbf{x}'(n)\}$ entering the i-th cell, and the conditional probability $p_{ji}^{(t)} (i = 1, \cdots, m, \ j = 1, \cdots, m)$ that $\{\mathbf{x}'(n)\}$ enters the i-th cell at time $k+t$ under the condition of $\{\mathbf{y}(n)\}$ entering the j-th cell at time k. Then, one can define the time-dependent mutual information [Matsumoto and Tsuda 1985, 1987, 1988] as follows:

$$I(t) = -\sum_{i=1}^{m} p_i \log p_i + \sum_{j=1}^{m} \sum_{i=1}^{m} p_j p_{ji}^{(t)} \log p_{ji}^{(t)}. \tag{6.34}$$

This quantity indicates a time course of shared information between two data sets $\{\mathbf{x}'(n)\}$ and $\{\mathbf{y}(n)\}$, in other words, information transmitted from $\{\mathbf{y}(n)\}$ to $\{\mathbf{x}'(n)\}$, since the coupling is unidirectional in the present case.

As the simplest case, let us suppose that $c_{kl} = c\delta_{kl}$, where c is a constant and the probabilities p_i and $p_{ji}^{(t)}$ are calculated in terms of only the first component of both $\{\mathbf{x}'(n)\}$ and $\{\mathbf{y}(n)\}$. To see the relevance of the algorithm (6.33), the mutual information is calculated in the Lorenz chaos forced by another Lorenz chaos in two ways, i.e., by means of the above algorithm and by the equations of motion. The computed system is given as follows:

$$dx_1/dt = -\sigma_1 x_1 + \sigma_1 x_2 + cy_1,$$
$$dx_2/dt = -x_2 + r_1 x_1 - x_1 x_3,$$
$$dx_3/dt = -b_1 x_3 + x_1 x_2, \tag{6.35}$$
$$dy_1/dt = -\sigma_2 y_1 + \sigma_2 y_2,$$
$$dy_2/dt = -y_2 + r_2 y_1 - y_1 y_3,$$
$$dy_3/dt = -b_2 y_3 + y_1 y_2,$$

where $\sigma_1 = \sigma_2 = 10$, $r_1 = r_2 = 28$, $b_1 = b_2 = 8/3$.

The results are shown in Fig. 6.15. The reason why $I(t)$ is almost invariant for a change of the coupling constant in the case of using the algorithm (6.33) is that the orbits are simply renumbered and new values can never be added by forcing. In spite of this weak point in the algorithm, as is seen in the figure, several cases of the coupling strength give even quantitatively good correspondence. Thus, one can adopt our algorithm as the first approximation

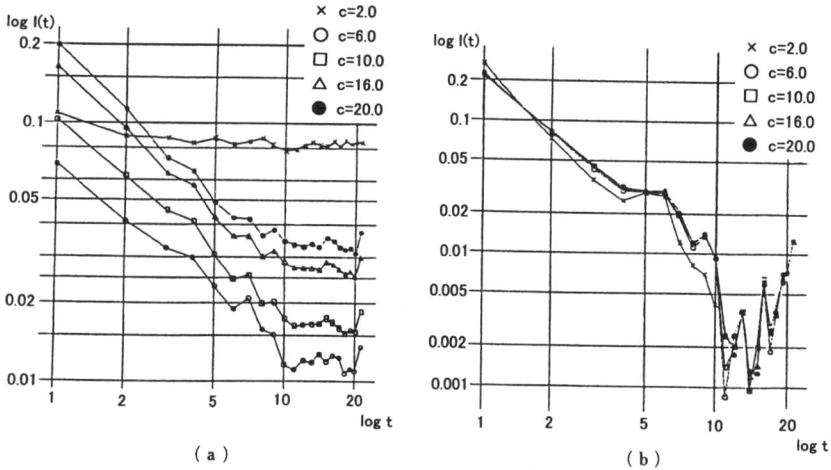

Fig. 6.15 a,b. Information quantity flowed from the Lorenz chaos to another Lorenz chaos. In the model, the coupling is unidirectional from the former to the latter. (a) Calculations by (6.36). (b) Calculations by the algorithm (6.34). (From I. Tsuda *et al. Int. J. Bifurcation and Chaos* **2** (1992) 313, with the permission of the publishers.)

for computing the information transmission to the experimental data from a known dynamical system or from other experimental data.

As is shown in Fig. 6.15a, the steepness of the decrease in a log–log plot is almost the same in all values of coupling strength except $c = 2$. This implies that the decay of information measured on a logarithmic scale is at a constant rate which is independent of the coupling strength, as far as the coupling strength is not too small. Moreover, the larger the coupling constant is, the larger is the transmitted information. Hence, in an information channel consisting of chaotic dynamical systems, the coupling strength determines the information content of the input, and its decay rate with time represents a channel capacity which is an inherent quantity of chaos. In Fig. 6.15b, the initial value of the information is constant for any coupling strength. This stems from the fact that in this algorithm no new data set is added.

We are here interested in chaos obtained in experiment as a receiver of information. This capability is determined by the way of decay of mutual information in time [Matsumoto and Tsuda 1987]. The capability is low in the case of linear decay, and high in the cases of exponential and power decay. The above algorithm is useful for the purpose of studying the decay rate.

The capillary chaos driven by the Lorenz chaos is shown in Fig. 6.16a. By this calculation, one can see the transmitted information to the capillary chaos from the Lorenz chaos. Thus, this quantity can be used to discriminate the ability of the capillary chaos to receive information fed from outside. One example of an embedded attractor of capillary chaos driven by the Lorenz chaos is shown in Fig. 6.16b. The power decay in the capillary chaos of

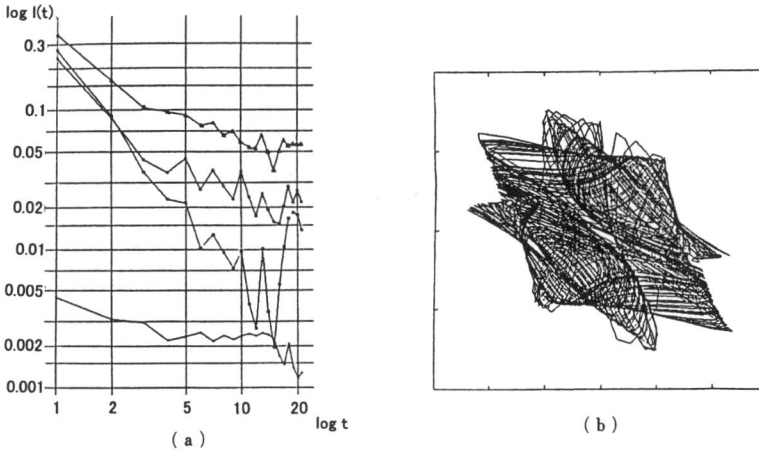

Fig. 6.16. (a) Information quantity flowed from the Lorenz chaos to the capillary chaos. From the above, the patient of senile dementia M. S., the neurotic patient S. S., the normal subject H. T. reading, and H. T. resting. (b) Attractor of capillary chaos connected by the Lorenz chaos (the case of H. T. reading). (From I. Tsuda *et al. Int. J. Bifurcation and Chaos* **2** (1992) 313, with the permission of the publishers.)

information created in the Lorenz chaos indicates that a network of capillary vessels can *dynamically* sustain external information [Matsumoto and Tsuda 1985; Tsuda 1991b, 1992a, 1995]. As mentioned earlier, such a network can perform a rapid nonlinear separation of patterns, acceleration of learning, and even provides a metaphor of deutero-learning (learning to learn [Bateson 1972]).

The choice of the Lorenz chaos is not essential for seeing the information-processing ability of the capillary chaos. One can choose other chaotic systems or quasirandom noise generators as a driving system. The calculated information quantity indicates a communication ability of the capillary chaos with the Lorenz chaos. By the present algorithm, one can know, in general, the information-storage capacity of any experimental data. Furthermore, by applying this algorithm to the various kinds of data sets, it will be possible to classify them in terms of their communication ability.

6.11.9 Implications of Capillary Chaos

Two crucial hypotheses have been proposed on metabolism. The first hypothesis is derived from the difference of the condition-sensitivity between chaos in the capillary vessels and those in the heart. Both the systems have been classified into the same category for autonomic nerve innervation. According to the observations by Tsuda *et al.*, however, it is plausible to think that there are at least two kinds of gates in the spinal cord. One is the gate with plasticity for innervating organs sensitive to mental or physical conditions, and the other is controlled rather automatically for innervating organs insensitive to those conditions. Only the latter should be called the autonomic nervous systems, whereas the former might be addressed as chaotically modulated autonomic nervous systems.

The cardiovascular systems exhibited deterministic chaos in their healthy conditions. This implies that at least these systems among organs innervated by the autonomic nervous systems need chaos to achieve a dynamic intelligent control [Tsuda *et al.* 1987; Tsuda 1991a, b], where chaos can buffer unexpected stimuli. Relating to this notion, the notion of *homeostasis* might be extended. By this extension, the second hypothesis is obtained.

Goldberger *et al.* [1988] and Tsuda *et al.* [1992] proposed a notion of *homeodynamics* in the sense that the autonomic control system acquires intelligent and flexible behaviors by generating deterministic chaos in its normal states. This notion was derived from the observation of chaos in the cardiovascular system, but it could be easily extended to other biological systems.

A similar notion has also been proposed by several authors. As discussed in Sect. 5.6.5, Kaneko and Ikegami [1992] introduced the notion of *homeochaos*, based on the computational results of their symbiotic-network model. Iberall [1978] and Yates [1980] proposed the notion of *homeokinesis* to capture the dynamic regulations and interactions essential for the self-maintenance of biological organisms. In a similar sense, Rössler and Hudson [1990] emphasized the significance of a metabolic chaos in living systems. To denote the dynamic state achieved by chaos in the metabolic control systems, they used the notion of chaotic maintenance.

The notion of homeochaos can be used for a network like the brain, immune systems, ecosystems and chemical networks in metabolic systems, where the network should be adaptive, flexible and intelligent. The hypothesis states that weakly chaotic states are naturally selected to acquire the flexibility in growing and evolving systems; thereby homeochaotic states become invariant, and among their systems the dynamically controlled system like cardiovascular systems is stabilized to maintain the chaotic states, namely an achievement of homeodynamic control.

7. Conversations with Authors

7.1 Concluding Discussions

In this final chapter we would like to review the current book by organizing the issues and novel viewpoints in a series of question and answer sessions with the editors.

The essence of this book is to deepen our understanding of the reality underlying artificial and natural systems from the viewpoint of chaos. In doing so, we pay attention to a new formalization of an observation process, for example.

Because it is almost impossible to understand chaotic dynamics by analytical methods, chaos is often studied experimentally on computers. However, the simulation of chaos has a fundamental difficulty as digital computers can only deal with a finite number of bits in their computation. Therefore, simulated chaos is generally different from naked chaos, except for a few types of chaos.[1] The properties of the intractability of chaos have yet to be fully addressed.

From Chap. 2 to the end of this book, we attempt to highlight some essential aspects of chaos with regard to the simulation-ability of chaotic systems. The difficulties in observing and describing chaos are discussed, by trying to understand its complexities at the same time. Indeed, the act of observing chaos eventually perturbs chaos itself due to the intractable nature of chaos.

A first step in formalizing the observation process (Chap. 2) is to study the effects of perturbing the chaos by adding noise. Observation here is "to have stable descriptions of the behavior of the system". If the system is in a chaotic state, the results of the observation may depend on the observation process itself (i.e. the structure of the noise). Indeed, the idea to take chaos as an information source originated here. In Sect. 2.4, BZ chaos, which becomes periodic when adding white noise, is given as an example of a system which behaves chaotically without observations and which demonstrates ordered behavior with observations. This characteristic of chaos, which is influenced by observation (noise), is called "descriptive instability".

---

[1] Those types of chaos are often said to have a pseudo-orbit tracing property.

Indeed, the concept of descriptive instability has brought the perspectives of information theory back into physics. In the 1960s, an information theory by Shannon became known to and spread out over many other fields. Especially in chemistry and biology, people tried to apply the information theory in their theories and experiments. The advantage of introducing the information theory lies in the possibility to describe systems by directions or strength of information flow between elements instead of describing them merely in terms of the excitability and/or inhibitory nature of their elements. The nature of chaos, which amplifies tiny differences, can be understood as a flow of information from the lower-bit level to the higher-bit level. However, the information does not flow uniformly from the lower to the higher level bits. In some chaotic systems, information can flow back and forth inhomogenously in a bit space. The way information flows in a bit space can be taken as another characterization of chaos. Stabilization of BZ chaos, described in Chap. 2, can be well characterized by the underlying information flow.

Furthermore, by taking a viewpoint of information flow, an observed system can be viewed as a machine which inputs information into a target system from which the information can be collected afterwards. It may be natural to think that an even more complex machine can be constructed then which suppresses the descriptive instability and yields a stable observation. For example, consider that the observing system is also expressed as a (chaotic) dynamical system. That is, we couple chaotic dynamical systems. By doing so, we have a larger dynamical system which embeds the observation process as a dynamical system. The nature of the descriptive instability can only be observed when chaos is perturbed externally or coupled with other chaos. In a globally coupled chaotic system, important bits move temporally in a bit space. This is understood as a cascading process occurring in a bit space (Chap. 4). By identifying a part of a whole system as an observation machine, we can investigate the disturbance caused by the machine into the information flow and the information flow into the machine separately. In contrast to the case where noise is applied to a system, the dynamics of the observation process can be positively studied in such coupled chaotic systems.

A system with the descriptive instability found in chaotic dynamics can be described as "a system in which the dynamics of the observation process is inseparable from the dynamics of the object itself".[2] It has been claimed that such inseparability does not often exist in nature. However, we now understand that chaos exists in many natural phenomena, and we thus have to ask paradoxically why we perceive our own existence and why we recognize the separation between subject and object. When we reflect on the world including ourselves, we can no longer be an external observer. Despite this,

---

[2] The importance of such inseparability is discussed in depth from a viewpoint other than chaos by Yukio Gunji. To search for this inseparability up to the final stage implies a redirection from a state-oriented to an observation-oriented approach. ("Life and time, and primordiality – calculation and ontological observation", Gendai Shiso **23** (1995) 253 (in Japanese).)

we are able to recognize the concept of an outside and separate ourselves from it. This may be related to the fact that our brain functions as a system with descriptive instability. All outside information is delivered to the brain through sensory organs and it is used to reconstruct the world. However, the world inside the brain includes us who construct it. This paradox is not limited to problems of the brain. It is necessary to reinterpret the copying of genes, mutation, cell differentiation and ecological systems, among others, from this viewpoint. This issue is also discussed by Kaneko and Ikegami [1998].

The study of complex systems might serve as a grand theory for unifying the inside and outside viewpoints.

7.2 Questions and Answers

This section highlights several issues in this book in the form of a question and answer session with the authors. Comments by Kaneko are marked with **K**, and those by Tsuda are marked with **T**.

7.2.1 The Significance of Models in Complex Systems Research

(**Q**) While emphasizing the relativism between the real and the artificial world, you also have insisted that the unique structure of the brain underlies the many chaotic processes which occur in it (Chap. 6). Does this mean that it is necessary to have a (brain) model which reflects the anatomical structure of the brain and that the real world is superior to an artificial world by providing a basis for an artificial world?

T: The world which we live in provides a playing field for reality. The world has changed, triggered by the development of human society, namely the increase of engineering products and apparatus, the acquisition of scientific and artistic viewpoints, and the extension of imagination and conception according to these developments. Thus, the world surrounding us is viewed as an artificial world rather than a naked world. The brain is a device to observe such a world and to create an internal model of it, and consequently to respond to signals coming from the internal world as well as the external world.

I would like to understand the neural correlates of the higher functions of the brain which can be considered a being under such circumstances. We are concerned with the question of what reality is and to what extent the "real" world reflects a naked world. There could be several ways of modeling. One way is to make a model based on the neuroanatomical structure. If the neural nets do not display any unstable dynamics such as chaos or chaotic itinerancy, no serious issues arise, but if such dynamic behaviors do appear, then we cannot clarify the origin of what we are observing without specifying the way the observations or measurements are made. Then, one solution which I actually used to clarify the relation between the real and the artificial worlds is to extract a dynamic concept and make an abstract model from the dynamic behaviors of a network model which possesses a neuroanatomical basis.

K: A model is generally constructed by abstracting "reality". As discussed in Chap. 1, the difference between reality and model cannot be neglected, especially in complex systems. Due to the inseparability of a complex 'system' from the outside, a model alone is not sufficient for representing reality, even if it is very detailed. Conversely, a model cannot completely escape from the shadow of reality, even if it is at a highly abstract level. In this respect, models at several levels of abstraction are necessary, and indeed, the construction of models with different levels is essential for understanding the complex reality of nature.

In Chap. 6, a modeling method of the brain is presented which takes physiological data into account. Of course, one cannot say that the model sufficiently reflects real data. At the same time, one cannot necessarily conclude that the abstract GCM model in Chap. 4, which is structurally much farther distant from a real brain, represents the brain less. Indeed, the discussions in Chaps. 4 and 6 are interrelated by chaotic itinerancy and hidden coherence. "Itinerancy over models at different levels" is necessary for approaching a complex "reality".

This discussion on modeling is not necessarily limited to issues of the brain and living organisms. In the examples in Chap. 3, there are different levels of CML models depending on the degree of abstraction of each CML process. For example, as with regard to the spatiotemporal chaos in thermal convection, the model in Sect. 3.3 is rather abstract, while the one in Sect. 3.5 is constructed at a rather concrete level.

Then, how can we relate the various levels of modeling? In the case of thermal convection such relationships can relatively easily be understood. One simply needs to understand how the abstract level of local chaos, discussed in Sect. 4.3, is derived from the concrete processes of real convection discussed in Sect. 4.5. In the same way, possible physiological mechanisms leading to chaos in neural systems are discussed in Sect. 5.5. However, in the case of neural information processing systems, the explanation on the connection between the two levels may be insufficient because there, in contrast to the case of convection, the question of "how the neural information processing system is organized through development and/or evolution" is essential. This is why we have discussed the necessity of a synthesis with the engineering approach in Chap. 1, and why we have included Chap. 5 for linking Chaps. 4 and 6. Unfortunately, we have to admit that the discussion in Chap. 5 still remains at a rather preliminary stage since the information-processing system discussed in Sect. 5.5 has not yet been constructed. So far we have only presented some guiding principles for future study, without succeeding in showing a clear, concrete example. In this sense the discussion in Chap. 5 might be considered rather incomplete.

Needless to say, the organization process of a biological system is fundamentally different from the designs adopted in conventional engineering. In the latter case, the design is carried out with a clear technological purpose by combining parts which have fixed, one-to-one correspondences with the desired functions. In fact one can say that the so-called genetic algorithms adopt a 'miniaturized' or 'caricaturized' evolutionary viewpoint, based on such a conventional engineering standpoint as a one-to-one correspondence between a 'gene' and a 'function'. As is discussed in Kaneko and Ikegami [1998], we need an alternative viewpoint other than this kind of conventional viewpoint for understanding the organization of a biological network system. In Sect. 5.6, we have briefly discussed a novel picture for biological organization in the context of complex systems studies.

T: As we already mentioned in Chap. 1, one of the major characteristics of complex systems is that various kinds of modes can appear depending on the way observation is made. Our basic idea is that the large effect of observations must stem from chaos involved in systems. This idea leads to the notion of "descriptive instability". Could the "real" world exist stably, or could it be some specific projection of a supposed entire world which is unstable for descriptions due to embedded chaos?

These questions are not merely philosophical but neuroscientific. For the latter questions, realistic models could be far from the truth since the descriptive instability is inherited in the real world. Hence, constructive or operational models at abstract levels could be more appropriate than "realistic" models for studying predictability and universality in the real world. The phenomenological models discussed in Sect. 3.3 are a successful example. On the other hand, for the former questions, one might say that the neuroanatomical structure may provide a real structure on which neuroactivities happen. Even in such a case, however, an interpretation is necessary for the physiological data, because a single physiological experiment is some projection of entire brain activities, which can be still unstable for observations.

There seems to be a stage at which a model for interpreting the data is necessary. This is different from the stage at which a model is made based on the experimental data. Then, it is probably undecidable to specify which model is appropriate, the realistic model or the abstract model. I personally think that the realistic model is necessary for a consistent interpretation at some stage, where "realistic" means the absence of descriptive instability. A realistic model in this sense should be based on logics and/or anatomical structures.

Not a duality (i.e. theory and experiment) but a triality (i.e. theory, experiment and engineering construction) can provide us with a new scientific method for studying complex systems like the brain which have descriptive instability. The reason why I started from a structural model is that an anatomical structure probably provides a certain "real" world, and thus can provide a design base for a constructive model.

(**Q**) It is interesting to ask why we perceive a stable reality, despite the fact that the world is full of chaos. Is this related to the suppression of chaos by observation (Chap. 2)? In addition, when we construct a model for complex systems, is the model required to have some universality? For example, would the coupled map lattice (CML) be as good a metaphor for complex systems as the Ising model is for magnetism? Then again, isn't there a paradox in the fact that such universality exists on the one hand, while on the other hand, the importance of individuality, history and the variety of complex systems is stressed in Chap. 1? How do you reconcile these aspects? Interpretative processes are proposed as a means for understanding the brain in the first part of Chap. 6. Can we think that constructive models for complex systems embody such interpretative processes? If so, I think that the CML should be one such powerful model.

T: The Ising model is a metaphorical model for magnetism, and the CML can be considered a metaphorical model for complex systems. But, the level of the metaphor is different. I think that the CML provides a higher-level metaphor than the Ising model, because the two states of each element in the Ising model represent the spin state, whereas in a CML the infinitely many states of the chaotic elements do not directly represent any specific physical state. Here we see a characteristic of complex systems in the fact that a universal description is obtained by raising the level of the metaphor.

Even if a stable real world exists, I do not think that it is the result of erasing chaos in the process of observation like in, e.g., noise-induced order. Rather, it seems to me that a stable real world is a "frozen state" of one of the unstable states. Therefore, it can become unstable by observations and be formed in various ways. Here we see the necessity of interpretation. For interpretation, a "plausible" metaphorical model is inevitable. I think the CML is one such plausible model.

Let me briefly comment on my model for the brain (Chap. 6). From the observations of the dynamical behaviors that the model exhibits, the notion of chaotic itinerancy was proposed. This notion has brought about dynamical interpretations of various brain functions such as dynamic associative memory in the neocortex and in the hippocampal CA3, perception in the olfactory system, and binding problems by dynamic cell assemblies in the prefrontal cortex. I believe further interpretations of brain dynamics in terms of chaotic itinerancy will also be found in other areas.

K: I would like to add a few remarks to the above statement of Tsuda-san and the discussion in Chap. 1. The Ising model was proposed as a minimal and universal model for a clear set of phenomena, that is, a phase transition represented by a single order parameter. On the other hand, the CML model corresponds to spatiotemporal chaos, which itself has diverse manifestations in nature. In this case, the "correspondence" is much looser than the one-to-one correspondence of a model for a specific class of phenomena like the one in the Ising model.

Such a difference in the level of abstraction of a model can also be seen in the Lorenz model for chaos. Phenomena which can be well described by the Lorenz equation do not exist except under extremely special circumstances. However, this model has implications for a wide range of chaotic phenomena. In this sense, the correspondence of the model with the phenomena is looser than the one in the Ising model. In the CML case, the coupling between the model and a class of phenomena may be more subtle.

Although a CML model or the Lorenz equation forms a 'qualitative universality class', the individuality of each specific phenomenon remains. In the case of the Ising model, since the universality class is well defined through the renormalization group theory, the universal features corresponding to the Ising model are clearly separated from other specific features. In the case of a 'qualitative universality class', however, the abstraction of universality remains more vague. It is hoped that this separation can be clarified in future studies on 'qualitative universality classes', but, on the other hand, some inseparability between universality and individuality may remain in complex systems, as discussed in Sect. 1.9.1.

7.2.2 Chaotic Itinerancy

(**Q**) I would like you to summarize the concept of chaotic itinerancy again. This seems to be a key concept throughout this book. The notion of chaotic itinerancy seems to be concerned more with the extremely long transient before reaching an attractor than with the final attractor itself.

For example, when spatially extended systems have several distinct domain structures, we say that the entire dynamics has two macroscopic time scales: one time scale which is related to the behaviors inside the domains and one which is related to the behaviors among the domains. If chaotic itinerancy is viewed as an extension of such domain structures into an abstract phase space, is there a nontrivial mechanism which causes chaotic itinerancy (after all, it sounds rather trivial to me to state that the generation of domain structures leads to chaotic itinerancy)? In a system with conserved energy, periodic motions are separated by "walls" in a phase space. It is known that these periodic motions merge through emerging holes in those walls when increasing the nonlinearity. Can we expect that even in dissipative systems "walls and their subsequent collapses" can be observed?

K: The phenomenon of chaotic itinerancy is characteristic for dynamical systems with many degrees of freedom. There a state itinerates over several ordered states through chaos. Although the dynamics, including all the itinerant states, as a whole is regarded as an attractor of a dynamic system with many degrees of freedom, it is understood as an itinerancy over "attractor ruins" with a fairly small number of degrees of freedom from the viewpoint of low-dimensional dynamical systems. Indeed there seem to be several types of "chaotic itinerancy" covered by this general definition for it. It can roughly be classified according to the degree of correlation between the ordered states visited successively. The correlation is high if the paths for the transitions between the ordered states are narrow, and the probabilities for visiting the next ordered state are rather low. On the other hand, the correlation is low when the memory of the previous state is lost due to high-dimensional chaos during the transition.

(1) Cases with strong correlation: For the traveling wave in the CML of Sect. 3.3.1, transitions to states with different velocities occur through a chaotic state. In this case, the coupling of the elements is local. However, since the traveling wave is a global phenomenon covering all lattice points, the wave can be interpreted as an example of chaotic itinerancy. Since the transition occurs through the creation or annihilation of a phase slip, the velocity difference between successive states is $\pm v_p$. Although the transition occurs through a chaotic state, its dimension is too low to allow for a variety of destinations. In this sense, the transition is rather limited.

(2) Cases with weak correlation: For the chaotic itinerancy in the GCM of Sect. 4.5, there are $N!/[(N/2)!(N/2)!]$ degenerate states, even if we consider only two-cluster states with equal-sized partitions (i.e., with $N/2$ elements for each cluster). The chaotic itinerancy occurs among such a huge number of possible attractor ruins that there is little correlation between two successive states. For example, the probability that a given pair of elements remains synchronized (up to a high precision) decays exponentially with time (although a power-law decay has also been observed in some rare cases). In this case, the high-dimensional chaotic motion lasts relatively long, and the memory of the previous state decays during the transition.

(3) Cases with medium correlation: let us recall Nozawa's model, discussed in Chap. 1. It is a kind of GCM where the coupling strength among the elements is not uniform but distributed. Depending on the embedded memory patterns, some couplings are strong, and others are weak. As a result, the number of "attractor ruins" is greatly reduced and the paths between such attractor ruins are limited. Hence there is correlation between the patterns before and after a transition. As for the chaotic itinerancy of the brain model in Sect. 6.8, again, some correlation among retrieval patterns exists. In both cases, the chaos during the transition has a relatively low dimension. Indeed this is why behavior close to that of a low-dimensional map such as the one-dimensional circle map was found in Fig. 6.4. The coupled maps on hypercubes, briefly discussed in Sect. 5.6.4, also belong to this medium-correlation case, as the number of couplings lies between the CML and GCM cases.

T: In the chaotic itinerancy observed in the model shown in Sect. 6.8, the unstable manifold of the attractor ruins has a relatively low dimension. In this model, the correlation between memory states may determine the dimensionality of the transition in such a way that a strong correlation generates a low-dimensional transition, whereas a high-dimensional transition is generated in a weak correlation case.

K: In cases (1) to (3), the escape from an ordered state follows some path (which may be along the unstable manifold of a saddle for a simple case), while the differences in the cases stem from the dimensionality of the exit path. It should be stressed that this chaotic itinerancy is (A) a general mechanism providing the formation and collapse of some order and (B) a universal phenomenon that characterizes the phase space of high-dimensional dynamical systems. As discussed in Chap. 1, a new principle, different from the slaving mode principle or from the principle of self-organization in dissipative structures, is inferred from chaotic itinerancy, allowing for the resolution of the antithesis between the top-down and bottom-up approaches.

It will be important to reinterpret many dynamical phenomena, such as those observed in physical systems like water and glass, chemical-reaction processes, physiological systems, neural and other information-processing systems, etc., from the viewpoint of chaotic itinerancy. For this purpose, the development of a dynamical systems theory for chaotic itinerancy is required. Also, I believe that chaotic itinerancy is essential to the problem of demons (Sect. 1.9) which underlie the functioning of biopolymers.

T: The applicability of the theory of dissipative structures and the slaving mode principle is restricted to those situations where chaos does not appear. Once chaos occurs, we have to look for another theory or principle. Interestingly, however, chaos itself can often be viewed as a nonthermodynamic branch, stemming from other nonthermodynamic branches which are covered by these theories. On the other hand, chaotic itinerancy cannot be explained in terms of these frameworks. This is a completely new concept. Recently, several mathematical studies on chaotic itinerancy have started in Japan, and it is hoped that a rigorous framework will appear in the near future.

The Milnor attractor may mathematically be related to the above-mentioned "attractor ruins". In my neural-network model too, the Milnor attractor was found, though I did not use this terminology since I was not familiar with it. By controlling the noise level, Kaneko-san has tried to determine whether the ruins in his GCM can be described as the Milnor attractors [Kaneko 1997, 1998b]. I have also investigated whether a coupled Milnor-attractor system can exhibit chaotic itinerancy; the results will be published in the near future.

K: I would like to mention one more feature of chaotic itinerancy in the terminology of dynamical systems. The systems with chaotic itinerancy studied so far commonly have a small number of positive Lyapunov exponents and many exponents close to zero. As a result, the dimension of the dynamical system is high, while the path in phase space is restricted.

7.2.3 New Information Theory and Internal Observation

(**Q**) In the information theory developed by Shannon, the issues are how accurately can an original message be transmitted in a noisy environment, and how can such accuracy be realized. With respect to the same issues, how can one construct an information theory for chaos channels or open-flow chaotic systems? Also, an information theory has resulted in an intriguing dilemma for thermodynamics, that is the Maxwell's demon. Can you say something about a new Maxwell's demon based on the information theory of chaotic systems?

T: Chaos produces a variety of dynamical orbits, hence it can be viewed as an information source. On the other hand, if you are concerned with the information content contained in the initial distribution, you observe an information decay in the course of time development. We investigated the detailed structure of such a decay, building on the pioneering works of Yoshitsugu Oono, Rob Shaw, Jim Crutchfield and Norman Packard [Oono 1978; Shaw 1981; Crutchfield and Packard 1982].

The Kullback divergence of one time step defines the information flow. If an absolutely continuous invariant measure (with respect to the Lesbegue measure) exists, the amount of the information flow is equivalent to the largest Lyapunov exponent. We introduced an observation window in a bit space and calculated the bit information going out of the window and thus determined the fluctuations in the information flow. If such fluctuations are small, the transition term of the mutual information is approximated by the information flow. On the other hand, if the fluctuations are large enough, the mutual information provides a detailed structure of the information flow. Then, we introduced a bit-wise mutual information and found that a slow decay of mutual information indicates an information mixing over the digits. This property of information mixing assures the transmission of input information without decay in a coupled chaotic system.

In this sense, chaotic networks can play a role of the information channel. The channel capacity is determined by the degree of mixing, and the transmission speed by the coupling strengths. Even in a linear chain of chaotic elements, multiple connections in the bit space of the chaotic elements are formed. In the binary expansion of each chaotic variable, each digit can be viewed as a formal neuron of the McCulloch–Pitts type. Because of the mixing property of information, virtual connections of digits between all neighboring chaotic elements in the linear chain are formed. Suppose we have N chaotic elements and M observed digits, that is, the code length for each element is M. Then, a linear chain of N chaotic elements is equivalent to an N-layered neural network, each layer of which having M formal neurons.

K: In a coupled chaotic system, information is generated by each chaotic element and transmitted by the coupling. The flow in bit space was discussed in Sects. 2.7 and 4.6.2. In a CML, the mutual relationship between the flows in real space and the flows in the bit space becomes an important issue. In fact, in a CML, information is selectively amplified or attenuated before transmission, depending on the relationships between the phases of oscillation of the elements. This issue is discussed for the one-way coupling model in Sect. 5.3. I have also worked on the information theory of CMLs with Crutchfield [1987, unpublished]. We studied the selective transmission of information created by active units (i.e., with information creation by chaos), in contrast with the information transmission of passive units (i.e., without information creation) in noisy media, as discussed by Shannon. In the CML, the transmission rate of information depends on the propagation velocity which is related with the error amplification rate as measured by the co-moving Lyapunov exponent (see Sect. 3.3.3). Information transmission, to some extent, can be discussed in connection with the co-moving Lyapunov exponent and the local Kolmogorov–Sinai entropy.

As discussed in Sects. 3.3.3 and 5.3, a one-way coupling system allows for the selective amplification of inputs based on their periodicity. During the transmission process, information is sometimes kept in the oscillation of a unit for a while before it is transmitted further down-flow. This is observed as a resonance-like structure in the information flow [Kaneko 1986b]. It will be necessary to construct a theory for the computation and information-processing abilities of nonlinear network systems by extending the study of the information theory of coupled chaotic systems.

K: I would also like to make a few additional remarks on demons from the viewpoint of coupled chaotic systems. In the present book, we have tried to outline a new framework for going beyond the conventional method of simplifying high-dimensional problems by separating them into systems with a small number of degrees of freedom and remnant noise. In coupled chaotic systems, chaotic itinerancy and hidden coherence have provided typical examples for phenomena which go beyond such a separation (see Chaps. 3 and 4). In these examples we have the following key characteristics; (a) systems with internal dynamics interact with each other, (b) the time scales of the internal dynamics and of the macroscopic interactions are of the same order (i.e., not clearly separated), (c) a bidirectional information flow between the microscopic motions of the individual elements and the macroscopic motions such as the mean-field dynamics is created. In Chap. 1, we have discussed the problem of energy transformation, storage and retrieval in a biomotor required for muscle contraction with regard to a possible relationship with a demon. I personally believe that the answer to this problem is given by a system satisfying (a), (b) and (c). In fact Nakagawa and I have succeeded in constructing a simple coupled pendulum system with chaotic itinerancy which allows for energy absorption and storage [Nakagawa and Kaneko, 1999]. Selective transmission of input, as observed in the one-way coupled map system, may also provide a key to answer the problem of energy transformation and retrieval. In the future, it will be important to construct an ideal system within the framework of (a)–(c), similarly to the "Carnot cycle" for the second law of thermodynamics.

T: When we discuss a dynamic demon, the starting point would do to explicitly introduce an "observer". This is the standpoint which Brillouin took for the explanation of the Maxwell's demon, and similar to the "endo" standpoint which Otto Rössler recently proposed. Although an answer to the question whether or not some dynamical demon is at work in chaotic itinerancy belongs to the future study, I believe a somewhat demon-like mechanism is embedded in chaotic itinerancy, as Kaneko-san pointed out. The significance of the demon can be seen in the relation of chaotic itinerancy with the mechanisms of enzymatic reactions, molecular motors, muscle contractions, and the threshold of excitable membranes like neurons (see also Tsuda and Tadaki [1997]).

In order to prove the relations explicitly, it seems that we need to study the way of representing the phenomena in the form of statements under some logic. According to our recent study [Tsuda and Tadaki 1997], it is possible to transform inference processes into dynamical systems. Furthermore we show that the propagation of the uncertainty generated via an observation of the truth value is related to a set of solutions of functional equations with dynamical variables. I would like to emphasize that this appears to be a very promising approach for analysing the dynamical demon in relation to the notions of dynamical observations and descriptive instability.

(**Q**) I would like to ask about the formalization of observation processes. When a phenomenon exists in front of us, it seems to be extremely subjective. However, it can be objective at the same time as we can share our experience with other people. How can one resolve this paradox? This question can only be addressed if we have a system which includes an internal observer. The important thing in the hermeneutic circle, discussed in the first part of Chap. 6, is: "how are we attached" to a model. The answer to this question seems to be closely related with the problem of how internal observations can be formalized. However, it is extremely difficult to simulate or model this issue because the act of observation for understanding does not always seem to be deterministic or computable. What does it mean to understand? We may need to surpass the Church–Turing thesis and discuss how an observer can be attached to a system.

T: We still have no clear ideas on how an observer can be attached to a system. However, I do hope that the problem of "care and cure", discussed in the last part of Chap. 6, can provide us with some hints as to how we can approach this problem in the future.

In the last section, we added the results of an experimental study and its dynamical system's analysis concerning patients undergoing mental treatment. The study was started with the motivation that this problem could provide us with an actual indicator on how nurses or medical doctors as an observer should deal with patients. It was made in collaboration with psychiatrist Takashi Tahara and engineer Hiroaki Iwanaga. Although the data analysis as such is not new, its aim is. This study also led us to consider what the data actually imply, and what the relation between the obtained data and the observers obtaining them is.

Furthermore, in medical treatments, the notion of care or nursing is quite important as well as the notion of cure. The notion of care leads to the relationship between nurse and patient. Here the nurse is a participant (in other words an observer) in a system, consisting of the patient and the nurse herself. This notion of participation is difficult to be felt and conceptualized by doctors who manage "cure". Our study could lead to one possible method for guiding nurses on how they should relate to patients. An essential point is that it could be dangerous if nurses are completely involved in the patient's world through the care process since it could be the so-called 'double-blind' state. In order to escape from this dangerous state, a certain objective indicator is necessary. The data of the chaotic pulsations of the capillary vessels could be used for this purpose. The Church–Turing thesis cannot solve this problem. The study of chaos in mental treatment shows that our human brain cannot be described by recursive functions.

Reality may manifest itself in the interface between the internal and the external world. Reality is a sensation accompanied by a manifestation of consciousness. My personal view on consciousness is that it is generated in the interface when the inside world self-referentially confronts the outside world. A collapse of the interface is often seen in patients with mental illness. Medical treatment for such patients usually consists of prompting the reconstruction of the patient's interface, and this can only be accomplished through interfacing with a nurse. Here we see the essence of the observation problem treated in this book.

In the transformation of an inference process to a dynamical system, the so-called self-referential paradox corresponds to a periodic orbit of the truth value. Chaos appears in those cases where the dynamics of the truth value of a self-referential statement has a fixed-point solution when adopting classical logic, but becomes unstable when adopting continuous logic. If the self-reference in the internal world is paradoxical, a sufficient interface cannot be generated. On the other hand, if the self-reference is consistent with the logics of the external world, an interface can be generated. Then, if we demand consistency among the various logics appearing in the external world, the determination process of the truth value must be unstable. Here I see the significance of chaos study for the treatment of mental illness. According to the experiments by Tahara, the capillary chaos obtained from schizophrenic patients cannot be distinguished from the capillary chaos of normal subjects. On the other hand, patients with senile dementia and neurotic patients show a relatively low degree of chaos. From these observations, I anticipate that a low degree of capillary chaos can be an indicator for the collapse of the interface in the above sense.

[Takashi Ikegami]

References

1. Aertsen, A., ed., Brain Theory (Elsevier, Amsterdam, 1993).
2. Aertsen, A. and E. Vaadia, 'Coding and computation in the cortex: single-neuron activity and cooperative phenomena', Information Processing in the Cortex: Experiments and Theory (eds. A. Aertsen and A.V. Braitenberg, Springer, Berlin, Heidelberg, New York, Tokyo, 1992).
3. Aertsen, A., M. Erb and G. Palm, 'Dynamics of functional coupling in the cerebral cortex: an attempt at a model-based interpretation', *Physica D* **75** (1994) 103.
4. Aihara, K., 'Chaotic neural networks', Bifurcation Phenomena in Nonlinear Systems and Theory of Dynamical Systems (ed. H. Kawakami, World Scientific, Singapore, 1990a) 143.
5. Aihara, K. (ed.), Chaos (Science-sha, Tokyo, 1990b, in Japanese).
6. Aihara, K., G. Matsumoto and M. Ichikawa, 'An alternating periodic–chaotic sequence observed in neural oscillators', *Phys. Lett.* **111A** (1985) 251.
7. Aihara, K., M. Kotani and G. Matsumoto, 'Chaos and bifurcations in dynamical systems of nerve membranes', Structure, Coherence and Chaos in Dynamical Systems (eds. P. Christiansen and R.D. Parmentier, Manchester University Press, 1989) 613.
8. Aihara, K., T. Tanabe and M. Toyoda, 'Chaotic neural networks', *Phys. Lett.* **144A** (1990) 333.
9. Aizawa, Y., 'The law of multiple-causation', *SuriKagaku* **368** (1994) 5 (in Japanese).
10. Amari, S., The Mathematical Theory of Neural Networks (Sangyo Tosho, Tokyo, 1978, in Japanese).
11. Amit, D.J., 'Neural networks – achievement, prospects, difficulties', Int. Symp. on The Physics of Structure Formation (Tübingen, Oct. 1986).
12. Araki, K., Master's Thesis (1994) University of Tokyo, *Bussei Kenkyu* **62** (1994) 793 (in Japanese).
13. Aranson, I.S., A.V. Gaponov-Grekhov and M.I. Rabinovich, 'The onset and spatial development of turbulence in flow systems', *Physica D* **33** (1988) 1.
14. Arbib, M. A., Érdi, P., and Szentágothai, J., Neural Organization (A Bradford Book, The MIT Press, Cambridge, 1998).
15. Arecchi, F.T., 'Rate processes in nonlinear optical dynamics with many attractors', *Chaos* **1** (1991) 357.
16. Arnold, V.I., 'Small denominators and problems of stability of motion in classical and celestial mechanics', *Russian Math. Surveys* **18** (1963) 85.
17. Arnold, V.I., Geometrical Methods in the Theory of Ordinary Differential Equations (Springer, New York, 1982).
18. Arnold, V.I. and A. Avez, Problèmes Ergodiques de la Mécanique Classique (Gauthier-Villars, Paris 1967; Japanese translation by Kosaku Yoshida, published by Yoshioka Shoten Publ. Inc., Kyoto, 1972).

19. Aubry, S., 'The new concept of transitions by breaking of analyticity in a crystallographic model', Solitons and Condensed Matter Physics (eds. A.R. Bishop and T. Schneider, Springer, Berlin, Heidelberg, 1979).

20. Babloyantz, A., 'Evidence of chaotic dynamics of brain activity during the sleep cycle', Dimension and Entropies in Chaotic Systems (ed. G. Mayer-Kress, Springer, Berlin, Heidelberg, 1986) 241.

21. Babloyantz, A. and A. Destexhe, 'Is the normal heart a periodic oscillator?', *Biological Cybernetics* **58** (1988) 203.

22. Bak, P., C. Tang and K. Wiesenfeld, 'Self-organized criticality – An explanation of $1/f$ noise', *Phys. Rev. Lett.* **59** (1987) 381.

23. Barna, G. and I. Tsuda, 'A new method for computing Lyapunov exponents', *Phys. Lett.* **A 175** (1993) 421.

24. Barnsley, M., Fractal Everywhere (Academic Press, Orlando, 1988).

25. Basar, E., 'Theoretical approaches to brain function (linear and nonlinear)', Brain Dynamics (eds. E. Basar and T.H. Bullock, Springer, Berlin, Heidelberg, 1990) 109.

26. Bascompte J. and R.V. Solé ed., Modeling Spatiotemporal Dynamics in Ecology, (Springer, Berlin, Heidelberg and Landes Bioscience Georgetown, 1998)

27. Bateson, G., Steps to an Ecology of Mind (Chandler Publishing Company, Ballantine Books, New York 1972).

28. Bauer, M., H. Heng and W. Martinessen, 'Characterization of spatiotemporal chaos from time series', *Phys. Rev. Lett.* **71** (1993) 521.

29. Bers, A., 'Space–time evolution of plasma instabilities – absolute and convective', Handbook of Plasma Physics 1 (ed. A.A. Galeev, Elsevier, Amsterdam, 1983) 451.

30. Blum, L., M. Shub and S. Smale, 'On a theory of computation and complexity over the real numbers: NP-completeness, recursive functions and universal machine', *Bull. Amer. Math. Soc.* **21** (1989) 1.

31. Blum, L., F. Cucker, M. Shub and S. Smale, Complexity and Real Computation (Springer, New York, 1998)

32. Bowen, R., 'Invariant measures for Markov maps of the interval', *Comm. Math. Phys.* **81** (1979) 1.

33. Bowen, R. and D. Ruelle, 'The ergodic theory of Axiom A flows', *Inventiones Math.* **29** (1975) 181.

34. Bracikowski, C. and R. Roy, 'Chaos in a multimode solid-state laser system', *Chaos* **1** (1991) 49.

35. Briggs, R.J., Electron-stream Interaction with Plasmas (MIT Press, Cambridge, MA, 1964).

36. Bunimovich, L.A. and Y.G. Sinai, 'Spacetime Chaos in coupled map lattices', *Nonlinearity* **1** (1989) 491.

37. Campbell, D.K., J.P. Crutchfield, J.D. Farmer and E. Jen, "Experimental mathematics: the role of computation in nonlinear science,' *Comm. ACM* **28** (1984) 374.

38. Chaitin, G.J., Algorithmic Information Theory (Cambridge University Press, 1987a).

39. Chaitin, G.J., Information, Randomness and Incompleteness (World Scientific, Singapore, 1987b).

40. Chaté, H. and P. Manneville, 'Transition to turbulence via spatiotemporal intermittency', *Phys. Rev. Lett.* **58** (1987) 112.

41. Chaté, H. and P. Manneville, 'Spatiotemporal intermittency in coupled map lattices', *Physica D* **32** (1988) 409.

42. Chaté, H. and P. Manneville, 'Collective behaviors in spatially extended systems with local interactions and synchronous updating', *Prog. Theor. Phys.* **87** (1992) 1.
43. Chaté, H. and M. Courbage eds., Lattice Dynamics, special issue of *Physica D* (vol. 103, 1997).
44. Chawanya, T. and S. Morita, 'On the bifurcation structure of the mean-field fluctuation in the globally coupled tent map systems', *Physica D* **116** (1998) 44.
45. Chirikov, B.V., 'A universal instability of many-dimensional oscillator systems', *Phys. Rep.* **52** (1979) 263.
46. Ciliberto, S. and P. Bigazzi, 'Spatiotemporal intermittency in Rayleigh–Benard convection', *Phys. Rev. Lett.* **60** (1988) 286.
47. Cole, B., 'Is animal behaviour chaotic? Evidence from the activity of ants', *Proc. Royal Soc. London B* **244** (1991) 253.
48. Collins, R.J. and D.R. Jefferson, 'Ant farm: towards simulated evolution', Artificial Life II (eds. C. Langton *et al.*, Addison-Wesley, Reading, MA, 1991).
49. Connel, J.H., 'Diversity in tropical rain forests and coral reefs', *Science* **199** (1978) 1302.
50. Cornfeld, I.P., S.V. Fomin and Y.G. Sinai, Ergodic Theory (Springer, New York, 1982).
51. Crick, F. and C. Asanuma, 'Certain aspects of the anatomy and physiology of the cerebral cortex', Parallel Distributed Processing (eds. J.L. McClelland, D.E. Rumelhart and the PDP Research Group, 1986), vol. 2, 333.
52. Cross, M.C. and P.C. Hohenberg, 'Pattern formation outside of equilibrium', *Rev. Mod. Phys.* **65** (1993) 851.
53. Crutchfield, J.P., 'The calculi of emergence: computation, dynamics and induction', *Physica D* **75** (1994) 11.
54. Crutchfield, J.P. and N.H. Packard, 'Symbolic dynamics of one-dimensional maps: entropies, finite precision, and noise', *Int. J. Theor. Phys.* **21** (1982) 433.
55. Crutchfield, J.P. and K. Kaneko, 'Phenomenology of spatiotemporal chaos', Directions in Chaos (World Scientific, Singapore, 1987).
56. Crutchfield, J.P. and K. Kaneko, 'Are attractors relevant to turbulence?', *Phys. Rev. Lett.* **60** (1988) 2715.
57. Crutchfield, J.P. and K. Young, 'Inferring statistical complexity', *Phys. Rev. Lett.* **63** (1989) 105.
58. Daviaud, F., M. Dubois and P. Berge, 'Spatio-temporal intermittency in quasi 1-dimensional Rayleigh–Benard convection', *Europhys. Lett.* **9** (1989) 441.
59. Daviaud, F., M. Bonetti and M. Dubois, 'Transition to turbulence via spatio-temporal intermittency in 1-d Rayleigh–Benard convection', *Phys. Rev. A* **42** (1990) 3388.
60. Davis, P., 'Application of optical chaos to temporal pattern search in a nonlinear optical resonator', *Jap. J. Appl. Phys.* **29** (1990) L1238.
61. Deissler, R.J. and K. Kaneko, 'Velocity dependent Lyapunov exponent as a measure of chaos for open flows', *Phys. Lett.* **119A** (1987) 397.
62. Derrida, B. and H. Flyvbjerg, 'The random map model: a disordered model with deterministic dynamics', *J. Physique* **48** (1988) 971.
63. Destexhe, A., J.A. Sepulchre and A. Babloyantz, 'A comparative study of the experimental quantification of deterministic chaos', *Phys. Lett.* **132A** (1988) 101.
64. Diltey, W., Wilhelm Dilthey, Gesammelte Schriften **8** (1931, Leipzig and Berlin); **5** (1924); **7** (1927) in H.O. Pöggeler, Hermeneutische Philosophie

(Nymphenburger Verlagshandlung GmbH, München, 1972) (Japanese translation: Kaishaku-Gaku no Konpon-Mondai, M. Tsukamoto, Koyo-shobo, Tokyo, 1977).

65. Dinse, H.R., K. Kruger and J. Best, 'A temporal structure of cortical information processing', *Concepts in Neuroscience* **1** (1990) 199.

66. Dominguez, D. and H.A. Cerdeira, 'Order and turbulence in rf-driven Josephson junction series arrays', *Phys. Rev. Lett.* **71** (1993) 3359.

67. Eckhorn, R., R. Bauer, W. Jordan, M. Brosch, W. Kruse, M. Munk and H.J. Reitboeck, 'Coherent oscillations: A mechanism of feature linking in the visual cortex?', *Biol. Cybernetics* **60** (1988) 121.

68. Eckmann, J.P., S.O. Kamphorst, D. Ruelle and S. Ciliberto, 'Lyapunov exponents from time series', *Phys. Rev.* **A34** (1986) 4971.

69. Edelman, G.M., Neural Darwinism (Basic Books, New York, 1987).

70. Eigen, M. and P. Schuster, The Hypercycle (Springer, Heidelberg, 1979).

71. Ellner, S. and P. Turchin, 'Chaos in a noisy world: new methods and evidence from time series analysis', *American Naturalist* **145** (1995) 343.

72. Elton, C.S., The Pattern of Animal Communities (Methuen and Co. Ltd., London, 1966).

73. Érdi, P. and Tsuda, I., Hermeneutic approach to the brain: Process versus Device?, *Theoria et Historia Scientiarium* (in press, 2000)

74. Ershov, S.V. and A.B. Potapov, 'On mean field fluctuations in globally coupled maps', *Physica D* **86** (1995) 532.

75. Ershov, S.V. and A.B. Potapov, 'On mean field fluctuations in globally coupled logistic-type maps', *Physica D* **106** (1997) 9.

76. Falconer, K.J., 'Random fractals', *Proc. Cambridge Philos. Soc.* **100** (1986) 559.

77. Farmer, J.D., N.H. Packard and A.S. Perelson, 'The immune system, adaptation and machine learning', *Physica D* **22** (1986) 187.

78. Feigenbaum, M.J., 'The universal metric properties of nonlinear transformations', *J. Stat. Phys.* **21** (1979) 669.

79. Finkelstein, D., 'Holistic method in quantum logic', Quantum Theory and the Structures of Time and Space (eds. L. Castell *et al.*, Carl Hanser, Munich, 1979) 37.

80. Flesselles, J-M., V. Croquette, S. Jucquois and B. Janiaud, 'Behavior of a one-dimensional chain of nonlinear oscillators', Spatiotemporal Patterns in Nonequilibirum Complex Systems (eds. P.E. Cladis and P. Palffy-Muhoray, Addison-Wesley, Reading, MA, 1995).

81. Freeman, W.J., 'Simulation of chaotic EEG patterns with a dynamic model of the olfactory system', *Biol. Cybern.* **56** (1987) 139.

82. Freeman, W.J., 'Controlled chaos in the basal forebrain: bifurcation during learning, itinerancy during perception', Proc. of the 2nd Int. Conf. on Fuzzy Logic and Neural Networks (Fuzzy Logic Systems Institute, Iizuka, 1992) 933.

83. Freeman, W.J., 'Neural mechanisms underlying destabilization of cortex by sensory input', *Physica D* **75** (1994) 151.

84. Freeman, W.J., Societies of Brains: A Study in the Neuroscience of Love and Hate (Lawrence Erlbaum, Hillsdale, NJ, 1995).

85. Freeman, W.J. and C.A. Skarda, 'Spatial EEG patterns, nonlinear dynamics and perception: the neo-Sherringtonian view', *Brain Res. Rev.* **10** (1985) 147.

86. Frisch, U., P.L. Sulem and M. Nelkin, 'A simple dynamical model of intermittent fully developed turbulence', *J. Fluid Mech.* **87** (1978) 719.

87. Fujii, H., H. Ito, K. Aihara and M. Tsukada, 'Dynamic cell assembly hypothesis – theoretical possibility of spatio-temporal coding in the cortex', *Neural Networks* **9** (1996) 1303.

88. Fujimoto, K. and K. Kaneko (1998) 'Noise Induced Boundary Dependence Through Convective Instability', *Physica D* **129** (1999) 203.
89. Fujisaka, H. and T. Yamada, 'Stability theory of synchronized motion in coupled-oscillator systems', *Prog. Theor. Phys.* **69** (1983) 32.
90. Fujisaka, H. and T. Yamada, 'Stability theory of synchronized motion in coupled-oscillator systems IV', *Prog. Theor. Phys.* **75** (1986) 1087.
91. Fukushima, K., Neural Networks and Self-Organization (Kyoritsu Publ., Tokyo, 1979, in Japanese).
92. Furusawa, C. and K. Kaneko 'Emergence of rules in cell society: differentiation, hierarchy, and stability', *Bull. Math. Biol.* **60** (1998) 659.
93. Gadamer, H. G., Philosophical Hermeneutics (translated and edited by D.E. Linge, University of California Press, 1976).
94. Gerbel, K. and P. Weibel eds., The World from Within – ENDO · NANO (PVS Verleger, Linz, 1992).
95. Giacomelli, G. and A. Politi, 'Spatiotemporal chaos and localization', *Europhys. Lett.* **15** (1991) 387.
96. Goldberger, H.L., D.R. Rigney, J. Mietus, E.M. Antman and S. Greenwald, 'Nonlinear dynamics in sudden cardiac death syndrome: heart rate oscillations and bifurcations', *Experientia* **44** (1988) 983.
97. Gollub, J. and R. Ramshankar, 'Spatiotemporal chaos in interfacial waves', New Perspectives in Turbulence (eds. S. Orszag and L. Sirovich, Springer, Berlin, Heidelberg, 1991).
98. Goodwin, B., 'How the leopard changed its spots – The evolution of complexity' (Charles Scribner's Sons, New York, 1994).
99. Grassberger, P., 'Towards a quantitative theory of self-generated complexity', *Int. J. Theor. Phys.* **25** (1986) 907.
100. Grassberger, P., 'Information content and predictability of lumped and distributed dynamical systems', *Physica Scripta* **40** (1989) 346.
101. Grassberger, P. and I. Procaccia, 'Characterization of strange attractors', *Phys. Rev. Lett.* **50** (1983a) 346.
102. Grassberger, P. and I. Procaccia, 'Measuring the strangeness of strange attractors', *Physica D* **9** (1983b) 189.
103. Grassberger, P. and T. Schreiber, 'Phase transitions in coupled map lattices', *Physica D* **50** (1991) 177.
104. Gray, C.M., P. Koenig, A.K. Engel and W. Singer, 'Oscillatory responses in cat visual cortex exhibit inter-columnar synchronization which reflects global stimulus properties', *Nature* **338** (1989) 334.
105. Hadley, P. and K. Wiesenfeld, 'Attractor crowding in oscillator arrays', *Phys. Rev. Lett.* **62** (1989) 1335.
106. Haken, H., Synergetics (Springer, Berlin, Heidelberg, 1979).
107. Hakim, V. and W.J. Rappel, 'Dynamics of the globally coupled complex Ginzberg–Landau equation', *Phys. Rev.* **A 46** (1992) 7347.
108. Hassell, M.P., H.N. Comins and R.M. May, 'Spatial structure and chaos in insect population dynamics', *Nature* **353** (1991) 255.
109. Hastings, A. and K. Higgins, 'Persistence of transients in spatially structured ecological models', *Science* **263** (1994) 1133.
110. Hata, H., S. Oku and K. Yabe, 'Structural criticality to dynamics glass state in spatially coupled map', *Prog. Theor. Phys.* **95** (1996) 45.
111. Hata, M., 'Dynamics of Caianiello's equation', *J. Math. Kyoto Univ.* **22** (1982) 155.
112. Hata, M., Chaos in Neural Network Model (Asakura Publ., Tokyo, 1998, in Japanese).

113. Hata, M. and M. Yamaguti, 'Takagi function and its generalization', *Japan J. of Applied Math.* **1** (1984) 186.

114. Hayashi, H. and S. Ishizuka, 'Chaos in molluscan neuron', Chaos in Biological Systems (eds. H. Degn, A.V. Holden and L.F. Olsen, NATO ASI Series, 1987) 157.

115. Hayashi, H. and S. Ishizuka, 'Chaotic activity in hippocampus neural network and intracranial self-stimulation', Proc. Int. Conf. Fuzzy Logic and Neural Networks, *Iizuka* **2** (1990) 583.

116. Hayashi, H. and S. Ishizuka (private communication, 1992).

117. Hayashi, H. and S. Ishizuka, 'Chaotic responses of the hippocampal CA3 region to a mossy fiber stimulation *in vitro*', *Brain Res.* **686** (1995) 194.

118. Hayashi, H., S. Ishizuka, M. Ohta and K. Hirakawa, 'Chaotic behavior in the onchidium giant neuron under sinusoidal stimulation', *Phys. Lett.* **88A** (1982) 435.

119. Heidegger, M., Sein und Zeit (Tübingen, 1927), sections 32 & 33 in H.O. Pöggeler, ibid.

120. Hilbert, D. and P. Bernays, Grundlagen der Mathematik I, II (Springer, Berlin, Heidelberg, New York 1970; Japanese Translation by N. Yoshida and S. Fuchino by Springer, Tokyo, 1993).

121. Hinton, G.E., T.J. Sejnowski and D.H. Ackley, 'Boltzmann machines: Constraint satisfaction networks that learn', *Cognitive Science* **9** 147.

122. Hofstadter, D.R., 'Gödel, Escher, Bach' (Basic Books Inc., London, New York, Victoria, Ontario, Aukland, 1979).

123. Holland, J.H., 'Escaping brittleness: the possibilities of general purpose learning algorithms applied to parallel rule-based systems', Machine Learning II (eds. R.S. Mishalski, J.G. Carbonell and T.M. Mitchell, Kaufmann, Los Altos, 1986).

124. Hopfield, J.J., 'Computing with neural circuits: a model', *Science* **233** (1986) 625.

125. Horai, M., 'Experimental studies on the Belousov–Zhabotinsky reactions' (Master Thesis Osaka University, 1983).

126. Houlrik, J.M., I. Webman and M.H. Jensen, 'Mean field theory and critical behavior of coupled map lattices', *Phys. Rev.* **A41** (1990) 4210.

127. Hudson, J.L. and J.C. Mankin, 'Chaos in the Belousov–Zhabotinsky reaction', *J. Chem. Phys.* **74** (1981) 6171.

128. Hudson, J.L., M. Hart and D. Marinko, 'An experimental study of multiple peak periodic and nonperiodic oscillations in the Belousov–Zhabotinsky reaction', *J. Chem. Phys.* **71** (1979) 1601.

129. Iberall, A.S., 'A field and thermodynamics for integrative physiology', *Amer. J. Physiology* **233** (1978) 171.

130. Ikeda, K., H. Daido and O. Akimoto, 'Optical turbulence: chaotic behavior of transmitted light from a ring cavity', *Phys. Rev. Lett.* **45** (1980) 709.

131. Ikeda, K., K. Otsuka and K. Matsumoto, 'Maxwell–Bloch turbulence', *Prog. Theor. Phys. Suppl.* **99** (1989) 295.

132. Ikegami, T. and I. Tsuda, 'Measuring complexity: a relative aspect of pattern recognition', *Int. J. Mod. Phys. C* **3** (1992) 447.

133. Ikegami, T. and K. Kaneko, 'Evolution of host–parasite network through homeochaotic dynamics', *Chaos* **2** (1992) 397.

134. Ito, M., Design of Brain (Chuo Koron Sha, Tokyo, 1980, in Japanese).

135. Ito, S. and Y. Takahashi, 'Markov subshifts and realization of β-expansions', *J. Math. Soc. Japan* **26** (1974) 33.

136. Jerne, N.K., 'The immune system', *Sci. Am.* **229** (1973) 52.

137. Jerne, N.K., 'Towards a network theory of the immune system', *Ann. Immunol.* **125C** (1974) 373.
138. Johnson-Laird, P.N., Mental Models (Cambridge University Press, 1983).
139. Just, W., 'Bifurcations in globally coupled map lattices', *J. Stat. Phys.* **79** (1995) 429.
140. Kaneko, K., 'Fractalization of torus', *Prog. Theor. Phys.* **71** (1984a) 1112.
141. Kaneko, K., 'Period-doubling of kink-antikink patterns, quasi-periodicity in antiferro-like structures and spatial intermittency in coupled map lattices – toward a prelude to a 'field theory of chaos', *Prog. Theor. Phys.* **72** (1984b) 480.
142. Kaneko, K., 'Spatiotemporal intermittency in coupled map lattices', *Prog. Theor. Phys.* **74** (1985a) 1033.
143. Kaneko, K., 'Spatial period-doubling in open flow', *Phys. Lett.* **111A** (1985b) 321.
144. Kaneko, K., Ph. D. Thesis 'Collapse of Tori and Genesis of Chaos in Dissipative Systems', 1983 (enlarged version is published by World Scientific, Singapore, 1986a).
145. Kaneko, K., 'Lyapunov analysis and information flow in coupled map lattices', *Physica D* **23** (1986b) 436.
146. Kaneko, K., 'Pattern dynamics in spatiotemporal chaos', *Physica D* **34** (1989a) 1.
147. Kaneko, K., 'Spatiotemporal chaos in one- and two-dimensional coupled map lattices', *Physica D* **37** (1989b) 60.
148. Kaneko, K., 'Towards thermodynamics of spatiotemporal chaos', *Prog. Theor. Suppl.* **99** (1989c) 263.
149. Kaneko, K., 'Self-consistent Perron–Frobenius operator for spatiotemporal chaos', *Phys. Lett.* **139 A** (1989d) 47.
150. Kaneko, K., 'Chaotic but regular posi-nega switch among coded attractors by cluster size variation', *Phys. Rev. Lett.* **63** (1989e) 219.
151. Kaneko, K., 'Simulating physics with coupled map lattices – pattern dynamics, information flow, and thermodynamics of spatiotemporal chaos', Formation, Dynamics, and Statistics of Patterns (eds. K. Kawasaki, A. Onuki and M. Suzuki, World Scientific, Singapore, 1990a).
152. Kaneko, K., 'Supertransients, spatiotemporal intermittency, and stability of fully developed spatiotemporal chaos', *Phys. Lett.* **149 A** (1990b) 105.
153. Kaneko, K., 'Clustering, coding, switching, hierarchical ordering, and control in network of chaotic elements', *Physica D* **41** (1990c) 137.
154. Kaneko, K., 'Globally coupled chaos violates law of large numbers', *Phys. Rev. Lett.* **65** (1990d) 1391.
155. Kaneko, K., 'Partition complexity in network of chaotic elements', *J. Phys.* **A24** (1991a) 2107.
156. Kaneko, K., 'Globally coupled circle maps', *Physica D* **54** (1991b) 5.
157. Kaneko, K., 'Global traveling wave triggered by local phase slips', *Phys. Rev. Lett.* **69** (1992a) 905.
158. Kaneko, K., 'Mean field fluctuation in network of chaotic elements', *Physica D* **55** (1992b) 368.
159. Kaneko, K., 'Propagation of disturbance, co-moving Lyapunov exponents, and path summation', *Phys. Lett.* **170A** (1992c) 210.
160. Kaneko, K. ed., 'Chaos focus issue on coupled map lattices', *Chaos* **2** (1992d) 279.
161. Kaneko, K., 'Chaotic traveling wave in coupled map lattices', *Physica D* **68** (1993a) 299.

162. Kaneko, K. ed., Theory and Applications of Coupled Map Lattices (John Wiley & Sons, Chichester, 1993b).
163. Kaneko, K., 'Relevance of clustering to biological networks', *Physica D* **75** (1994a) 55.
164. Kaneko, K., 'Information cascade with marginal stability in network of chaotic elements', *Physica D* **77** (1994b) 456.
165. Kaneko, K., 'Chaos as a source of complexity and diversity in evolution', *Artificial Life* **1** (1994c) 163.
166. Kaneko, K., 'Diversity induced by chaos', *Nikkei-Science* May (1994d) 34 (in Japanese)
167. Kaneko, K., 'Remarks on the mean field dynamics of network of chaotic elements', *Physica D* **86** (1995) 158.
168. Kaneko, K., 'Dominance of Milnor attractors and noise-induced selection in a multi-attractor system', *Phys. Rev. Lett.* **78** (1997) 2736.
169. Kaneko, K., 'Diversity, stability, recursivity, hierarchy, and rule generation in a biological system studied as intra–inter dynamics', *Int. J. Mod. Phys. B.* **12** (1998a) 285.
170. Kaneko, K., 'Life as complex systems: viewpoint from intra–inter dynamics', *Complexity* **3** (1998b) 53.
171. Kaneko, K., 'On the strength of attractors in a high-dimensional system: Milnor attractor network, robust global attraction, and noise-induced selection', *Physica D* **124** (1998c) 308.
172. Kaneko, K. and T. Ikegami, 'Homeochaos: dynamic stability of a symbiotic network with population dynamics and evolving mutation rates', *Physica D* **56** (1992) 406.
173. Kaneko, K. and T. Ikegami, Evolutionary Scenario of Complex Systems (Asakura 1998; in Japanese)
174. Kaneko, K. and T. Konishi, 'Peeling the onion of order and chaos in a high-dimensional Hamiltonian system', *Physica D* **71** (1994) 146.
175. Kaneko, K. and I. Tsuda, 'Constructive complexity and artificial reality: an introduction', *Physica D* **75** (1994) 1.
176. Kaneko, K. and T. Yomo, 'Cell division, differentiation, and dynamic clustering', *Physica D* **75** (1994) 89.
177. Kaneko, K. and T. Yomo, 'A theory of differentiation with dynamic clustering', Advances in Artificial Life (eds. E. Moran *et al.*, Springer, Berlin, Heidelberg, 1995) 329.
178. Kaneko, K. and T. Yomo, 'Isologous diversification: a theory of cell differentiation', *Bull. Math. Biol.* **59** (1997) 139.
179. Kaneko, K. and T. Yomo, 'Isologous Diversification for Robust Development of Cell Society' *J. Theor. Biol.* **199** (1999) 243.
180. Kapral, R., 'Chemical waves and coupled map lattices', Theory and Applications of Coupled Map Lattices (ed. K. Kaneko, John Wiley & Sons, Chichester, 1993).
181. Kauffman, S.A., 'Metabolic stability and epigenesis in randomly constructed genetic nets', *J. Theor. Biol.* **22** (1969) 437.
182. Kauffman, S.A., The Origin of Order (Oxford University Press, 1993).
183. Kay, L., K. Shimoide and W.J. Freeman, 'Comparison of EEG time series from rat olfactory system with model composed of nonlinear coupled oscillators', *Int. J. Bifurcation and Chaos* **5** (1995) 849.
184. Keeler, J.D. and J.D. Farmer, 'Robust space–time intermittency and $1/f$ noise', *Physica D* **23** (1986) 413.
185. Kessler, D.A., H. Levine and W.N. Reynolds, 'Coupled-map lattice model for crystal growth', *Phys. Rev.* **A42** (1990) 6125.

186. Kikkawa, J., The current status of the studies in tropical rain forest, *Kagaku* **60** (1990) 603 (in Japanese).

187. Kirkpatrick, S., C.D. Gelatt Jr. and M.P. Vecchi, 'Optimization by simulated annealing', *Science* **220** (1983) 671.

188. Koerner, E. and H.-J. Boehme, 'A neural network model for control and stabilization of reverberating pattern sequences', Proc. of ICANN–91 (Helsinki, 1991).

189. Koerner, E., I. Tsuda, and H. Shimizu, 'Parallel in sequence – towards the architecture of an elementary cortical processor', Mathematical Research: Parallel Algorithms and Architectures (eds. Albrecht, A., Jung, H., and Mehlhorn, K., Akademie-Verlag, Berlin, 1987) 37.

190. Kolmogorov, A.N., 'The local structure of turbulence in incompressible viscous fluid for very large Reynolds numbers', *Dokl. Akad. Nauk. SSSR* **30** (1941) 301.

191. Kolmogorov, A.N., 'Quantity of information defined (algorithmic approach which uses recursive functions for defining concept, and quantity of information)', *Prob. Info. Trans.* **1** (1965) 1.

192. Kolodner, P., J.A. Glazier and H. Williams, 'Dispersive chaos in one-dimensional traveling-wave convection', *Phys. Rev. Lett.* **65** (1990) 1579.

193. Konishi, T. and K. Kaneko, 'Clustered motion in symplectic coupled map system', *J. Phys.* **A25** (1992) 6283.

194. Kuramoto, Y., Chemical Oscillations, Waves, and Turbulence (Springer, Berlin, Heidelberg, 1984).

195. Kuznetsov, S.P., 'Renormalization group, universality, and scaling in dynamics of one-dimensional dissipative medium', *Radiofizika* **29** (1986) 888.

196. Kuznetsov, S.P., 'Renormalization group, universality, and scaling in dynamics of coupled map lattices', Theory and Applications of Coupled Map Lattices (ed. K. Kaneko, John Wiley & Sons, Chichester, 1993).

197. Layne, S.P., G. Mayer-Kress and J. Holzfuss, 'Problems associated with dimensional analysis of electroencephalogram data', Dimensions and Entropies in Chaotic Systems (ed. G. Mayer-Kress, Springer, Berlin, Heidelberg, 1986).

198. Lichtenberg, A.J. and M.A. Lieberman, Regular and Stochastic Motion (Springer, Berlin, 1983).

199. Lloyd, S. and H. Pagel, 'Complexity as thermodynamic depth', *Ann. Phys.* **188** (1988) 186.

200. Lorenz, E. N., 'Deterministic nonperiodic flow', *J. Atmos. Sci.* **20** (1963) 130.

201. MacKay, R.S. and J. Meiss eds., Hamiltonian Dynamical Systems (Adam Hilger, Bristol, 1987).

202. Malsburg, C. von der, 'A neural cocktail party processor', *Biol. Cybern.* **54** (1986) 29.

203. Mandelbrot, B.B., 'Intermittent turbulence in self-similar cascades: divergence of high moments and dimension of the carrier', *J. Fluid Mech.* **62** (1974) 331.

204. Mandelbrot, B.B., Fractals – Form, Chance and Dimension (Freeman, San Francisco, 1977).

205. Mandell, A.J., 'Dynamical complexity and pathological order in the cardiac monitoring problem', *Physica D* **27** (1987) 235.

206. Marr, D., Vision (W.H. Freeman and Company, San Francisco, 1982).

207. Matsumoto, K. and I. Tsuda, 'Noise-induced order', *J. Stat. Phys.* **31** (1983) 87.

208. Matsumoto, K. and I. Tsuda, 'Information theoretical approach to noisy dynamics', *J. Phys. A: Math. Gen.* **18** (1985) 3561.

209. Matsumoto, K. and I. Tsuda, 'Extended information in one-dimensional maps', *Physica D* **26** (1987) 347.

210. Matsumoto, K. and I. Tsuda, 'Calculation of information flow rate from mutual information', *J. Phys. A: Math. Gen.* **21** (1988) 1405.
211. Matsuoka, S., Nomadology (Yamato Shobo, Tokyo, 1986, in Japanese).
212. May, R., Stability and Complexity in Model Ecosystems (Princeton University Press, 1973).
213. Mayer-Kress, G. and K. Kaneko, 'Spatiotemporal chaos and noise', *J. Stat. Phys.* **54** (1989) 1489.
214. Mezard, M., G. Parisi and M.A. Virasoro eds., Spin Glass Theory and Beyond (World Scientific, Singapore, 1987).
215. Michalland, S., M. Rabaud and Y. Couder, 'Transition to chaos by spatiotemporal intermittency in directional viscous fingering', *Europhys. Lett.* **22** (1993) 17.
216. Milnor, J., 'On the concept of attractor', *Comm. Math. Phys.* **99** (1985) 177; **102** (1985) 517.
217. Miyashita, Y., 'Inferior temporal cortex: where visual perception meets memory', *Annual Rev. Neuroscience* **16** (1993) 245.
218. Moore, C., 'Recursion theory on the reals and continuous-time computation', *Theor. Comp. Sci.*, **162** (1996) 23.
219. Morita, S., 'Bifurcation in globally coupled chaotic maps', *Phys. Lett. A* **211** (1996) 258.
220. Moser, J., 'On a theorem of Anosov', *J. Differ. Eqns.* **5** (1969) 411.
221. Nagel, E. and J.R. Newman, Gödel's Proof (New York University Press, 1964).
222. Nagumo, J. and S. Sato, 'On a response characteristic of a mathematical neuron model', *Kybernetik* **10** (1972) 155.
223. Nakagawa, N. and K. Kaneko, 'Energy storage in a Hamiltonian system in partial contact with a heat bath', *J. Phys. Soc. Japan* **69** (2000) 1255.
224. Nakagawa, N. and Y. Kuramoto, 'Collective chaos in a population of globally coupled oscillators', *Prog. Theor. Phys.* **89** (1993) 313.
225. Nakagawa, N. and Y. Kuramoto, 'From collective oscillations to collective chaos in a globally coupled oscillator system', *Physica D* **75** (1994a) 74.
226. Nakagawa, N. and Y. Kuramoto, 'Anomalous Lyapunov spectrum in globally coupled oscillators', *Physica D* **80** (1994b) 307.
227. Nakagawa, N. and T. Komatsu, 'Collective motion occurs inevitably in a class of populations of globally coupled chaotic elements', *Phys. Rev. E* **57** (1998) 1570.
228. Nara, S. and P. Davis, 'Chaotic wandering and search in a cycle-memory neural network', *Prog. Theor. Phys.* **88** (1992) 845.
229. Nara, S. and P. Davis, Chaos and Intelligent Information Processings (Just System, Tokushima, 1994, in Japanese).
230. Nara, S., P. Davis and H. Totsuji, 'Memory search using complex dynamics in a recurrent neural network model', *Neural Networks* **6** (1993) 963.
231. Nara, S., P. Davis, M. Kawachi and H. Totsuji, 'Chaotic memory dynamics in a recurrent neural network with cycle memories embedded by pseudo-inverse method', *Int. J. Bifurcation and Chaos* **5** (1995) 1205.
232. Nasuno, S. and S. Kai, 'Instabilities and transition to defect turbulence in electrohydrodynamic convection to nematics', *Europhys. Lett.* **14** (1991) 779.
233. Nasuno, S., M. Sano and Y. Sawada, 'Defect turbulence in EHD convection of liquid crystal', Cooperative Dynamics in Complex Systems (ed. H. Takayama, Springer, Berlin, Heidelberg, 1989).
234. Neumann, J. von, 'Probabilistic logics and the synthesis of reliable organisms from unreliable components', Automata Studies (eds. C. Shannon and J. McCarthy, Princeton University Press, 1956).

235. Neumann, J. von, Theory of Self-Reproducing Automata (ed. A.W. Burks, University of Illinois Press, 1966).
236. Nicolis, J.S., 'Should a reliable information processor be chaotic?', *Kybernets* **11** (1982) 269.
237. Nicolis, J.S., Chaotic Information Processing (World Scientific, Singapore, 1991).
238. Nicolis, G. and I. Prigogine, Self-Organization in Nonequilibrium Systems (John Wiley & Sons, Chichester, 1977)
239. Nicolis, J.S. and I. Tsuda, 'Chaotic dynamics of information processing: the magic number seven plus–minus two revisited', *Bull. Math. Biol.* **47** (1985) 343.
240. Nicolis, J.S. and I. Tsuda, 'On the parallel between Zipf's law and $1/f$ processes in chaotic systems possessing coexisting attractors', *Prog. Theor. Phys.* **82** (1989) 254.
241. Nishimori, H. and N. Ouchi, 'Formation of ripple patterns and dunes by wind-blown sand', *Phys. Rev. Lett.* **71** (1993) 197.
242. Noé, K., Hermeneutics of Science (Shin'Yo-sha, Tokyo, 1993, in Japanese).
243. Nozawa, H., 'A neural network model as a globally coupled map and applications based on chaos', *Chaos* **2** (1992) 377.
244. Obukhov, S.P., 'The problem of directed percolation', *Physica A* **101** (1980) 145.
245. Ohmine, I. and H. Tanaka, 'Potential-energy surfaces for water dynamics. 2. vibrational-mode excitations, mixing, and relaxations', *J. Chem. Phys.* **93** (1990) 8138.
246. Oono, Y., 'A heuristic approach to the Kolmogorov entropy as a disorder parameter' *Prog. Theor. Phys.* **60** (1978) 1944.
247. Oono, Y. and S. Puri, 'Computationally efficient modelling of ordering of quenched phases', *Phys. Rev. Lett.* **58** (1986) 836.
248. Oono, Y. and S. Puri, 'Study of phase separation dynamics by use of cell dynamical systems', *Phys. Rev.* **A 38** (1988) 1542.
249. Oono, Y. and Y. Takahashi, 'Chaos, external noise and Fredholm theory', *Prog. Theor. Phys.* **63** (1980) 1804.
250. Otsuka, K. and K. Ikeda, 'Cooperative dynamics and functions in a collective nonlinear optical element system', *Phys. Rev.* **A 39** (1989) 5209.
251. Packard, N.H., J.P. Crutchfield, J.D. Farmer and R.S. Shaw, 'Geometry from a time series', *Phys. Rev. Lett.* **45** (1980) 712.
252. Paladin, G., M. Serva and A. Vulpiani, 'Complexity in dynamical systems with noise', *Phys. Rev. Lett.* **74** (1995) 66.
253. Parisi, G., 'Asymmetric neural networks and the process of learning', *J. of Phys.: Math. and Gen.* **19** (1986) L675.
254. Pecora, L.M. and T. Carroll, 'Synchronization in chaotic systems', *Phys. Rev. Lett.* **64** (1990) 821.
255. Perez, G., C. Pando-Lambruschini, S. Sinha and H.A. Cerdeira, 'Nonstatistical behavior of coupled optical systems', *Phys. Rev.* **A 45** (1992) 5469.
256. Perez, G. and H.A. Cerdeira, 'Instabilities and nonstatistical behavior in globally coupled systems', *Phys. Rev.* **A 46** (1992) 7492.
257. Perez, G., S. Sinha and H.A. Cerdeira, 'Order in the turbulent phase of globally coupled maps', *Physica D* **63** (1993) 341.
258. Pikovsky, A.S., 'Spatial development of chaos in nonlinear media', *Phys. Lett.* **156 A** (1991) 223.
259. Pikovsky, A.S. and J. Kurths, 'Do globally coupled maps really violate the law of large numbers?', *Phys. Rev. Lett.* **72** (1994) 1644.

260. Pöggeler, H.O. ed., Hermeneutische Philosophie (Nymphenburger Verlagshandlung GmbH, München, 1972; Japanese translation published by Koyo shobo, Tokyo, 1977).

261. Poincaré, H., Les Méthodes Nouvelles de la Mécanique Céleste III (Ganthier-Villars, Du Bureau des Longitudes, De L'école Polytechnique, Paris, 1899) (Japanese translation by M. Fukuhara and T. Ura, Kyoritsu Publ., Tokyo, 1970).

262. Polanyi, M., Personal Knowledge (The University of Chicago Press, 1958).

263. Pomeau, Y., 'Front motion, metastability and subcritical bifurcations in hydrodynamics', *Physica D* **23** (1986) 3.

264. Rapp, P.E., T.R. Bashore, J.M. Martinerie, A.M. Albano, I.D. Zimmerman and A.I. Mees, 'Oscillations and chaos in cellular metabolism and physiological systems,' Chaos (ed. A.V. Holden, Manchester University Press, 1980) 179.

265. Rapp, P.E., T.R. Bashore, J.M. Martinerie, A.M. Albano, I.D. Zimmerman and A.I. Mees, 'Dynamics of brain electrical activity,' *Brain Topography* **2** (1989) 99.

266. Reitboeck, H.J., R. Eckhorn, M. Arndt and P. Dicke, 'A model for feature linking via correlated neural activity', Synergetics of Cognition (ed. H. Haken, Springer, Berlin, Heidelberg, 1990) 112.

267. Richmond, B.J., L.M. Optican, M. Prdell and H. Spitzer, 'Temporal encoding of two-dimensional patterns by single units in primate interior temporal cortex', *J. Neurophysiology* **57** (1987a) 132.

268. Richmond, B.J. and L.M. Optican, 'Temporal encoding of two-dimensional patterns by single units in primate interior temporal cortex, II. Quantification of response wave form', *J. Neurophysiology* **57** (1987b) 147.

269. Richmond, B.J., L.M. Optican and H. Spitzer, 'Temporal encoding of two-dimensional patterns by single units in primate interior temporal cortex, I. Stimulus–response relation', *J. Neurophysiology* **64** (1990a) 351.

270. Richmond, B.J. and L.M. Optican, 'Temporal encoding of two-dimensional patterns by single units in primate primary visual cortex, II. Information transmission', *J. Neurophysiology* **64** (1990b) 370.

271. Rössler, O.E., 'Recursive evolution', *Biosystem* **13** (1979) 193.

272. Rössler, O.E., 'The Chaotic Hierarchy', *Z. Naturforsch.* **38a** (1983) 788.

273. Rössler, O.E., 'Endophysics', Real Brains and Artificial Minds (eds. J.L. Casti and A. Karlqvist, North-Holland, New York, 1987) 25.

274. Rössler, O.E., 'Micro constructivism', *Physica D* **75** (1994) 438.

275. Rössler, O. E. and J.L. Hudson, 'Self-similarity in hyperchaotic data', Chaos in Brain Function (ed. E. Basar, Springer, Berlin, Heidelberg, 1990) 74.

276. Rössler, O.E., C. Knudsen, J.L. Hudson and I. Tsuda, 'Nowhere-differentiable attractors', *Int. J. Intell. Sys.* **10** (1995) 15.

277. Roux, J.C., A. Rossi, B. Bachelart and C. Vidal, 'Representation of a strange attractor from an experimental study of chemical turbulence', *Phys. Lett.* **77A** (1980) 391.

278. Ruelle, D. and F. Takens, 'On the nature of turbulence', *Comm. Math. Phys.* **20** (1971) 167.

279. Ruelle, D., Thermodynamic Formalism (Addison-Wesley, Reading, MA, 1978).

280. Sano, M. and Y. Sawada, 'Measurement of the Lyapunov spectrum from a chaotic time series', *Phys. Rev. Lett.* **55** (1985) 1082.

281. Schaffer, W.M., L.F. Olsen, C.L. Truty, S.L. Fulmer and D.J. Graser, 'Periodic and chaotic dynamics in childhood infection', From Chemical to Biological Organization (eds. M. Markus, S.C. Muller and G. Nicolis, Springer, Berlin, Heidelberg, 1988) 331.

282. Schmitz, R.A., K.R. Graziani and J.L. Hudson, 'Experimental evidence of chaotic states in the Belousov–Zhabotinsky reaction', *J. Chem. Phys.* **67** (1977) 3040.

283. Seelen, W. von, H.A. Mallot and F. Giannakopoulos, 'Computation in cortical nets', Neural and Synergetic Computers (ed. H. Haken, Springer, Berlin, Heidelberg, 1988) 123.

284. Shaw, R., 'Strange attractors, chaotic behavior, and information flow', *Z. Naturf.* **36a** (1981) 80.

285. Shaw, R., The Dripping Faucet as a Model Chaotic System (Aerial Press, Santa Cruz, 1984).

286. Shibata, T. and K. Kaneko, 'Coupled map gas', unpublished [1995].

287. Shibata, T. and K. Kaneko, 'Heterogeneity induced order in globally coupled chaotic systems', *Europhys. Lett.* **38** (1997) 417.

288. Shibata, T. and K. Kaneko, 'Tongue-like bifurcation structures of the mean-field dynamics in a network of chaotic elements', *Physica D* **124** (1998a) 163.

289. Shibata, T. and K. Kaneko, 'Collective chaos', *Phys. Rev. Lett.* **81** (1998b) 4116.

290. Shibata, T., T. Chawanya and K. Kaneko, 'Noiseless collective motion out of noisy chaos', *Phys. Rev. Lett.* **82** (1999) 4424.

291. Shimada, I., 'Gibbsian distribution on the Lorenz attractor', *Prog. Theor. Phys.* **62** (1979) 61.

292. Shimada, I., private communication.

293. Shimizu, H., Life and Place (NTT Publ., Tokyo 1992, in Japanese).

294. Shimizu, H. and Y. Yamaguchi, 'Synergetic computer and holonics – information dynamics of a semantic computer', *Physica Scripta* **36** (1987) 970.

295. Shimizu, H., Y. Yamaguchi, I. Tsuda and M. Yano, 'Pattern recognition based on holonic information dynamics: towards synergetic computers', Complex Systems – Operational Approaches (ed. H. Haken, Springer, Berlin, Heidelberg, 1985) 225.

296. Shinjo, K., 'Formation of a glassy solid by computer-simulation', *Phys. Rev.* **B40** (1989) 9167.

297. Shinjo, K. and T. Sasada, 'Commensurate, incommensurate and random structures in one-dimensional lattice gas model', *Prog. Theor. Phys.* **78** (1987) 573.

298. Showalter, K., R.M. Noyes and K. Bar-Eli, 'A modified Oregonator model exhibiting complicated limit cycle behavior in a flow system', *J. Chem. Phys.* **69** (1978) 2514.

299. Siegelmann, H. and E. Sontag, 'Analog computation via neural networks', *Theoretical Computer Science* **131** (1994) 331.

300. Sinha, S., D. Biswas, M. Azam and S.V. Lawande, 'Nonstatistical behavior of higher-dimensional coupled systems', *Phys. Rev.* **A 46** (1992) 3193.

301. Skarda, C.A. and W.J. Freeman, 'How brains make chaos in order to make sense of the world', *Behavioral and Brain Sciences* **10** (1987) 161.

302. Sole, R.V., J. Bascompte and J. Vallis, 'Nonequilibrium dynamics in lattice ecosystems', *Chaos* **2** (1992) 387.

303. Sommerer, J.C. and E. Ott., 'A physical system with qualitatively uncertain dynamics', *Nature* **365** (1993) 138.

304. Sompolinsky, H., A. Crisanti and H.J. Sommers, 'Chaos in random neural networks', *Phys. Rev. Lett.* **61** (1988) 259.

305. Szentagothai, J., 'The 'module-concept' in cerebral cortex architecture', *Brain Research* **95** (1975) 475.

306. Szentagothai, J., 'The modular architectonic principle of neural centers', *Reviews of Physiology, Biochemistry, and Pharmacology* **98** (1983) 11.

307. Takagi, Y., Lifespan of an Organism and Cell (Heibon-sha, Tokyo, 1993, in Japanese).
308. Takagi, Y., 'Clonal life cycle of *Paramecium* in the context of evolutionally acquired mortality', Cell Immortalization (ed. A. Macieira-Coelho, Springer, Berlin, Heidelberg, 1999).
309. Takens, F., 'Detecting strange attractors in turbulence', Lect. Notes in Math. (eds. D.A. Rand and L.S. Young, Springer, Berlin, Heidelberg) **898** (1981) 366.
310. Tanaka, S. and S. Ito, 'Random iteration of unimodal linear transformations', *Tokyo J. Math.* **58** (1982) 463.
311. Terada, T., Collected essays 1–5 (Iwanami shoten, Tokyo, 1946, in Japanese).
312. Tomita, K., 'Chaotic response of nonlinear oscillators', *Phys. Rep.* **86** (1982) 113.
313. Tomita, K., 'Significance of chaos', *Trans. Physical Society of Japan* **40** (1985) 99 (in Japanese).
314. Tomita, K. and I. Tsuda, 'Chaos in the Belousov–Zhabotinsky reaction in a flow system', *Phys. Lett.* **71A** (1979) 489.
315. Tomita, K. and I. Tsuda, 'Towards the interpretation of Hudson's experiment on the Belousov–Zhabotinsky reaction – Chaos due to delocalization', *Prog. Theor. Phys.* **64** (1980) 1138.
316. Tomita, K. and T. Kai, 'Chaotic behavior of deterministic orbits: the problem of turbulent phase', *Prog. Theor. Phys. Suppl.* **64** (1978) 280.
317. Tomita, K., A. Ito and T. Ohta, 'Simplified model for Belousov–Zhabotinsky reaction', *J. Theor. Biol.* **68** (1977) 459.
318. Torcini, A., A. Politi, G.P. Puccioni and G. D'Alessandro, 'Fractal dimension of spatially extended systems', *Physica D* **53** (1991) 85.
319. Tsuda, I., 'A hermeneutic process of the brain', *Prog. Theor. Phys. Suppl.* **79** (1984) 241.
320. Tsuda, I., 'Roles of chaos in information dynamics of the brain', II–2 (1985) 1 (in Japanese).
321. Tsuda, I., 'Hermeneutics of the brain', Adventure in Interpretation (ed. by H. Shimizu, NTT Publ., Tokyo, 1987a, in Japanese).
322. Tsuda, I., 'Deus ex machina – from chaos theory to brain theory', *Trans. Japan Society of System and Control: System and Control* **31** (1987b) 36 (in Japanese).
323. Tsuda, I., E. Koerner and H. Shimizu, 'Memory dynamics in asynchronous neural networks', *Prog. Theor. Phys.* **78** (1987) 51.
324. Tsuda, I., Chaos Aspect of Brain (Science-sha Publ. Inc., Tokyo, 1990a, in Japanese).
325. Tsuda, I., 'Possible biological and cognitive functions of neural networks probabilistically driven by an influence of probabilistic release of synaptic vesicles', Proc. 12th Ann. Int. Conf. IEEE/EMBS (1990b) 1772.
326. Tsuda, I., 'Chaotic neural networks and thesaurus', Neurocomputers and Attention (Manchester University Press, 1991a) 430.
327. Tsuda, I., 'Chaotic itinerancy as a dynamical basis of hermeneutics in brain and mind', *World Futures* **32** (1991b) 167.
328. Tsuda, I., 'Dynamic link of memory – chaotic memory map in nonequilibrium neural networks', *Neural Networks* **5** (1992a) 313.
329. Tsuda, I., 'Nonlinear dynamical systems theory and engineering neural network: can each afford plausible interpretation of 'how' and 'what'? ', *Behavioral and Brain Sciences* **15** (1992b) 802.
330. Tsuda, I., 'Chaotic dynamical systems in neural networks', Neurocomputing (ed. K. Matsuoka, Asakura shoten, Tokyo, 1992c, in Japanese).

331. Tsuda, I., T. Tahara and H. Iwanaga, 'An observation of a chaotic pulsation in human capillary vessels and its dependence on the mental and physical conditions', *Int. J. Bifurcation and Chaos* **2** (1992) 313.

332. Tsuda, I., 'Dynamic-binding theory is not plausible without chaotic oscillation', *Behavioral and Brain Sciences* **16** (1993) 475.

333. Tsuda, I., 'Can stochastic renewal of maps be a model for cerebral cortex?', *Physica D* **75** (1994a) 165.

334. Tsuda, I., 'From micro-chaos to macro-chaos: chaos can survive even in macroscopic states of neural activities', Psycoloquy. 94.5.12. eeg–chaos.3. tsuda (ISSN 1055–0143, American Psychological Association, New York, 1994b).

335. Tsuda, I., 'A new type of self-organization associated with chaotic dynamics in neural systems', *Int. J. Neural Systems* **7** (1996) 451.

336. Tsuda, I. and A. Yamaguchi, 'Singular-continuous nowhere-differentiable attractors in neural systems', *Neural Networks* **11** (1998) 927.

337. Tsuda, I. and H. Shimizu, 'Self-organization of the dynamical channel', Complex Systems – Operational Approaches in Neurobiology, Physics and Computers (ed. H. Haken, Springer, Berlin, Heidelberg, 1985) 240.

338. Tsuda, I. and K. Tadaki, 'A logic-based dynamical theory for a genesis of biological threshold', *BioSystems* **42** (1997) 45.

339. Tsukada, M., 'Theoretical model of the hippocampal–cortical memory system motivated by physiological functions in the hippocampus', *Concepts in Neuroscience* **3** (1992) 213, and M. Tsukada's papers cited therein.

340. Tsukada, M., M. Terasawa and G. Hauske, 'Temporal pattern discrimination in the cat's retinal cells and Markov system models', IEEE Trans. Systems, Man. Cybern. SMC **13** (1983) 953.

341. Tyson, J.J., 'Belousov–Zhabotinsky reaction', Lect. Notes in Biomath. **10** (Springer, Berlin, Heidelberg, New York, 1976).

342. Tyson, J.J., 'Appearance of chaos in a model of Belousov–Zhabotinsky reaction', *J. Math. Biol.* **5** (1978) 351.

343. Ueda, Y., The Road to Chaos (Aerial Press, Santa Cruz, 1994).

344. Waller, I. and R. Kapral, 'Spatial and temporal structure in systems of coupled nonlinear oscillators', *Phys. Rev.* **A30** (1984) 2047.

345. Wiener, N., Cybernetics, or Control and Communication in the Animal and the Machine (John Wiley and Sons, Inc., New York, 1948, 2nd edition, The MIT press, 1961).

346. Willaime, H., O. Cardoso and P. Tabeling, 'Spatiotemporal intermittency in lines of vortices', *Phys. Rev.* **E48** (1993) 288.

347. Willeboordse, F.H. and K. Kaneko, 'Bifurcations and spatial chaos in an open flow model', *Phys. Rev. Lett.* **73** (1994) 533.

348. Willeboordse, F.H. and K. Kaneko, 'Pattern dynamics of a coupled map lattice for open flow', *Physica D* **86** (1995) 428.

349. Wilson, E.O., The Diversity of Life (W.W. Norton and Company Inc., New York, 1992).

350. Winfree, A.T., The Geometry of Biological Time (Springer, Berlin, Heidelberg, 1980).

351. Wolf, A., J.B. Swift, H.L. Swinney and J.A. Vastano, 'Determining Lyapunov exponents from a time series', *Physica D* **16** (1985) 285.

352. Wolfram, S. ed., Theory and Applications of Cellular Automata (World Scientific, Singapore, 1986).

353. Yamada, M. and K. Ohkitani, 'Lyapunov spectrum of a model of two-dimensional turbulence', *Phys. Rev. Lett.* **60** (1988) 983.

354. Yamaguchi, A., On the mechanism of spatial bifurcations in the open flow system, *Int. J. Bifurcation and Chaos* **7** (1997) 1529.

355. Yamaguti, M., Chaos and Fractal (Kodansha, Tokyo, 1986, in Japanese).
356. Yamaguti, M., M. Hata, and J. Kigami, Mathematical Theory of Fractal (Iwanami shoten, Iwanami Lecture Series on Applied Mathematics 7, 1993, in Japanese).
357. Yanagita, T., 'Coupled map lattice model for boiling', *Phys. Lett.* **165 A** (1992) 405.
358. Yanagita, T. and K. Kaneko, 'Coupled map lattice model for convection', *Phys. Lett.* **175 A** (1993) 415.
359. Yanagita, T. and K. Kaneko, 'Rayleigh–Bénard convection', *Physica D* **82** (1995) 288.
360. Yanagita, T. and K. Kaneko, 'Modeling and Characterization of Cloud Dynamics', Phys. Rev. Lett. **78** (1997) 4297.
361. Yao, Y. and W.J. Freeman, 'Model of biological pattern recognition with spatially chaotic dynamics', *Neural Networks* **3** (1990) 153.
362. Yates, F.E., 'Physical causality and brain theories', *Amer. J. Physiology* **238** (1980) 277.
363. Yu, L., E. Ott and Q. Chen, 'Transition to chaos for random dynamical systems', *Phys. Rev. Lett.* **65** (1990) 2935.
364. Zhuangzhi, Philosophy of Zhuangzhi (Iwanami shoten, 1973). An English translation: The complete works of Chuang Tzu (translated by B. Watson, Columbia University Press, New York and London, 1968). Zhuangzhi (or Chuang Tzu in pronunciation) lived around 400–300 B.C.

Index

Printing: Saladruck, Berlin
Binding: H. Stürtz AG, Würzburg